Natural Degradation: Polymer Degradation under Different Conditions

Natural Degradation: Polymer Degradation under Different Conditions

Editors

Alexandre Vetcher
Alexey Iordanskii

MDPI • Basel • Beijing • Wuhan • Barcelona • Belgrade • Manchester • Tokyo • Cluj • Tianjin

Editors
Alexandre Vetcher
Peoples' Friendship
University of Russia
Complementary and
Integrative Health Clinic of
Dr. Shishonin
Russia

Alexey Iordanskii
Russian Academy of Sciences
Russia

Editorial Office
MDPI
St. Alban-Anlage 66
4052 Basel, Switzerland

This is a reprint of articles from the Special Issue published online in the open access journal *Polymers* (ISSN 2073-4360) (available at: https://www.mdpi.com/journal/polymers/special_issues/Natural_Degradation_Polymer_Degradation_under_Different_Conditions).

For citation purposes, cite each article independently as indicated on the article page online and as indicated below:

LastName, A.A.; LastName, B.B.; LastName, C.C. Article Title. *Journal Name* **Year**, *Volume Number*, Page Range.

ISBN 978-3-0365-5133-3 (Hbk)
ISBN 978-3-0365-5134-0 (PDF)

© 2022 by the authors. Articles in this book are Open Access and distributed under the Creative Commons Attribution (CC BY) license, which allows users to download, copy and build upon published articles, as long as the author and publisher are properly credited, which ensures maximum dissemination and a wider impact of our publications.

The book as a whole is distributed by MDPI under the terms and conditions of the Creative Commons license CC BY-NC-ND.

Contents

About the Editors vii

Preface to "Natural Degradation: Polymer Degradation under Different Conditions" ix

Alexandre A. Vetcher and Alexey L. Iordanskii
Natural Degradation: Polymer Degradation under Different Conditions
Reprinted from: *polymers* 2022, *14*, 3595, doi:10.3390/polym14173595 1

Kirill V. Zhukov, Alexandre A. Vetcher, Bagrat A. Gasparuan and Alexander Y. Shishonin
Alteration of Relative Rates of Biodegradation and Regeneration of Cervical Spine Cartilage through the Restoration of Arterial Blood Flow Access to Rhomboid Fossa: A Hypothesis
Reprinted from: *Polymers* 2021, *13*, 4248, doi:10.3390/polym13234248 3

Arpaporn Teamsinsungvon, Chaiwat Ruksakulpiwat and Yupaporn Ruksakulpiwat
Effects of Titanium–Silica Oxide on Degradation Behavior and Antimicrobial Activity of Poly (Lactic Acid) Composites
Reprinted from: *Polymers* 2022, *14*, 3310, doi:10.3390/polym14163310 13

Johanna Langwieser, Andrea Schweighuber, Alexander Felgel-Farnholz, Christian Marschik, Wolfgang Buchberger and Joerg Fischer
Determination of the Influence of Multiple Closed Recycling Loops on the Property Profile of Different Polyolefins
Reprinted from: *Polymers* 2022, *14*, 2429, doi:10.3390/polym14122429 39

Svetlana G. Karpova, Natalia A. Chumakova, Anton V. Lobanov, Anatoly A. Olkhov, Alexandre A. Vetcher and Alexey L. Iordanskii
Evaluation and Characterization of Ultrathin Poly(3-hydroxybutyrate) Fibers Loaded with Tetraphenylporphyrin and Its Complexes with Fe(III) and Sn(IV)
Reprinted from: *Polymers* 2022, *14*, 610, doi:10.3390/polym14030610 57

Olivia A. Attallah, Muhammad Azeem, Efstratios Nikolaivits, Evangelos Topakas and Margaret Brennan Fournet
Progressing Ultragreen, Energy-Efficient Biobased Depolymerization of Poly(ethylene terephthalate) via Microwave-Assisted Green Deep Eutectic Solvent and Enzymatic Treatment
Reprinted from: *Polymers* 2022, *14*, 109, doi:10.3390/polym14010109 73

Anatoly A. Olkhov, Elena E. Mastalygina, Vasily A. Ovchinnikov, Tatiana V. Monakhova, Alexandre A. Vetcher and Alexey L. Iordanskii
Thermo-Oxidative Destruction and Biodegradation of Nanomaterials from Composites of Poly(3-hydroxybutyrate) and Chitosan
Reprinted from: *Polymers* 2021, *13*, 3528, doi:10.3390/polym13203528 89

Ji-Bong Choi, Yu-Kyoung Kim, Seon-Mi Byeon, Jung-Eun Park, Tae-Sung Bae, Yong-Seok Jang and Min-Ho Lee
Characteristics of Biodegradable Gelatin Methacrylate Hydrogel Designed to Improve Osteoinduction and Effect of Additional Binding of Tannic Acid on Hydrogel
Reprinted from: *Polymers* 2021, *13*, 2535, doi:10.3390/polym13152535 105

Ping He, Haoda Ruan, Congyang Wang and Hao Lu
Mechanical Properties and Thermal Conductivity of Thermal Insulation Board Containing Recycled Thermosetting Polyurethane and Thermoplastic
Reprinted from: *Polymers* 2021, *13*, 4411, doi:10.3390/polym13244411 119

Gloria Amo-Duodu, Emmanuel Kweinor Tetteh, Sudesh Rathilal, Edward Kwaku Armah, Jeremiah Adedeji, Martha Noro Chollom and Maggie Chetty
Effect of Engineered Biomaterials and Magnetite on Wastewater Treatment: Biogas and Kinetic Evaluation
Reprinted from: *Polymers* **2021**, *13*, 4323, doi:10.3390/polym13244323 . . . . **137**

Mohammed Al-Yaari and Ibrahim Dubdub
Pyrolytic Behavior of Polyvinyl Chloride: Kinetics, Mechanisms, Thermodynamics, and Artificial Neural Network Application
Reprinted from: *Polymers* **2021**, *13*, 4359, doi:10.3390/polym13244359 . . . . **151**

Camille Gillet, Ferhat Tamssaouet, Bouchra Hassoune-Rhabbour, Tatiana Tchalla, Valérie Nassiet
Parameters Influencing Moisture Diffusion in Epoxy-Based Materials during Hygrothermal Ageing—A Review by Statistical Analysis
Reprinted from: *Polymers* **2022**, *14*, 2832, doi:10.3390/polym14142832 . . . . **169**

Andrey A. Vodyashkin, Parfait Kezimana, Alexandre A. Vetcher and Yaroslav M. Stanishevskiy
Biopolymeric Nanoparticles–Multifunctional Materials of the Future
Reprinted from: *Polymers* **2022**, *14*, 2287, doi:10.3390/polym14112287 . . . . **201**

About the Editors

Alexandre Vetcher

Deputy Director (Science) at scientific and educational center "Nanotechnology" at the Institute of Biochemical Technology and Nanotechnology of the Peoples' Friendship University of Russia (RUDN), and Associate Director (Science) at the Complementary and Integrative Health Clinic of Dr. Shishonin.

Education:

1988—Ph.D. in Biophysics, Institute of Chemical Physics, Academy of Sciences, USSR

1983—M.S. in Molecular, Physical and Chemical Biology, M.V. Lomonosov Moscow State University branch, Pustchino-na-Oke

1981—B.S. and M.S. in Chemistry, V.I. Lenin Belorussian State University

Scientific Interests:

Polymers natural decay and its management and monitoring; nucleic acids properties, their characterization (PAGE, AG, PCR, AFM, TEM, etc.), and construction of nanoobjects on their basis; thermodynamics of irreversible processes in biodegradation in vivo.

Awards and achievements:

2010—The manuscript "Gel mobilities of linking-number topoisomers and their dependence on DNA helical repeat and elasticity" (2010), *Biophysical Chemistry*, has been selected and evaluated by Dr. Maxwell, a Member of the Faculty of 1000 (F1000), which places it in F1000 library in the top 2% of published articles in biology and medicine.

2010—UTD service award

2007—The GMU Apprenticeship scholarship award was devoted to a student from my research group, MS Gearheart, R.

1991—Soviet Union Business Convention Award for the development and production of gel electrophoresis equipment

1990—Byelorussian National Academy of Sciences Award in Molecular Biology

1990—Byelorussian YCL Scientific Award in Biological Sciences

1982—USSR Ministry of Higher Education Year Junior Scientist Award in Chemistry

Alexey Iordanskii

Principal Investigator at N.N. Semenov Federal Research Center for Chemical Physics, Russian Academy of Sciences, Moscow.

1998—PT Professor, M.V. Lomonosov Moscow Academy of Fine Chemical Technology

Education:

1975—Ph.D., N.N. Semenov Institute of Chemical Physics, Russian Academy of Sciences, Moscow

1968—M.S., M.V. Lomonosov Moscow State University

Scientific Interests:

Biopolymers characterization; transport phenomena in polymers; nanofibers; controlled release evaluation; biodegradation and hydrolysis in polymers and composites.

Awards and achievements:

1997—The Russian Federation medal in honor of the 850th anniversary of Moscow's founding (for contribution to science development in Moscow).

1990—DSc (Polymer Chemistry), N.N. Semenov Institute of Chemical Physics, Russian Academy of Sciences.

1988—Gold medal for biomedical research at the Soviet Union Business Convention Award for the creation of biomedical materials for an artificial heart.

Professional membership(s)

1975—D. I. Mendeleev Society Membership

Preface to "Natural Degradation: Polymer Degradation under Different Conditions"

Nowadays, different issues of natural degradation attract the attention of the research community. This growing interest is mostly associated to:

- the consideration of at least in vivo biodegradation as an interaction of both synthesis and decay;

- the development of novel ways to manage natural degradation through the creation of special composites;

- the nano-level structural control of composites which allow either increased or decreased decay speed, taking into account that different applications often require achievements in both directions;

- the application of centralized aerobic–anaerobic energy balance compensation (CAAEBC) theory to medical issues of biodegradation of the human body.

All these points are reflected at certain extent in the current Special Issue.

Alexandre Vetcher and Alexey Iordanskii
Editors

Editorial

Natural Degradation: Polymer Degradation under Different Conditions

Alexandre A. Vetcher [1,2,*] and Alexey L. Iordanskii [3]

1. Institute of Biochemical Technology and Nanotechnology (IBTN) of the Peoples' Friendship University of Russia (RUDN), 6 Miklukho-Maklaya St., 117198 Moscow, Russia
2. Complementary and Integrative Health Clinic of Dr. Shishonin, 5 Yasnogorskaya St., 117588 Moscow, Russia
3. N.N. Semenov Federal Research Center for Chemical Physics, Russian Academy of Sciences, 4 Kosygin St. 4, 119334 Moscow, Russia

* Correspondence: avetcher@gmail.com

Natural degradation (ND) is currently one of the main directions of polymer research. This direction is important for the technological applications of both natural and synthetic polymers and their composites. In some cases, we are interested in increasing their stability, e.g., if we are talking about the decay of intervertebral cartilage, teeth, or the polymer coatings of vehicles and buildings. In others, we are interested in accelerating their decay. The most prominent objects of such applications are polymer carriers for targeted drug delivery, used packaging materials, or non-recoverable car tires. There is a third direction of managed degradation, also. It could be employed in controlled release applications. In all these cases, even ND could occur inside cells or inside a living organism or in the environment under the influence of the degradation-causing factors inherent in it (e.g., ozone, oxygen, and ultraviolet, etc.). It is also obvious that polymers and composites are involved in dissipative synthesis decay type structures that are at least inside a living organism. The contributions to this Special Issue, although not covering the full range of the ND field, give a good overview of some important topics that are required for a sustainable future.

The possibility to apply the theory of centralized aerobic-anaerobic energy balance compensation to the balance of the cervical spine cartilage's polymeric part in the system of biodegradation and regeneration is the focus of [1]. This contribution employs the thermodynamics of irreversible processes to the above-mentioned system.

Another important topic is the influence of composite components and their structure on ND dynamics. In this Special Issue, we have reports devoted to the influence of Ti_xSi_y oxides on ND of PLA [2], the peculiarities of polyolefins' ND [3], the effect of small additions (1–5 wt.%) of tetraphenylporphyrin and its complexes with Fe (III) and Sn (IV) on the structure, and subsequently, the ND-associated properties of ultrathin fibers that are based on poly(3-hydroxybutyrate) [4], the combined ultragreen chemical and biocatalytic depolymerization of polyethylene terephthalate (PET) using a deep eutectic solvent-based low-energy microwave treatment followed by an enzymatic hydrolysis to facilitate PET biodepolymerisation [5], the biodegradation and thermo-destruction of films from blends of poly(3-hydroxybutyrate) and chitosan by pouring them from a solution, and nonwoven fibrous materials that are obtained by electrospinning [6], and in the rat-transplanted biodegradation of a single and double cross-linked GelMA hydrogel [7].

Some research studies are devoted directly to waste utilization/recycling, e.g., the application of a mechanochemical method to analyze the recycling mechanism of polyurethane foam and optimize the recycling process, and the regeneration the polyurethane foam powder breaks the C–O bond of the polyurethane foam (in order to greatly enhance the activity of the powder) [8], and also to the wastewater and utilizing its energy potential via the anaerobic digestion of municipality wastewater [9].

Citation: Vetcher, A.A.; Iordanskii, A.L. Natural Degradation: Polymer Degradation under Different Conditions. *Polymers* 2022, 14, 3595. https://doi.org/10.3390/polym14173595

Received: 26 August 2022
Accepted: 26 August 2022
Published: 31 August 2022

Publisher's Note: MDPI stays neutral with regard to jurisdictional claims in published maps and institutional affiliations.

Copyright: © 2022 by the authors. Licensee MDPI, Basel, Switzerland. This article is an open access article distributed under the terms and conditions of the Creative Commons Attribution (CC BY) license (https://creativecommons.org/licenses/by/4.0/).

The contemporary research is hard to imagine without the creation of a working mathematical model. We have two papers; one is devoted to the development of novel models of the pyrolysis of waste polyvinyl chloride [10]; another is devoted to the modelling of the hygrothermal ageing of epoxy resins and epoxy matrix composite materials [11].

The Special Issue ends with the review on biopolymeric nanoparticles and highlights their various synthesis methods as well as the modulation of their abilities to degrade at different conditions [12].

Author Contributions: Conceptualization, A.A.V. and A.L.I.; methodology, software, validation, formal analysis, investigation, resources, A.A.V.; data curation, writing—original draft preparation, writing—review and editing, A.A.V. and A.L.I.; visualization, supervision, project administration, funding acquisition, A.A.V. All authors have read and agreed to the published version of the manuscript.

Funding: This paper has been supported by the RUDN University Strategic Academic Leadership Program.

Institutional Review Board Statement: Not applicable.

Informed Consent Statement: Not applicable.

Data Availability Statement: Not applicable.

Conflicts of Interest: The authors declare no conflict of interests.

References

1. Zhukov, K.V.; Vetcher, A.A.; Gasparuan, B.A.; Shishonin, A.Y. Alteration of Relative Rates of Biodegradation and Regeneration of Cervical Spine Cartilage through the Restoration of Arterial Blood Flow Access to Rhomboid Fossa: A Hypothesis. *Polymers* **2021**, *13*, 4248. [CrossRef] [PubMed]
2. Teamsinsungvon, A.; Ruksakulpiwat, C.; Ruksakulpiwat, Y. Effects of Titanium–Silica Oxide on Degradation Behavior and Antimicrobial Activity of Poly (Lactic Acid) Composites. *Polymers* **2022**, *14*, 3310. [CrossRef] [PubMed]
3. Langwieser, J.; Schweighuber, A.; Felgel-Farnholz, A.; Marschik, C.; Buchberger, W.; Fischer, J. Determination of the Influence of Multiple Closed Recycling Loops on the Property Profile of Different Polyolefins. *Polymers* **2022**, *14*, 2429. [CrossRef] [PubMed]
4. Karpova, S.G.; Chumakova, N.A.; Lobanov, A.V.; Olkhov, A.A.; Vetcher, A.A.; Iordanskii, A.L. Evaluation and Characterization of Ultrathin Poly(3-hydroxybutyrate) Fibers Loaded with Tetraphenylporphyrin and Its Complexes with Fe(III) and Sn(IV). *Polymers* **2022**, *14*, 610. [CrossRef] [PubMed]
5. Attallah, O.A.; Azeem, M.; Nikolaivits, E.; Topakas, E.; Fournet, M.B. Progressing Ultragreen, Energy-Efficient Biobased Depolymerization of Poly(ethylene terephthalate) via Microwave-Assisted Green Deep Eutectic Solvent and Enzymatic Treatment. *Polymers* **2022**, *14*, 109. [CrossRef] [PubMed]
6. Olkhov, A.A.; Mastalygina, E.E.; Ovchinnikov, V.A.; Monakhova, T.V.; Vetcher, A.A.; Iordanskii, A.L. Thermo-Oxidative Destruction and Biodegradation of Nanomaterials from Composites of Poly(3-hydroxybutyrate) and Chitosan. *Polymers* **2021**, *13*, 3528. [CrossRef] [PubMed]
7. Choi, J.-B.; Kim, Y.-K.; Byeon, S.-M.; Park, J.-E.; Bae, T.-S.; Jang, Y.-S.; Lee, M.-H. Characteristics of Biodegradable Gelatin Methacrylate Hydrogel Designed to Improve Osteoinduction and Effect of Additional Binding of Tannic Acid on Hydrogel. *Polymers* **2021**, *13*, 2535. [CrossRef] [PubMed]
8. He, P.; Ruan, H.; Wang, C.; Lu, H. Mechanical Properties and Thermal Conductivity of Thermal Insulation Board Containing Recycled Thermosetting Polyurethane and Thermoplastic. *Polymers* **2021**, *13*, 4411. [CrossRef] [PubMed]
9. Amo-Duodu, G.; Tetteh, E.K.; Rathilal, S.; Armah, E.K.; Adedeji, J.; Chollom, M.N.; Chetty, M. Effect of Engineered Biomaterials and Magnetite on Wastewater Treatment: Biogas and Kinetic Evaluation. *Polymers* **2021**, *13*, 4323. [CrossRef] [PubMed]
10. Al-Yaari, M.; Dubdub, I. Pyrolytic Behavior of Polyvinyl Chloride: Kinetics, Mechanisms, Thermodynamics, and Artificial Neural Network Application. *Polymers* **2021**, *13*, 4359. [CrossRef] [PubMed]
11. Gillet, C.; Tamssaouet, F.; Hassoune-Rhabbour, B.; Tchalla, T.; Nassiet, V. Parameters Influencing Moisture Diffusion in Epoxy-Based Materials during Hygrothermal Ageing—A Review by Statistical Analysis. *Polymers* **2022**, *14*, 2832. [CrossRef] [PubMed]
12. Vodyashkin, A.A.; Kezimana, P.; Vetcher, A.A.; Stanishevskiy, Y.M. Biopolymeric Nanoparticles–Multifunctional Materials of the Future. *Polymers* **2022**, *14*, 2287. [CrossRef] [PubMed]

Communication

Alteration of Relative Rates of Biodegradation and Regeneration of Cervical Spine Cartilage through the Restoration of Arterial Blood Flow Access to Rhomboid Fossa: A Hypothesis

Kirill V. Zhukov [1], Alexandre A. Vetcher [1,2,*], Bagrat A. Gasparuan [1] and Alexander Y. Shishonin [1]

[1] Complementary and Integrative Health Clinic of Dr. Shishonin, 5 Yasnogorskaya Str., 117588 Moscow, Russia; kirizhuk@yandex.ru (K.V.Z.); b.gasparyan@shishonin.ru (B.A.G.); ashishonin@yahoo.com (A.Y.S.)
[2] Peoples' Friendship University of Russia (RUDN), 6 Miklukho-Maklaya Str., 117198 Moscow, Russia
* Correspondence: avetcher@gmail.com

Citation: Zhukov, K.V.; Vetcher, A.A.; Gasparuan, B.A.; Shishonin, A.Y. Alteration of Relative Rates of Biodegradation and Regeneration of Cervical Spine Cartilage through the Restoration of Arterial Blood Flow Access to Rhomboid Fossa: A Hypothesis. *Polymers* **2021**, *13*, 4248. https://doi.org/10.3390/polym13234248

Academic Editor: George Z. Papageorgiou

Received: 3 November 2021
Accepted: 27 November 2021
Published: 3 December 2021

Publisher's Note: MDPI stays neutral with regard to jurisdictional claims in published maps and institutional affiliations.

Copyright: © 2021 by the authors. Licensee MDPI, Basel, Switzerland. This article is an open access article distributed under the terms and conditions of the Creative Commons Attribution (CC BY) license (https://creativecommons.org/licenses/by/4.0/).

Abstract: We found the logical way to prove the existence of the mechanism that maintains the rates of biodegradation and regeneration of cervical spine cartilage. We demonstrate, that after we restore access to arterial blood flow through cervical vertebral arteries to rhomboid fossa it causes the prevalence of regeneration over biodegradation. This is in the frames of consideration of the human body as a dissipative structure. Then the recovery of the body should be considered as a reduction of the relative rates of decay below the regeneration ones. Then the recovery of cervical spine cartilage through redirecting of inner dissipative flow depends on the information about oxygen availability that is provided from oxygen detectors in the rhomboid fossa to the cerebellum. Our proposed approach explains already collected data, which satisfies all the scientific requirements. This allows us to draw conclusions that permit reconsidering the way of dealing with multiple chronic diseases

Keywords: biodegradation rates; arterial hypertension; vertebral cartilage; rhomboid fossa

1. Introduction

Cervical intervertebral disc damage has long been considered a major source of neck pain for quite a long time [1]. Generally, the major reason for such damage is attributed to osteochondrosis [2]. The analysis of accumulated data connects disc degeneration with multiple chronic diseases, including for instance diabetes mellitus type 2 [3]. Namely, for the reason of observed correlations, the recovery of damaged vertebral cartilage is one of the main challenges of modern medicine [4,5]. Therefore, the explanation of reasons for such correlations' existence plays great importance. The structure of cartilage (mostly extracellular matrix (65–80% liquid content) the dry part of which consists of collagens, proteoglycans, and in smaller amounts lipids, phospholipids, non-collagenous proteins, and glycoproteins [6]) allows the consideration of the process of its decay as biopolymers biodegradation [7]. For that reason, the process of non-invasive integrated recovery could be considered as events with cartilage positioned on the flow (Figure 1).

$$\text{Current Cartilage} = \text{Previous Cartilage} + \text{Total Cartilage Inflows}(V_i) - \text{Total Cartilage Outflows}(V_o)$$

Figure 1. Schematic representation of cartilage on the flow, where V_i is the rate of regeneration and V_o is the rate of biodegradation.

Since biopolymers are the main part of the dry extracellular matrix, therefore, the cartilage condition could be described in terms of ($V_i - V_o$). In the outcome of positive value, we observe recovery, while negative—decay. The observation of the recovery of the cartilage on the molecular level confirms that Vi becomes greater than Vo.

In general, there are two approaches to recover damaged vertebral cartilage—surgery [4,8] and integrated treatment [9–11]. Nowadays there is a wide variety of techniques for recovery from gene-based and cell-based therapies [12] to a multitude of polymer-based structures [12,13] to replace or support the damaged part of the cartilage. This is a so-called symptomatic treatment that does not change Vi—Vo.

Therefore, the issue of regeneration as an alternative to continuous degeneration eventually attracts the attention of the scientific community [14,15]. The experiments are usually conducted on model animals (e.g., rats [15,16]) that make it possible to measure some macro and molecular parameters of cartilages, analyzing separated tissues at different stages on the level of histological preparations. Such an experimental set usually compares to the averaged data since we could use the same animal only once, not the data from the same object on the different stages of recovery.

The absence of the paired data (before→after) does not provide the ability to conclude about events at the same object, which is of maximal interest. Indeed, only in this case, we will compare apples to apples. Thus, we need to collect data from the same area of the same cartilage just on different stages of the cartilage recovery process without the separation of the cartilage for the histological staining. Moreover, the data, which has been obtained from humans becomes significantly more valuable for Medicine.

What methods give us the best opportunity to observe the level of cartilage deterioration? At present, the most acceptable methods for such data collection are X-ray, CT, and MRI [17]. The analysis of the accumulated data confirms that comparison of the images taken from the same area by X-ray and MRI can exhibit cartilage growth [18].

Below we will connect contemporary observations of cervical cartilage balancing on generation→biodegradation flow [19] with the ideas of homeostasis [20] from the point of view of biological applications of the thermodynamics of irreversible processes [21–23].

2. Materials and Methods

As a method, we employed a retrospective analysis. As a sample, we used the already collected data from medical records of 2622 patients treated in our Clinic for arterial hypertension (AHT) so far in 2021. The sample consists of 1163 male (66.3 ± 10.2 y/o) and 1459 female (64.1 ± 9.3 y/o) patients. An ultrasound examination of the brachiocephalic arteries was performed twice during the course of therapy in our clinic at the preliminary and final stages. The restoration of the access to arterial blood flow through cervical vertebral arteries to rhomboid fossa was performed according to [24,25]. Therapy started from the manual correction of the cervical intervertebral discs to restore blood flow through the brachiocephalic arteries. The correction is followed by a cycle of 12 visits to corrective exercises leading to the strengthening of the muscular corset of the neck. These visits took from 15 to 41 days. By the end of the course of therapy, BP had been normalized. The measurements of the blood pressure (BP) and linear velocity of arterial blood flow (V_A) were conducted as described earlier [24] according to the regulations in Russia. The contemporary project, therefore, did not require any consent agreements from human beings. MRI images were taken on 1.5 T Siemens Magnetom Essenza (Siemens AG, Erlangen, Germany) at the MRI Expert Clinic (Moscow, Russia).

3. Formulation of the Problem

Contemporary thermodynamics of irreversible processes suggests that every living creature reminds a surfer of the energy flow, and is called a dissipative structure [22,23]. These structures require somehow organized feedback, which can adjust variations in different parameters to keep the structure in certain frames. In chemistry, it causes self-oscillation like Belousov [26] or Sel'kov [27] reactions. Namely, this way of organization of dissipative structure looks for the bystander as homeostasis of a living creature. We need to underline, that homeostasis is maintainable only if parameters do not cross the critical borders that keep the system (dissipative structure) inside the realm where it can still surf [28]. In biological systems the adjustment is centralized. The supreme center of the regulation for vertebrae is located in the cerebellum [29]. Now we would like to understand if it is possible to treat the patient's herniated cervical cartilage in a way that everything starts to work properly, how can we prove that homeostasis will no longer let cartilage balancing on generation→biodegradation flow, leave the borders, which restrain it from the way of irreversible decay.

4. The Explanation of the Hypothesis

We hypothesize that AHT appears as a reaction of oxygen detectors in the brain on the decrement of the oxygen availability in the blood flow to the detector because of the jam, caused by the cervical cartilage damage and the cascade of the processes described extensively below (Figure 2). This cascade resulted in the elevation of blood pressure (BP).

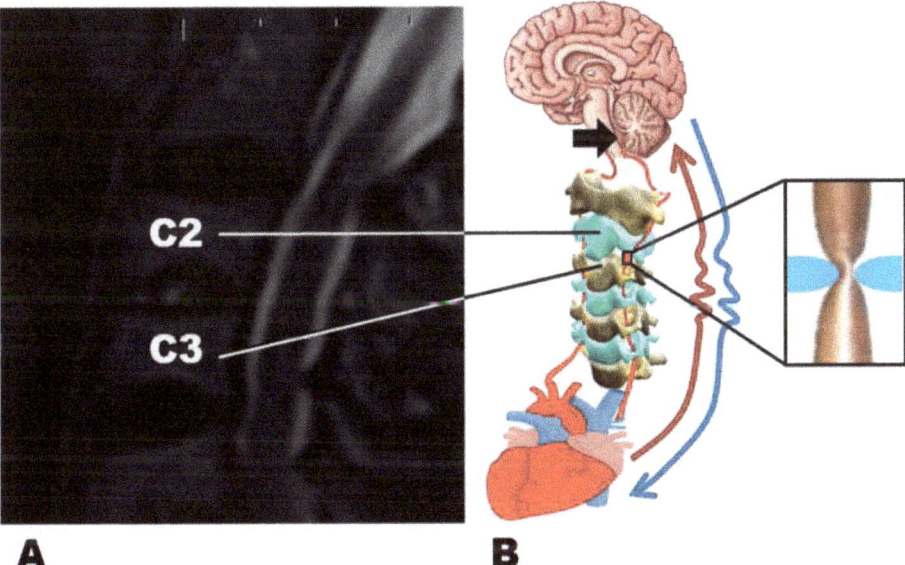

Figure 2. (**A**)—Cervical spine MRI sagittal image demonstrated damaged C2/3 intervertebral disc. The patient is 75.6 y/o. (**B**)—Schematic representation of the caused by the protrusion of intervertebral disc restriction of the access to arterial blood to rhomboid fossa (black arrow).

The elevation is a result to attempt to maintain the constant level of (E_{CONST}) energy metabolism in the brainstem through energy balance compensation between aerobic (AE) and anaerobic (AN) ways of glycolysis.

The microcirculatory and cellular levels of the AE (oxygen—E_{AE}) and AN (glucose, lipoproteins, etc., E_{AN}) molecular components of the metabolism are constantly monitored by special brain centers to fulfill

$$E_{CONST} = E_{AE} + E_{AN} \tag{1}$$

The decrease of E_{AE}, due to particular reasons (reduction in oxygen content in the microcirculatory bed and brain stem cells), two types of centralized adaptation reactions take place to maintain the overall unchanged E_{CONST} value. These are reactions of centralized aerobic-anaerobic energy balance compensation (CAAEBC) to maintain the unchanged level of E_{CONST}. Moreover, the reactions of AN compensation, as less energy-efficient [30], are triggered only with the complete depletion of the reserves of AE compensation reactions. For the very schematic representation of this energy flow, AE compensation reactions are neurogenic cardiovascular reactions, which are expressed in a steady rise in BP (an increase in the cardiac output force), a narrowing of the peripheral capillaries at rest, and an increase in cardiac rate. The goal of the AE compensation reaction is to increase the brain stem blood perfusion and hence restoration of the E_{AE} level.

AN compensation reactions are neurohumoral metabolic reactions that lead to an increase in the AN metabolism of sugars, phospholipids, and other energetically rich biochemical compounds. The purpose of this reaction is to increase E_{AN}, to maintain the balance of E_{CONST}, with a reduced E_{AE}.

Such reactions of the organism are manifestations of phenotypic adaptation (PA) according to [31]. The PA of the living organism to any changes in the environment first takes place by small forces along a simpler path. First of all, oxygen saturation of the brain occurs to a certain level, after which the reflex mechanism causing compensatory AHT is disconnected. If the brain encounters oxygen starvation for an extended period, then according to the PA theory, changes occur at the biochemical level, namely, the balance of biochemical process shifts, that is, the contribution to the energy balance of AN processes increases and the contribution of AE processes decreases.

The brain, lacking oxygen, determines its decrease, as a decrease in the level of oxygen in the atmosphere and thus, tries to adapt the work of the organism under AE conditions [32,33]. In other words, the brain tries to adapt to the already changed, according to the information from the oxygen detector, external environment, which has remained the same. Since the brain in such a situation begins to receive signals about the critical conditions of the heart, then, as a control center, to save the cardiac resource, it rebuilds the biochemical processes under conditions of a reduced partial pressure of oxygen [34]. There is a shift in AE $\rightleftarrows$ AN balance towards AN, thus preserving the overall balance of energy that is necessary to fulfill Bauer's universal principle of biology [35] and to balance the effect on the body of the second law of thermodynamics.

Considering the work of these compensatory mechanisms—"fast" and "slow", we could present a remarkable clinical example—squeezing the vessels of the neck to cause a person short-term hypoxia (oxygen deprivation) or tough physical exercise (sudden increment of oxygen consumption) will immediately result in the reflexive increase of BP and heart rate [32,33]. When the squeezing or exercise ends, all vital signs will quickly recover back to normal. This is an example of "quick" adaptation.

If a person already has a long-term occlusion of the vessels due to cervical osteochondrosis or there is a narrowing of the lumen of the vessels due to the atherosclerotic process, we will see the manifestations of the action of the "slow" adaptation with a shift in the AE $\rightleftarrows$ AN balance, namely, the development of the metabolic syndrome, type 2 diabetes [36], and cervical cartilage decay.

Therefore, if we restore the access of blood flow to the oxygen detector, it will lead stepwise, if it is not too late to:

1. Increment of arterial linear blood flow velocity(V_A) through cervical arteries, since namely they have access to the oxygen detector.
2. Decrement of BP.
3. Restoration of measurable body parameters like pulse, pH, [Fe], etc., to the normal values.
4. Restoration of cartilage, starting from its biopolymeric part, easily visible on MRI.

Let's see how such a hypothesis allows predicting the work of the behavior of cervical cartilage discs and what measurements should be presented to justify it.

5. Discussion on the Hypothesis Verification

Let's continue with AHT, caused by intervertebral discs compression with hernias and protrusions of the cervical spine. The anatomical features of the cervical vertebrae are such that veins and arteries pass through the holes in their transverse processes (*arteria vertebralis, venae vertebralis*). Due to the displacement of the vertebrae and the constant deep muscle spasm around them, the vessels are subsequently clamped, their lumen is narrowed (the arteries are thin and convoluted), and the blood flow is reduced up to ten times or may even be stopped and, as a result, the amount of delivered oxygen to the oxygen detectors in the brain is dramatically reduced. The systolic peak (PS), end diastolic velocity (ED), maximum and average speed in the cardiac cycle (TAMAX and TAMEAN, respectively), pulsatility index (PI), resistance index (RI) and the systole/diastole (S/D) ratio [37] characterize the recovery of the flow in different extent (Figure 3).

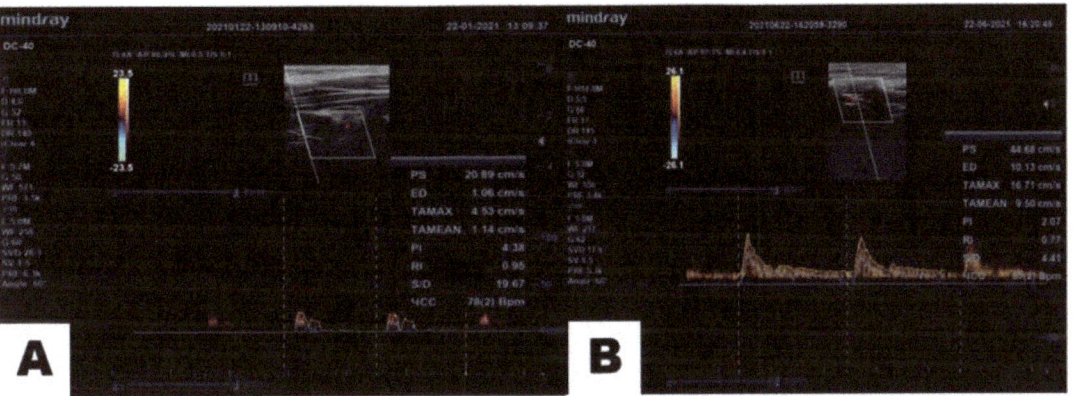

Figure 3. Restoration of V_A of a 75.6 y/o patient. (**A**)—V_A measurements before and (**B**)—5 months after the treatment according to [24,25]. The 10 fold increment of diastolic V_A (ED) and nine fold increment of TAMEAN confirm the opening of the access to rhomboid fossa.

Since the brain through the detectors' signals observes the lack of oxygen, it takes emergency measures and commands the heart to increase the strength and/or heart rate so that blood, through all the blocks and obstacles, is still able to reach the brain and provide much-needed oxygen. A stable compensatory increase in pressure and/or heart rate is developed—thus the brain is protected from hypoxia. Therefore, as soon as it is possible to unlock the vertebral arteries and veins, the pressure and heart rate should return to normal [32,33]. In this case, if we can heal the patients from AHT through the restoration of vertebral arteries patency and confirm that cervical spine cartilage discs are developing toward restoration, then our hypothesis should be considered as confirmed. The normalization of AHT through the restoration of the blood flow of the brain stem is easy to register by measurement of BP and arterial V_A (This preliminary data is collected and exhibited in Table 1).

Table 1. Results on the restoration of vertebral arteries V_A by cervical cartilage recovery according to [24,25]. The data was collected for the entire sample.

Parameter	Patients Number	% of Total
Lowering BP on 10–20 torr	740 *	28
Lowering BP on 20–40 torr	1604 *	61
Lowering BP for ≥40 torr	278 *	11
Pulse normalization	2342	89
Increment of $V_A \geq 25\%$	2622	100

* 740 + 1604 + 278 = 2622, so the BP-lowering experienced 100% of the patients.

In this case, the decompression of the vertebral arteries and veins during the correction of deep neck muscles leads to measurable restoration of V_A of *sinistra* and *dextra arteria vertebralis* to the normal [38]. The real truth is the fact that with age, there is a progression of osteochondrosis and complications arise that are associated with the gradual displacement of the cervical vertebrae. To prove this hypothesis, we will need to compare data before and after treatment. In Figure 4 the MRI axial images from the same C2/3 disk of a 75.6 y/o patient is demonstrated. The observation proves, that 5 months after the treatment the biodegradation becomes slower than regeneration and allows regeneration to start the restoration of cartilage. According to [39], such changes in MRI allow assuming that changes are cartilage recovery results.

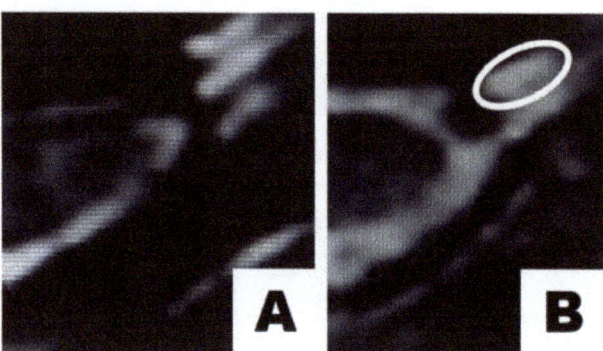

Figure 4. (**A**)—Cervical spine MRI axial image of C2/3 intervertebral disc before and (**B**)—5 months after the treatment according to [24,25]. The area of the prevalance of the regeneration over biodegradation is surrounded by the white line.

As a rule, for the majority of elderly people, the diagnosis of "essential" or "idiopathic" AHT simply implies nothing more than a compensatory increase in BP due to circulatory disturbances in the brain stem due to compression of the vessels at the level of the cervical spine [40–43]. Over time the patient develops an osteochondrosis process in the cervical region, and since this is where the vertebral arteries pass through the transverse processes of the vertebrae, their lumen naturally narrows blood flow into the brain stem and into the rhomboid fossa where the vascular center is located. Osteochondrosis is a disease that is directly related to the psychological state of the individual, namely, the accumulation of stress factors in the body. This happened evolutionarily so that any nervous shock manifests itself in the tension of the neck muscles. Therefore, it is necessary to "squeeze the head in the shoulders, hide it" to protect the cervical arteries, because the predator, attacks the neck first. At its core, this reaction is atavism. Evolution should continue and similar atavisms will eventually disappear. However, while there is such an effect, any stress, social or any other kind, produces an automatic reaction—the neck strains, and the head is drawn into the shoulders. Most people have special target muscles that harden when stressed [44]. This is a kind of vicious circle: stress causes the release of adrenaline, which in turn strains the muscles of the neck and upper back, making even more adrenaline, etc.

Soon enough this condition becomes habitual. There is a chronic spasm of deep muscles and in general a spasm of all muscles of the cervical and thoracic sections. Against this background the microcirculation and nutrition of the intervertebral discs and ligaments are disturbed, their weakening occurs, and the vertebrae begin to shift and clamp the vessels. The partial pressure of oxygen in the brain stem decreases because the flow of blood decreases. Due to the activation of the vascular center in the rhomboid fossa, the brain stem delivers an efferent signal to the heart and it starts increasing the pressure by increasing the heart force and the heart rate. According to Dobroborskiy's theory, this is the first stage of PA. According to the CAAEBC, it is an AN compensation of the energy balance.

That is to say that the central apparatus—the brain—regularly receives less oxygen than it should have received by the parameters that are genetically set as normal [45]. Accordingly, the control center proceeds to the processes of additional energy release from processes not related to breathing (AN).

The brain marks regular overloads and strains on the heart muscle and then activates the mechanism of spasm of the peripheral vessels, thus reducing the burden on the heart. Gradually, a slow centralization of blood circulation takes place. Therefore, the brain temporarily compensates for the state of hypoxia. With the unjustified administration of beta-blockers and vasodilators acting on the renin-angiotensin-aldosterone system, the brain receives an additional toxic effect that nullifies all its adaptive responses. The control center persistently continues the process of compensating the energy balance, due to which, over time, these drugs cease to function properly [46–48].

When the process of oxygen starvation is aggravated, the following adaptive mechanism is activated: the biochemical component. There is a need for extracting energy from AN source. There is a shift in AE $\rightleftarrows$ AN energy balance, and then a shift in the acid-base balance, resulting in body acidosis [49–51]. As a result, the picture of PA related to functioning in a medium without oxygen is traced. It looks like diabetes mellitus type 2 could also be developed this way in the form of impaired insulin metabolism and disruption of the normal functioning of the pancreas. We would like to emphasize, that this is only a suggestion and it requires additional data to draw a sturdy conclusion. That is, the organism tries to keep the energy level in the body with the help of these reactions.

If we can replenish the access oxygen to the detector, then the brain stem should reorganize the regulation of biochemical processes from the condition "lack of oxygen" to "normal". We should be able to observe this process of continuous recovery. In Figure 4 we demonstrated just one of many observations.

6. Conclusions

We demonstrated that the restoration of access to the arterial blood flow to the rhomboid fossa leads to:

1. The restoration of the arterial BP;
2. The restoration of the organism's homeostatic ability to self-repair the cartilage.

The confirmation of the central role of the rhomboid fossa in homeostasis (including BP restoration, cartilage regeneration, etc.).

To additionally prove this hypothesis with objective methods in addition to the data on BP, V_A and MRI we will need to provide the data on blood biochemical parameters, e.g., pH, [Fe], etc., to demonstrate the beginning of physiological functions restoration.

Author Contributions: Conceptualization, A.Y.S., A.A.V.; methodology, A.Y.S.; software, B.A.G.; validation, B.A.G., K.V.Z.; formal analysis, A.Y.S., A.A.V.; investigation, A.Y.S., K.V.Z.; resources, A.Y.S.; data curation, B.A.G.; writing—original draft preparation, A.Y.S., A.A.V.; writing—review and editing, A.Y.S., A.A.V.; visualization, A.A.V.; supervision, A.Y.S. and A.A.V.; project administration K.V.Z.; funding acquisition, A.Y.S. All authors have read and agreed to the published version of the manuscript.

Funding: This paper has been supported by the RUDN University Strategic Academic Leadership Program (recipient A.A.V.).

Institutional Review Board Statement: Since we employed a retrospective analysis as a method, and, therefore, analysed already collected data, then, there is no any necessity to obtain any Review Board Statement, according to both Swiss [52] and Russian [53] regulations.

Informed Consent Statement: Since we worked only with the already collected data (we applied a retrospective analysis as a method) and all patients undersighned their consent for the treatment in Clinic, therefore, this retrospective study is exempt from the necessity to provide any Informed Consent Statement according to both Swiss [52] and Russian [53] regulations.

Data Availability Statement: As a method, we employed a retrospective analysis and we are able to provide the employed data with the exception of the part, that is covered by the The Russian Federal Law on Personal Data (No. 152-FZ).

Acknowledgments: The authors wish to thank Vasilsa D. Bystrykh for her assistance with the edition of the submission's final version. Alexandre A. Vetcher expresses acknowledgments to the RUDN University Strategic Academic Leadership Program for the obtained support.

Conflicts of Interest: The authors declare no conflict of interest.

Ethical Approval: Since we worked only with the already collected data without any involvement of human being and sharing their personal data, therefore, this study is exempt from the necessity to provide it according to both Swiss [52] and Russian [53] regulations.

References

1. Liu, T.H.; Liu, Y.Q.; Peng, B.G. Cervical intervertebral disc degeneration and dizziness. *World J. Clin. Cases* **2021**, *9*, 2146–2152. [CrossRef] [PubMed]
2. Armbrecht, G.; European Vertebral Osteoporosis Study; European Prospective Osteoporosis Study Groups. Degenerative intervertebral disc disease osteochondrosis intervertebralis in Europe: Prevalence, geographic variation and radiological correlates in men and women aged 50 and over. *Rheumatology* **2017**, *56*, 1189–1199. [CrossRef] [PubMed]
3. Mahmoud, M. The Relationship between Diabetes Mellitus Type II and Intervertebral Disc Degeneration in Diabetic Rodent Models: A Systematic and Comprehensive Review. *Cells* **2020**, *9*, 2208. [CrossRef]
4. Choy, W.J. Annular closure device for disc herniation: Meta-analysis of clinical outcome and complications. *BMC Musculoskelet. Disord.* **2018**, *19*, 290. [CrossRef]
5. Kerr, D.; Zhao, W.; Lurie, J.D. What Are Long-term Predictors of Outcomes for Lumbar Disc Herniation? A Randomized and Observational Study. *Clin. Orthop. Relat. Res.* **2015**, *473*, 1920–1930. [CrossRef] [PubMed]
6. Sophia Fox, A.J.; Bedi, A.; Rodeo, S.A. The basic science of articular cartilage: Structure, composition, and function. *Sports Health* **2009**, *1*, 461–468. [CrossRef] [PubMed]
7. Reddy, M.S.B. A Comparative Review of Natural and Synthetic Biopolymer Composite Scaffolds. *Polymers* **2021**, *13*, 1105. [CrossRef] [PubMed]
8. Oosterhuis, T. Rehabilitation after lumbar disc surgery. *Cochrane Database Syst. Rev.* **2014**, *3*, CD003007. [CrossRef]
9. Spoto, M.M.; Dixon, G. An integrated approach to the examination and treatment of a patient with chronic low back pain. *Physiother. Theory Pract.* **2015**, *31*, 67–75. [CrossRef]
10. Azevedo, D.C. Movement System Impairment-Based Classification Versus General Exercise for Chronic Low Back Pain: Protocol of a Randomized Controlled Trial. *Phys. Ther* **2015**, *95*, 1287–1294. [CrossRef]
11. Robinson, M. Clinical diagnosis and treatment of a patient with low back pain using the patient response model: A case report. *Physiother. Theory Pract.* **2016**, *32*, 315–323. [CrossRef] [PubMed]
12. Tavakoli, J.; Diwan, A.D.; Tipper, J.L. Advanced Strategies for the Regeneration of Lumbar Disc Annulus Fibrosus. *Int. J. Mol. Sci.* **2020**, *21*, 4889. [CrossRef]
13. Choi, Y.; Park, M.H.; Lee, K. Tissue Engineering Strategies for Intervertebral Disc Treatment Using Functional Polymers. *Polymers* **2019**, *11*, 872. [CrossRef]
14. Chu, G. Biomechanics in Annulus Fibrosus Degeneration and Regeneration. *Adv. Exp. Med. Biol.* **2018**, *1078*, 409–420. [PubMed]
15. Moriguchi, Y. In vivo annular repair using high-density collagen gel seeded with annulus fibrosus cells. *Acta Biomater.* **2018**, *79*, 230–238. [CrossRef]
16. Bowles, R.D.; Setton, L.A. Biomaterials for intervertebral disc regeneration and repair. *Biomaterials* **2017**, *129*, 54–67. [CrossRef]
17. Benneker, L.M. Correlation of radiographic and MRI parameters to morphological and biochemical assessment of intervertebral disc degeneration. *Eur. Spine J.* **2005**, *14*, 27–35. [CrossRef] [PubMed]
18. Laor, T.; Clarke, J.; Yin, H. Development of the long bones in the hands and feet of children: Radiographic and MR imaging correlation. *Pediatr. Radiol.* **2016**, *46*, 551–561. [CrossRef]
19. Karsdal, M.A. Cartilage degradation is fully reversible in the presence of aggrecanase but not matrix metalloproteinase activity. *Arthritis Res. Ther.* **2008**, *10*, R63. [CrossRef]
20. Billman, G.E. Homeostasis: The Underappreciated and Far Too Often Ignored Central Organizing Principle of Physiology. *Front. Physiol.* **2020**, *11*, 200. [CrossRef] [PubMed]
21. Tlidi, M. Observation and modelling of vegetation spirals and arcs in isotropic environmental conditions: Dissipative structures in arid landscapes. *Philos. Trans. A Math. Phys. Eng. Sci.* **2018**, *376*, 2135–2146. [CrossRef]
22. Tlidi, M.; Clerc, M.G.; Panajotov, K. Dissipative structures in matter out of equilibrium: From chemistry, photonics and biology, the legacy of Ilya Prigogine (part 2). *Philos. Trans. A Math. Phys. Eng. Sci.* **2018**, *376*, 2147–2155.
23. Tlidi, M.; Clerc, M.G.; Panajotov, K. Dissipative structures in matter out of equilibrium: From chemistry, photonics and biology, the legacy of Ilya Prigogine (part 1). *Philos. Trans. A Math. Phys. Eng. Sci.* **2018**, *376*, 2124–2134.
24. Vetcher, A.A. The cervical blood flow parameters with the best correlation from arterial blood pressure in hypertension cases. *Int. J. Rec. Sci. Res.* **2021**, *12*, 42957–42958.
25. Shishonin, A. Method for Treating Cervical Osteochondrosis. Patent of RF RU 2 243 758C2, 2003.
26. Belousov, B.P. A periodic reaction and its mechanism. In *Collection of Short Papers on Radiation Medicine Conference for 1958*; Medgiz: Moscow, Russia, 1959.

27. Sel'kov, E.E. Self-Oscillations in Glycolysis.1. A Simple Kinetic Model. *Eur. J. Biochem.* **1968**, *4*, 79–86. [CrossRef] [PubMed]
28. Frederick, D.W. Loss of NAD Homeostasis Leads to Progressive and Reversible Degeneration of Skeletal Muscle. *Cell Metab.* **2016**, *24*, 269–282. [CrossRef] [PubMed]
29. Roh, E.; Kim, M.S. Brain Regulation of Energy Metabolism. *Endocrinol. Metab.* **2016**, *31*, 519–524. [CrossRef] [PubMed]
30. Nelson, D.L.; Cox, M.M. *Lehninger Principles of Biochemistry*, 5th ed.; W.H.Freeman: New York, NY, USA, 2008.
31. Dobroborsky, B.S. *Thermodynamics of Biological Systems*; North-Western State Medical University Press: Saint-Petersburg, Russia, 2006.
32. Curtelin, D. Cerebral blood flow, frontal lobe oxygenation and intra-arterial blood pressure during sprint exercise in normoxia and severe acute hypoxia in humans. *J. Cereb. Blood Flow Metab.* **2018**, *38*, 136–150. [CrossRef]
33. He, Z.B. Atlantoaxial Misalignment Causes High Blood Pressure in Rats: A Novel Hypertension Model. *Biomed. Res. Int.* **2017**, *2017*, 5986957. [CrossRef] [PubMed]
34. Silvani, A. Brain-heart interactions: Physiology and clinical implications. *Philos. Trans. A Math. Phys. Eng. Sci.* **2016**, *374*, 2067. [CrossRef]
35. Elek, G.; Muller, M. The living matter according to Ervin Bauer (1890–1938), (on the 75th anniversary of his tragic death) (History). *Acta Physiol. Hung.* **2013**, *100*, 124–132. [CrossRef]
36. Levit, S. Type 2 diabetes therapeutic strategies: Why don't we see the ELEPHANT in the room? *Diabetes Mellitus* **2016**, *19*, 341–349. [CrossRef]
37. Beltrame, R.T. Automatic and manual Doppler velocimetry measurements of the uterine artery in pregnant ewes. *Anim. Reprod. Sci.* **2017**, *181*, 103–107. [CrossRef]
38. Schoning, M.; Walter, J.; Scheel, P. Estimation of cerebral blood flow through color duplex sonography of the carotid and vertebral arteries in healthy adults. *Stroke* **1994**, *25*, 17–22. [CrossRef]
39. Eckstein, F. In vivo morphometry and functional analysis of human articular cartilage with quantitative magnetic resonance imaging–from image to data, from data to theory. *Anat. Embryol.* **2001**, *203*, 147–173. [CrossRef] [PubMed]
40. Gwadry-Sridhar, F.H. Impact of interventions on medication adherence and blood pressure control in patients with essential hypertension: A systematic review by the ISPOR medication adherence and persistence special interest group. *Value Health* **2013**, *16*, 863–871. [CrossRef] [PubMed]
41. Salvi, E. Genomewide association study using a high-density single nucleotide polymorphism array and case-control design identifies a novel essential hypertension susceptibility locus in the promoter region of endothelial NO synthase. *Hypertension* **2012**, *59*, 248–255. [CrossRef] [PubMed]
42. Morgan, T.O.; Anderson, A.; Bertram, D. Effect of indomethacin on blood pressure in elderly people with essential hypertension well controlled on amlodipine or enalapril. *Am. J. Hypertens.* **2000**, *13*, 1161–1167. [CrossRef]
43. Hoffmann, J. European headache federation guideline on idiopathic intracranial hypertension. *J. Headache Pain* **2018**, *19*, 93. [CrossRef]
44. Kroll, L.S. Level of physical activity, well-being, stress and self-rated health in persons with migraine and co-existing tension-type headache and neck pain. *J. Headache Pain* **2017**, *18*, 46. [CrossRef]
45. De Backer, J. A reliable set of reference genes to normalize oxygen-dependent cytoglobin gene expression levels in melanoma. *Sci. Rep.* **2021**, *11*, 10879. [CrossRef] [PubMed]
46. Lipworth, B. Beta-blockers in COPD: Time for reappraisal. *Eur. Respir. J.* **2016**, *48*, 880–888. [CrossRef] [PubMed]
47. Fu, M. Beta-blocker therapy in heart failure in the elderly. *Int. J. Cardiol.* **2008**, *125*, 149–153. [CrossRef]
48. Couffignal, C. Timing of beta-Blocker Reintroduction and the Occurrence of Postoperative Atrial Fibrillation after Cardiac Surgery: A Prospective Cohort Study. *Anesthesiology* **2020**, *132*, 267–279. [CrossRef]
49. Palmer, B.F.; Clegg, D.J. Hyperchloremic normal gap metabolic acidosis. *Minerva. Endocrinol.* **2019**, *44*, 363–377. [CrossRef]
50. Regolisti, G. Metabolic acidosis. *G. Ital. Nefrol.* **2016**, 33–72.
51. Sajan, A. Recurrent Anion Gap Metabolic Acidosis. *Am. J. Med. Case Rep.* **2019**, *7*, 200–202. [CrossRef]
52. Gloy, V. Uncertainties about the need for ethics approval in Swtzerland: A mixed method study. *Swiss. Med. Wkly.* **2020**, *150*, w20318.
53. Talantov, P. Unapproved clinical trials in Russia: Exception or norm? *BMC Medical. Ethics* **2021**, *22*, 46. [CrossRef]

Article

Effects of Titanium–Silica Oxide on Degradation Behavior and Antimicrobial Activity of Poly (Lactic Acid) Composites

Arpaporn Teamsinsungvon [1,2,3], Chaiwat Ruksakulpiwat [1,2,3] and Yupaporn Ruksakulpiwat [1,2,3,*]

[1] School of Polymer Engineering, Institute of Engineering, Suranaree University of Technology, Nakhon Ratchasima 30000, Thailand
[2] Center of Excellence on Petrochemical and Materials Technology, Chulalongkorn University, Bangkok 10330, Thailand
[3] Research Center for Biocomposite Materials for Medical Industry and Agricultural and Food Industry, Nakhon Ratchasima 30000, Thailand
* Correspondence: yupa@sut.ac.th; Tel.: +66-44-22-3033

Abstract: A mixed oxide of titania–silica oxides (Ti_xSi_y oxides) was successfully prepared via the sol–gel technique from our previous work. The use of Ti_xSi_y oxides to improve the mechanical properties, photocatalytic efficiency, antibacterial property, permeability tests, and biodegradability of polylactic acid (PLA) was demonstrated in this study. The influence of different types and contents of Ti_xSi_y oxides on crystallization behavior, mechanical properties, thermal properties, and morphological properties was presented. In addition, the effect of using Ti_xSi_y oxides as a filler in PLA composites on these properties was compared with the use of titanium dioxide (TiO_2), silicon dioxide (SiO_2), and TiO_2SiO_2. Among the prepared biocomposite films, the PLA/Ti_xSi_y films showed an improvement in the tensile strength and Young's modulus (up to 5% and 31%, respectively) in comparison to neat PLA films. Photocatalytic efficiency to degrade methylene blue (MB), hydrolytic degradation, and in vitro degradation of PLA are significantly improved with the addition of Ti_xSi_y oxides. Furthermore, PLA with the addition of Ti_xSi_y oxides exhibited an excellent antibacterial effect on Gram-negative bacteria (*Escherichia coli* or *E. coli*) and Gram-positive bacteria (*Staphylococcus aureus* or *S. aureus*), indicating the improved antimicrobial effectiveness of PLA composites. Importantly, up to 5% Ti_xSi_y loading could promote more PLA degradation via the water absorption ability of mixed oxides. According to the research results, the PLA composite films produced with Ti_xSi_y oxide were transparent, capable of screening UV radiation, and exhibited superior antibacterial efficacy, making them an excellent food packaging material.

Keywords: titanium silicon oxide; hydrolytic degradation; titania; silica; antimicrobial activity; photocatalytic degradation

1. Introduction

Higher usage rates of plastics all over the world are causing an increasing rate of disposal of this petroleum-based product. In addition, the limited obtainability of petrochemical resources has become a major global concern [1,2]. Poly (lactic acid) (PLA) is one of the most important biocompatible and biodegradable polymers, and it is a sustainable alternative to petrochemical-derived products [3]. PLA is a synthetic biodegradable polymer, made up of a repeated monomer unit: lactic acid (LA). It is derived from renewable and degradable resources such as corn and rice, and decomposes through simple hydrolysis into water and carbon dioxide. PLA has been viewed as one of the most promising materials because of its excellent biodegradability, biocompatibility, composability, renewability, transparency, high strength, and high modulus [4,5]. Moreover, PLA degradation products are non-toxic (at a lower composition) making it a natural choice for biomedical applications [5,6]. Therefore, this polymer has attracted a wide range of attention in various applications. However, PLA has a slow degradation rate and hydrophobicity, so it does not decompose fast enough for industrial decomposers [7–9].

To overcome such shortcoming, extra materials are added to fulfill the essential properties such as improved mechanical, heat stability, and barrier properties, and controlled degradation. In addition, the critical hydrolytic degradation rate often limits its application. The rate of degradation could be controlled by adding plasticizers and additives [10]. Although PLA can be degraded with microorganisms in the ground, it takes over two months to decompose, and does not degrade in air [11]. One method to solve this drawback is to add photodegradability filler into PLA, which helps improve degradability under any conditions [12].

Numerous modifications such as copolymerization, plasticization, polymer blends, and polymer composites have been applied to improve some PLA properties. An example of such modifications is to incorporate nanoparticles into the PLA matrix to enhance PLA properties and control the degradation process in various media [13]. The addition of selected nanofillers into PLA, such as organomodified layered silicates (OMLS) [14], carbon nanotubes (CNTs) [15], zinc oxide [16], silica nanofillers (SiO_2) [17,18], and titanium dioxide (TiO_2) [19], can enhance PLA's characteristic features. However, among the nanofillers mentioned above, TiO_2 is the most widely used. TiO_2 or titania is well recognized as a valuable material used in applications such as paints or filler in paper, polymers, textiles, photocatalysis, etc. Nano-TiO_2 particles hold many good properties such as good chemical resistance, high chemical stability [20], attractive photocatalytic activity, excellent photostability, biocompatibility, and antimicrobial activity [21]. Another most-popular nanofiller is silicon dioxide (SiO_2) or silica, a chemical compound that contains oxygen and silicon. Within inorganic oxide fillers, SiO_2 admits much concentration because of its well-defined ordered structure, the easy surface modification, high surface area, and cost-effective production [22]. SiO_2 helps improve the strength, modulus of elasticity, wear resistance, heat and fire resistances, and insulation of properties of polymer materials [23]. Moreover, SiO_2 has been widely applied in food additives, drug delivery, bioimaging, gene delivery, and engineering. Additionally, SiO_2 is classified by the FDA as a "generally regarded as safe" (GRAS) agent, thus making it an ultimate candidate for biomedical applications [24]. The binary oxides are prepared for many purposes: to expand the chemical properties, to develop specific textural properties, or to produce a particle with a personalized composition that is known to present unique characteristics (large surface areas, thermal stability, etc.) [25].

Furthermore, the antimicrobial activity of PLA is usually obtained by adding several metal particles and metal oxides such as silver (Ag) particles [26], zinc oxide (ZnO) [27,28], titanium dioxide (TiO_2) [29,30], and magnesium oxide (MgO) [31] as antibacterial agents.

The objective of the present research was to study PLA composites incorporated with Ti_xSi_y oxide of different concentrations in PLA. We have inspected the influence of TiO_2, SiO_2, and Ti_xSi_y oxide on the mechanical properties, thermal properties, morphological properties, and degradation of PLA in various media. In addition, the antimicrobial activity of the PLA composite was investigated. Furthermore, the addition of $Ti_{70}Si_{30}$ oxides into PLA significantly improved the photocatalytic efficiency of degrading methylene blue (MB), hydrolytic degradation, and the in vitro degradation of PLA. In addition, PLA, with the addition of $Ti_{70}Si_{30}$ oxides, exhibited an excellent antibacterial effect on Gram-negative bacteria (*E. coli*) and Gram-positive bacteria (*S. aureus*), indicating the improved antimicrobial effectiveness of PLA composites. As a result, $Ti_{70}Si_{30}$ oxide was used to study the influence of the contents of Ti_xSi_y oxides on the mechanical properties, thermal properties, morphological properties, and degradation of PLA in various media, and the antimicrobial activity of PLA.

2. Materials and Methods

2.1. Materials

Poly (lactic acid) (PLA, grade 4043D) was supplied from Nature Works LLC (Minnetonka, MN, USA), a commercial grade for 3D printing and film applications. Tetraethylorthosilicate (TEOS, 98%, AR grade) and Titanium (IV) isopropoxide (TTIP, 98%, AR grade) were purchased from Acros (Geel, Belgium). Absolute ethanol (C_2H_5OH, AR grade), hydrochloric acid (HCl, AR grade), and ammonium hydroxide (NH_4OH, AR grade) were supplied from Carlo Erba Reagents (Emmendingen, Germany). TiO_2, SiO_2, and Ti_xSi_y oxide prepared in-house were used as filler [32,33]. Particle size of SiO_2, $Ti_{70}Si_{30}$, $Ti_{50}Si_{50}$, and $Ti_{40}Si_{60}$ oxide used in this study were in the range of 130–150 nm. TiO_2 can be generally synthesized under high acid conditions to obtain a uniform spherical shape and small nanoparticles in the range of 25–50 nm [34]. However, the author could synthesise TixSiy under alkali conditions to control hydrolysis step of Ti-precursor to obtain a uniform shape and size in the range of 130–150 nm because Ti-precursor decreased the reactivity of the alkoxide, hence low concentration of Ti-precursor decreased reactivity of the alkoxide with the lower hydrolysis rate [35].

2.2. Preparation of PLA Composite Films

PLA and PLA composites were prepared by solvent film casting method. First, $Ti_{70}Si_{30}$, $Ti_{50}Si_{50}$, $Ti_{40}Si_{60}$, SiO_2, and TiO_2 were dispersed in chloroform with ultrasonic treatment for 1 day. After that, PLA was added to the Ti_xSi_y oxide and strictly stirred for 4 days. The dispersions of PLA composites were additionally ultrasonically treated for 1 h with frequency of 42 kHz for four times per day. Then, the treated dispersions were slightly poured onto Petri dishes, and the solvent was evaporated under room temperature. The films were dried to constant mass at room temperature for ~24 h and stored in oven at 40 °C for 4 h. Pure PLA and PLA composite films had uniform thickness of 250 ± 4.68 μm. After that, Ti_xSi_y oxides were synthesized using a modified Stöber method involving simultaneous hydrolysis and condensation of TEOS, and Ti_xSi_y mixed oxides' preparation was reported in details in Teamsinsungvon et al. [32].

2.3. Mechanical Properties

The tensile properties of PLA and PLA composites films were obtained in accordance with the ASTM standard method D882-18 using an Instron universal testing machine (UTM, model 5565, Norwood, MA, USA) with a load cell of 5 kN. Specimen samples were 10 cm × 2.54 cm. Crosshead speed was set at 50 cm/min. The values were presented as the average of seven measurements.

2.4. Thermal Properties

Thermal properties of PLA and PLA composites films were carried out using a differential scanning calorimeter (DSC204F1, Netzsch, Selb, Germany) equipped with a liquid nitrogen cooling system. The samples were heated from room temperature to 180 °C with a heating rate of 5 °C/min (1st heating scan) and stored for 5 min to erase previous thermal history. Then it was cooled to room temperature (25 °C) with a cooling rate of 5 °C/min. Finally, it was heated again to 180 °C with heating rate 5 °C/min (2nd heating scan). The degree of crystallinity (X_c) of PLA and PLA composites was estimated using Equation (1) [36,37]:

$$X_c = \frac{\Delta H_m}{\Delta H_m^0 \cdot (\mathcal{O}_{PLA})} \cdot 100 \quad (1)$$

where ΔH_m are the melting enthalpy in the second heating process, ΔH_m^0, which is the melting enthalpy of an infinitely large crystal, was taken as 93.6 J/g [38], and $\mathcal{O}_{PLA}$ is the PLA weight fraction in the composites.

Thermogravimetric analysis of PLA and PLA composite films were examined using thermogravimetric analyzer (TGA/DSC1, Mettler Toledo, Columbus, OH, USA). The temperature was raised from the room temperature to 650 °C under nitrogen and then heated to 800 °C under air atmosphere at heating rate of 10 °C/min. The weight change was recorded as a function of temperature.

2.5. Morphological Properties

Morphological properties of the film were examined through a scanning electron microscope (JEOL, JSM-6010LV, Tokyo, Japan). An EDAX Genesis 2000 energy dispersive spectrometer (AMETEK, Inc., Berwyn, PA, USA) was applied to determine the spatial distribution of Ti and Si in PLA composite. SEM images of the samples were collected using acceleration voltage 9–12 kV. The cross-sections of the films after tensile test and the freeze-fractured films in liquid nitrogen were sputtered with gold.

2.6. Water Vapor Transmission Rate (WVTR)

WVTR, a modified ASTM standard method E96, is a measure of the rate of water that passes through PLA and its composites film at a particular time interval. To analyze, the composite films were cut into circular discs of 1.5 cm diameter and placed on the top of the glass vial containing 5 mL water. Constant humidity was maintained by placing it in the desiccators maintained at room temperature and relative vapor pressure (RVP) = 0 by using silica gel. The vial assembly was weighted every hour on first day, and then the weighting was performed daily over a 20-day period. In terms of statistical approach, when the straight line adequately fit the weight change vs. time plot using linear regression with $r^2 \geq 0.99$, the constant rate of weight change was obtained. WVTR was calculated using the Equation (2) [39,40]. WVTR (g m^{-2} day^{-1}) for each type of film was determined using three individually prepared films as the replicated experimental units.

$$WVTR = \frac{G}{t \cdot A} \qquad (2)$$

where G = weight change of the vial with water and film (from the straight line) (g), T = the duration for the measurement (day), G/t = slope of the straight line (g day^{-1}), and A = the test area of the film (m^2).

2.7. Photocatalytic Degradation of Methylene Blue (MB)

The photocatalytic activity of PLA composite films was evaluated by degrading methylene blue (MB) under UV light according to Chinese standard GB/T 23762-2009. Specimens of pure PLA and PLA composite films were 5 cm × 15 cm. Five pre-wetted film of each pure PLA and PLA composite films were placed in a flask, then 200 mL of MB solution (10 mg/L) was added. The flask was placed on a mechanical shaker at 50 rpm in a UV chamber with 4 UV lamps (LP Hg lamps, 8 watts, main light emission at 245 nm). The 4 mL MB solution was collected every 60 min, and analyzed using UV-vis spectrophotometer (Cary300, Agilent Technology, Santa Clara, CA, USA). To maintain the volume of solution in the flask, the samples were placed back after each measurement. The maximum absorbance of MB occurs at 664 nm (Figure 1a). The spectrometer was calibrated with solution of MB at 1 mg/L, 3 mg/L, 5 mg/L, 7 mg/L, and 10 mg/L concentrations, respectively. Calibration curve of methylene blue aqueous solutions are shown in Figure 1b. The samples and the solution were stored in a black box to correct the possible decompositions of MB under UV light in absence of any photocatalyst. The concentration of MB was also measured every 60 min to evaluate the absorption of MB. The values were presented as the average of five measurements.

2.8. Light Transmittance and Opacity Measurements

The UV-vis transmission spectra of the film specimens were recorded in the range of 200–800 nm using a UV-vis spectrophotometer (Cary300, Agilent Technology, Santa Clara, CA, USA), according to a modified standard procedure of the British Standards Institution (BSI 1968). The film specimens were directly placed in cells, and empty cells were used as reference. The opacity of the film specimens was determined with well-controlled thicknesses. Equation (3) was used to calculate the *Opacity* (AU.nm·mm^{-1}) of the films:

$$Opacity = \frac{Abs_{600}}{b} \quad (3)$$

where Abs_{600} = the absorbance at 600 nm; the transmittance values were converted to absorbance values using the Lambert–Beer equation and b = the film thickness (mm). This test was triplicated for each type of film [41].

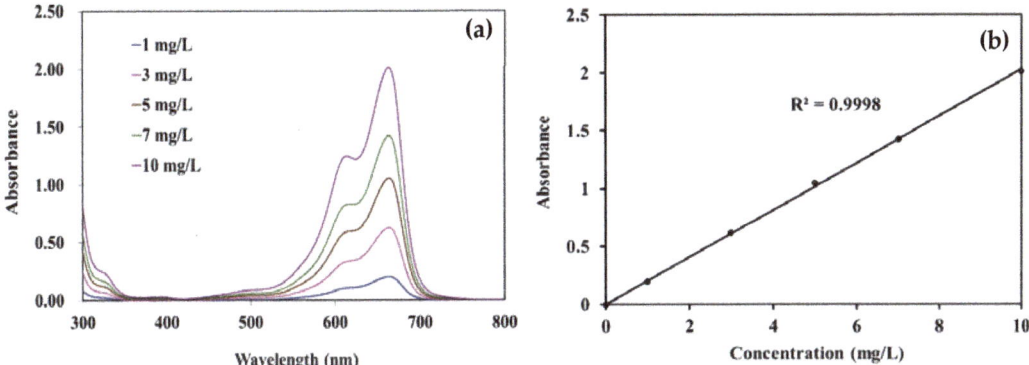

Figure 1. (a) Wavelength–absorbance curves and (b) calibration curve of methylene blue aqueous solutions.

2.9. Hydrolytic Degradation

Hydrolytic degradation of PLA and PLA composite films (10 × 10 mm^2) were carried out at 37 °C in small bottles containing 1 mol·dm^{-3} NaOH solutions (pH 13). Following the incubation for a given time (0, 60, 120, 180, 240, 300, 360, and 420 min), the films were periodically removed, washed with distilled water, and dried in oven at 40 °C for 48 h. The weight loss (W_{loss}) was estimated using following Equation (4):

$$W_{loss}(\%) = \frac{W_0 - W_t}{W_0} \cdot 100 \quad (4)$$

where W_0 = the initial weight of polymer film, and W_t = the weight of degraded sample measured at time t after drying in oven for 48 h [12].

2.10. In Vitro Degradation

The study of degradation of PLA and its composite was carried out following the standards specified by BS EN ISO 10993-13:2010 [42] and ASTM F1635-11 [43]. Each of PLA and PLA composite samples of (0.4–0.5 g) was accurately weighed. The samples were then immersed separately in 0.01 M phosphate-buffered saline (PBS) (pH = 7.4 ± 0.2) solution and maintained at 37 °C with different soaking times from 1 to 8 weeks (0, 2, 4, 6, 8). At various time points, the specimens were washed with deionized water to remove the salts, then oven dried at 40 °C for 48 h. Later, dry weights of the samples were recorded. The percentage mass change was determined using the following Equation (5);

$$W_{loss}(\%) = \frac{W_t - W_0}{W_0} \cdot 100 \quad (5)$$

where W_t = mass of degraded sample measured at time t after drying at 40 °C in oven for 48 h, and W_0 = the initial mass of the sample. At each time point, every sample was weighed and mechanically tested, allowing the degradation pathway of each individual sample to be followed with time.

The tensile properties of each specimen were measured by an Instron universal testing machine (UTM, model 5565) with a load cell of 5 kN and crosshead speed of 50 cm/min. The values were presented as the average of five measurements.

2.11. Antimicrobial Activity

Antimicrobial effects of the different samples were determined using the JIS Z 2801:2006 method. The ability of PLA and PLA composite films to restrain the growth of Escherichia coli and Staphylococcus aureus were investigated. The bacteria were incubated at 37 °C for 24 h. A plate containing a test sample was inoculated with 0.2 mL of an overnight culture of Escherichia coli and Staphylococcus aureus, while bacterial culture concentration was adjusted to 10^6 CFU/mL. All petri dishes were incubated at 37 °C for 24 h and colony-forming units (CFU) were counted. Percentage reduction of the colonies was calculated using Equations (6) and (7) below, which relates the number of colonies of neat PLA with that of the composites.

$$\% \ Reduction = \frac{(Log \ CFU \ at \ 0 \ h - Log \ CFU \ at \ 24 \ h\ )}{Log \ CFU \ at \ 0 \ h} \cdot 100 \qquad (6)$$

$$Antimicrobial \ activity(R) = Ut - At \qquad (7)$$

where Ut = average of CFU per milliliter after inoculation on untreated test pieces after 24 h; At = average of CFU per milliliter after inoculation on antibacterial test pieces after 24 h.

3. Results and Discussion

3.1. Mechanical Properties

The tensile properties of PLA and PLA composites with various $Ti_{70}Si_{30}$ oxide contents and different types of nanoparticles are listed in Table 1.

Table 1. Tensile properties of PLA, PLA/TiO_2, PLA/SiO_2, PLA/Ti_xSi_y, and PLA/TiO_2SiO_2 composites.

Sample	Tensile Strength (MPa)	Elongation at Break (%)	Young's Modulus (GPa)
PLA	33.94 ± 1.38	19.62 ± 8.64	0.98 ± 0.23
97PLA/3TiO_2	28.22 ± 1.47	18.1 ± 4.24	1.01 ± 0.04
97PLA/3SiO_2	29.75 ± 1.38	11.8 ± 4.96	1.07 ± 0.10
97PLA/3$Ti_{70}Si_{30}$	35.64 ± 3.21	11.12 ± 1.81	1.18 ± 0.06
95PLA/5$Ti_{70}Si_{30}$	32.06 ± 2.29	9.88 ± 2.81	1.01 ± 0.10
97PLA/3$Ti_{50}Si_{50}$	35.31 ± 2.01	9.61 ± 2.57	1.29 ± 0.12
97PLA/3$Ti_{40}Si_{60}$	30.67 ± 1.74	22.30 ± 2.53	1.02 ± 0.05
97PLA/3TiO_2SiO_2	29.00 ± 1.42	58.37 ± 5.00	0.96 ± 0.07

The addition of 3 wt.% of $Ti_{70}Si_{30}$ and $Ti_{50}Si_{50}$ slightly increased the tensile strength and Young's modulus and decreased the elongation at break of the PLA composite films compared to neat PLA (Figure 2). This may be due to higher interfacial adhesion between the oxide filler and the PLA matrix by the Van der Waals force or induction interactions, which decreased thereafter when adding a mixed oxide content of up to 5 wt.%, however, it was slightly higher than pure PLA. The result could be attributed to the increased filler quantity leading to a weaker filler–matrix interface and the agglomeration of filler particles, which consequently decreases the tensile strength. The Young's modulus of the composites insignificantly varied in correspondence with Ti_xSi_y oxide (Figure 2c). However, elongation at the break decreased with the addition of Ti_xSi_y oxide (Figure 2b) as a result of the addition of a rigid phase in the PLA composite, which contributed to a reduction of the PLA ductility. Nevertheless, the addition of TiO_2SiO_2 and $Ti_{40}Si_{60}$ in the PLA matrix increased elongation at the break of PLA.

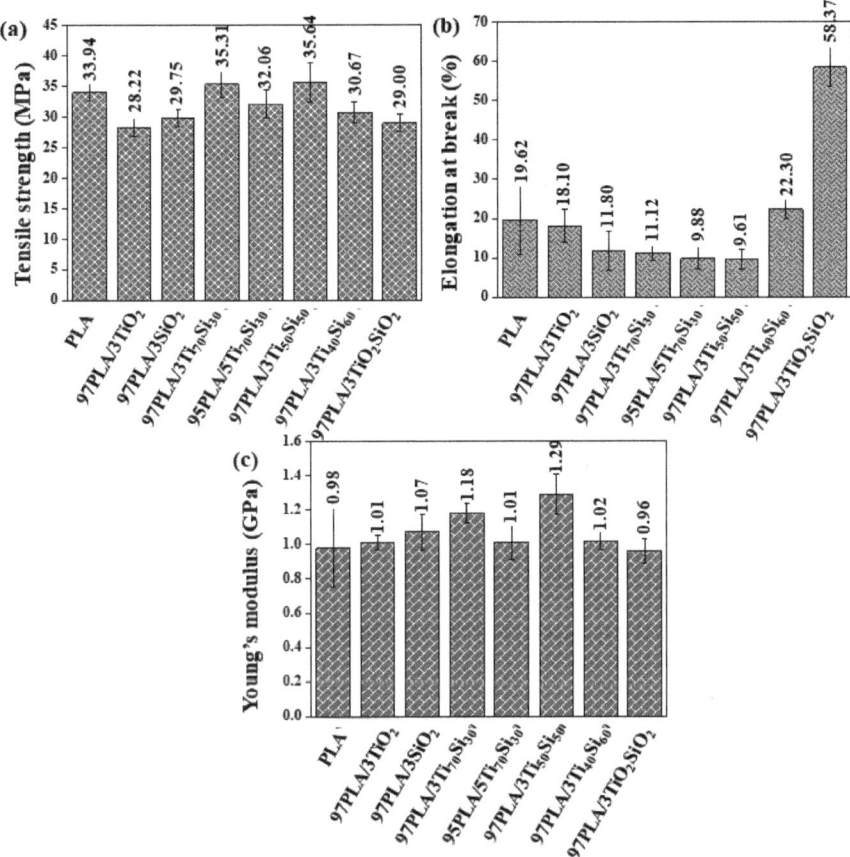

Figure 2. Comparison of the (**a**) Tensile strength, (**b**) Elongation at break, and (**c**) Young's modulus of PLA, PLA/TiO$_2$, PLA/SiO$_2$, PLA/Ti$_x$Si$_y$, and PLA/TiO$_2$SiO$_2$ composites.

During tensile testing, it was observed that the fracture behavior of the film changed for PLA, PLA/TiO$_2$, PLA/SiO$_2$, PLA/Ti$_x$Si$_y$, and PLA/TiO$_2$SiO$_2$ composites. This was demonstrated in the tensile stress–strain curves, as shown in Figure 3. With the addition of filler to the PLA matrix, the composite exhibits elastic behavior, enhancing the toughness of PLA. Furthermore, it was shown that the addition of Ti$_{70}$Si$_{30}$ oxide to the PLA polymer matrix resulted in a decrease in the ductile characteristics, while slightly increasing the tensile strength and decreasing elongation at the break. However, the addition of TiO$_2$SiO$_2$ and Ti$_{40}$Si$_{60}$ oxide increased elongation at the break of PLA. This may be due to the particles of TiO$_2$SiO$_2$ in the formulation being homogenously distributed in the polymer matrix, contributing to the occurrence of plastic deformations in the whole sample, allowing a higher elongation than that of the pure PLA.

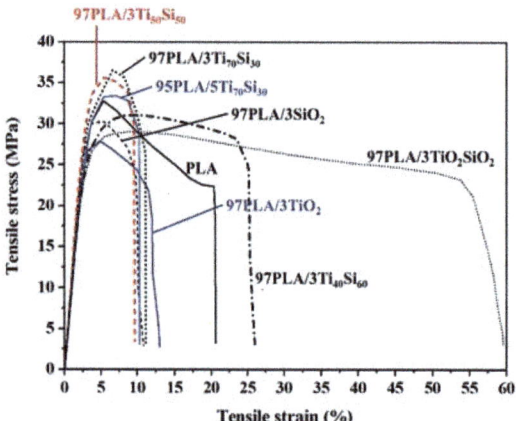

Figure 3. Stress–strain curves of PLA, PLA/TiO$_2$, PLA/SiO$_2$, PLA/Ti$_x$Si$_y$, and PLA/TiO$_2$SiO$_2$ composites.

3.2. Thermal Properties

Thermal behaviors of pure PLA and PLA composites with TiO$_2$, SiO$_2$, Ti$_{70}$Si$_{30}$, Ti$_{50}$Si$_{50}$, Ti$_{40}$Si$_{60}$, and TiO$_2$SiO$_2$ were investigated using differential scanning calorimetry (DSC), in which first heating scans (1st heating scans), cooling scans, and second heating scans (2nd heating scans) of PLA and PLA composites were performed. DSC thermograms of PLA and PLA composites are shown in Figure 4. The thermal properties of all samples are listed in Table 2. The glass transition temperature (T$_g$), cold crystallization temperature (T$_{cc}$), and melting temperature (T$_m$) of the PLA composites were observed.

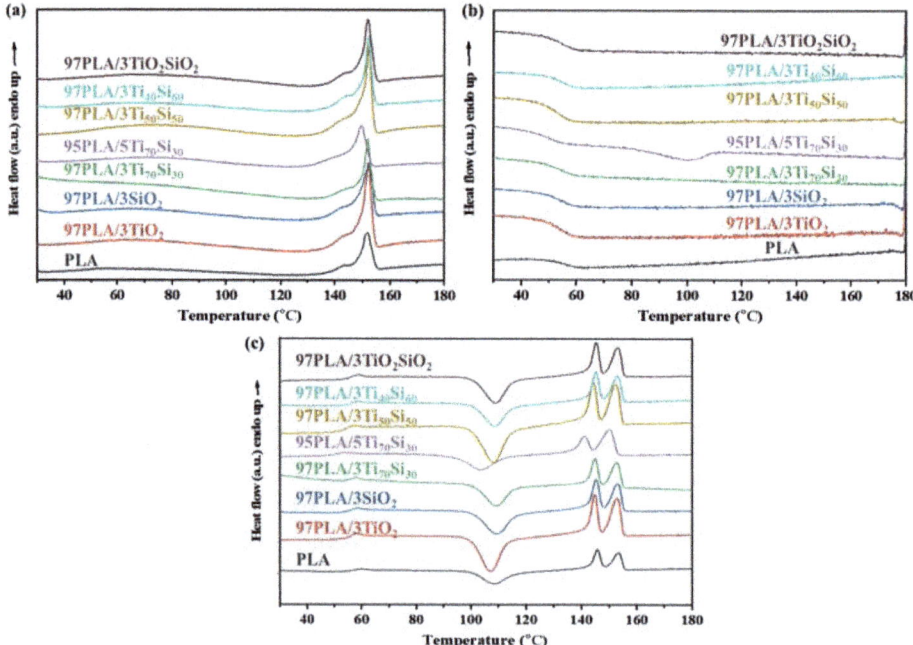

Figure 4. DSC thermogram of (**a**) first heating, heating rate 5 °C/min, (**b**) cooling, cooling rate 5 °C/min, (**c**) second heating, heating rate 5 °C/min of PLA, PLA/TiO$_2$, PLA/SiO$_2$, PLA/Ti$_x$Si$_y$, and PLA/TiO$_2$SiO$_2$ composites.

Table 2. Thermal characteristics of PLA, PLA/TiO$_2$, PLA/SiO$_2$, PLA/Ti$_x$Si$_y$, and PLA/TiO$_2$SiO$_2$ composites (the second heating, heating rate 5 °C/min).

Samples	T_g, °C	T_{cc}, °C	ΔH_c, J/g	T_{m1}, °C	T_{m2}, °C	ΔH_m, J/g	X_c,%
PLA	58.87	109.47	36.15	145.84	153.35	31.73	33.86
97PLA/3TiO$_2$	56.86	107.36	33.61	144.89	152.72	33.17	36.50
97PLA/3SiO$_2$	56.64	108.22	33.06	145.39	145.33	33.31	36.65
97PLA/3Ti$_{70}$Si$_{30}$	54.81	107.3	34.41	145.07	152.75	32.71	35.99
95PLA/5Ti$_{70}$Si$_{30}$	51.84	103.09	26.80	141.29	150.22	32.18	36.15
97PLA/3Ti$_{50}$Si$_{50}$	55.96	107.14	33.06	144.41	152.50	34.44	37.89
97PLA/3Ti$_{40}$Si$_{60}$	57.40	107.72	34.16	145.33	153.01	33.38	36.73
97PLA/3TiO$_2$SiO$_2$	57.46	107.79	30.54	145.39	153.07	32.31	35.55

T_g and T_{cc} of neat PLA occur at 58.87 °C and 109.47 °C, respectively, while T_m appeared at 145.84 °C and 152.35 °C, respectively. The double melting endotherms of neat PLA were explained by the melting and recrystallization. The peak at a low temperature was attributed to the melting of the crystals formed during the non-isothermal melt crystallization, while the peak at a high temperature corresponded to the re-melting of the newly-formed crystallite during melting, and recrystallization during the DSC heating scans [44]. However, T_c of the neat PLA did not appear in the cooling cycle. The addition of TiO$_2$, SiO$_2$, Ti$_{70}$Si$_{30}$, Ti$_{50}$Si$_{50}$, Ti$_{40}$Si$_{60}$, and TiO$_2$SiO$_2$ in the PLA matrix showed insignificant effects on the glass transition temp (T_g), cold crystallization temp (T_{cc}), and melting temp (T_m) of PLA. After the cooling scan (Figure 4b), no crystallization temperatures (T_c) were observed in all composites in the cooling cycle. However, by adding 5wt.% of Ti$_{70}$Si$_{30}$ oxide into PLA, T_c was only observed at 101 °C when cooling the sample in the DSC measurement at a cooling rate of 5 °C/min.

Cold crystallization is a phenomena that occurs from re-crystallization during the heating process of polymers [45]. Cold crystallization behaviors (Figure 4c) of PLA films containing 3 wt.% of TiO$_2$, SiO$_2$, Ti$_{70}$Si$_{30}$, Ti$_{50}$Si$_{50}$, Ti$_{40}$Si$_{60}$, and TiO$_2$SiO$_2$ nanoparticles and 5 wt.% of Ti$_{70}$Si$_{30}$ were observed. For neat PLA, there is a slightly cold crystallization peak around 109.47 °C. While adding 3 wt.% of TiO$_2$, SiO$_2$, Ti$_{70}$Si$_{30}$, Ti$_{50}$Si$_{50}$, Ti$_{40}$Si$_{60}$, and TiO$_2$SiO$_2$ into PLA, the cold crystallization peak of PLA composites shifted to a lower temperature by approximately 1–2 °C. Moreover, the T_{cc} of the PLA composite shifted to 103.09 °C with Ti$_{70}$Si$_{30}$-loading rising to 5 wt.%, showing that a 5 wt.% Ti$_{70}$Si$_{30}$ addition can promote PLA crystallization. This might be attributed to the increase of the chain mobility of PLA, and the Ti$_x$Si$_y$ oxide could act as an efficient cold crystal nuclei site, which consequently increased the crystallinity of PLA. These results suggest that Ti$_{70}$Si$_{30}$ oxide had a positive effect on the promotion of the crystallization of PLA and could act as a nucleating agent. Similar to other results, Chen et al. [45] found that by incorporating a composite nucleating agent (CNA) to PLA, the polymer could decrease T_{cc}, indicating that the crystallization ability of the PLA composite can be enhanced in such a way.

The addition of 5 wt.% Ti$_{70}$Si$_{30}$ is found to be able to enhance the T_m of PLA composites. This is possibly a result of the heterogeneous nucleation effects of Ti$_{70}$Si$_{30}$ nanoparticles on PLA during the crystallization process. The lamella formation of PLA was hindered by Ti$_{70}$Si$_{30}$ and led to less perfect crystals of PLA [46]. While the addition of a 3 wt.% of TiO$_2$, SiO$_2$, Ti$_{70}$Si$_{30}$, Ti$_{50}$Si$_{50}$, Ti$_{40}$Si$_{60}$, and TiO$_2$SiO$_2$ did not significantly affect the T_m of PLA, the degree of crystallinity (χ_c) of neat PLA significantly increased with incorporating TiO$_2$, SiO$_2$, Ti$_{70}$Si$_{30}$, Ti$_{50}$Si$_{50}$, Ti$_{40}$Si$_{60}$, and TiO$_2$SiO$_2$, indicating that TiO$_2$, SiO$_2$, Ti$_{70}$Si$_{30}$, Ti$_{50}$Si$_{50}$, Ti$_{40}$Si$_{60}$, and TiO$_2$SiO$_2$ can act as nucleating agents for PLA.

Thermal degradation at 5% weight loss ($T_{0.05}$), 50% weight loss ($T_{0.5}$), final degradation (T_f), and the char formation at 800 °C of PLA and PLA composites are listed in Table 3, respectively. TGA and DTG curves of PLA and PLA composites at a heating rate of 10 °C/min are shown in Figure S1a,b, respectively. The presence of TiO$_2$, SiO$_2$, Ti$_{70}$Si$_{30}$, Ti$_{50}$Si$_{50}$, Ti$_{40}$Si$_{60}$, and TiO$_2$SiO$_2$ did not change the thermal decomposition behavior of PLA, while the mass loss between 250–365 °C was observed, which corresponded to the decomposition of PLA. Then from 365 to 600 °C thermal analysis curves slowed down to complete the decomposition of the PLA matrix until a constant mass was reached. The constant mass remaining at the end of each TGA experiment corresponded to amounts of nanoparticles in PLA composites. In this study, the temperature at 5% weight loss ($T_{0.05}$) was defined as the onset degradation temperature for the evaluation of the TiO_2, SiO_2, and Ti_xSi_y oxide effects on the thermal stability of the PLA composites.

Table 3. Thermal degradation temperature of PLA, PLA/TiO_2, PLA/SiO_2, PLA/Ti_xSi_y, and PLA/TiO_2SiO_2 composites.

Samples	$T_{0.05}$, °C	$T_{0.5}$, °C	T_d, °C	T_f, °C	Residual, %
PLA	321.67	360.33	358.83	424.64	1.20
97PLA/3TiO_2	336.17	363.67	363.17	429.22	4.07
97PLA/3SiO_2	331.17	361.17	359.33	427.27	4.14
97PLA/3$Ti_{70}Si_{30}$	304.00	351.83	350.17	408.68	3.91
95PLA/5$Ti_{70}Si_{30}$	284.50	347.33	346.33	407.16	5.46
97PLA/3$Ti_{50}Si_{50}$	324.50	358.36	356.50	425.49	4.22
97PLA/3$Ti_{40}Si_{60}$	326.50	358.67	356.83	416.83	3.53
97PLA/3TiO_2SiO_2	330.67	362.33	361.33	418.69	4.50

It is obvious that the T_{onset} of the PLA composites shifted to a higher temperature with the presence of 3 wt.% of TiO_2, SiO_2, $Ti_{50}Si_{50}$, $Ti_{40}Si_{60}$, and TiO_2SiO_2. Consequently, the thermal stability of the PLA composites was improved. A possible reason to explain this behavior is that TiO_2, SiO_2, $Ti_{50}Si_{50}$, $Ti_{40}Si_{60}$, and TiO_2SiO_2 particles may act as a heat barrier in the early stage of thermal decomposition [12]. Similar data have been reported by Zhang et al., who studied PLA composites obtained by adding TiO_2 to poly (lactic acid) [47]. However, PLA with the addition of 3 wt.% of $Ti_{70}Si_{30}$ oxide was found to present a lower onset temperature than that of pure PLA, which resulted in a decrease in the PLA thermal stability. Moreover, it was found that the onset degradation temperature of the composites shifted to a lower temperature with increasing $Ti_{70}Si_{30}$ oxide nanoparticles loading from 3 to 5 wt.%. This suggests that there might be degradation due to the water absorption of the filler that would be associated with the cleavage of the chain of PLA at the ester group (–C–O–) by water molecules due to hydrolysis leading to decreased the thermal stability of PLA.

Moreover, the peak in the DTG curves represented the temperature maximum degradation rate (Figure S1b). PLA/3TiO_2 exhibited the fastest degradation rate at the highest temperature, compared to neat PLA and other PLA composites. However, the degradation temperature of the PLA/$Ti_{70}Si_{30}$ composite shifted to a lower temperature. This suggested that the thermal stability of PLA decreased with the incorporation of $Ti_{70}Si_{30}$ loading. In addition, when 3 wt.% TiO_2, SiO_2, $Ti_{70}Si_{30}$, $Ti_{50}Si_{50}$, $Ti_{40}Si_{60}$, and TiO_2SiO_2, and 5 wt.% $Ti_{70}Si_{30}$ mixed oxides were added to PLA, the composites left the char residual of fillers at 4.07, 4.14, 3.91, 4.22, 3.53, 4.50, and 5.46%, respectively, for the PLA composites. The char residual generally depended on the amount of added nanoparticles [48].

3.3. Morphological Properties

In order to investigate the dispersion and distribution of TiO_2, SiO_2, $Ti_{70}Si_{30}$, $Ti_{50}Si_{50}$, $Ti_{40}Si_{60}$, and TiO_2SiO_2 in the PLA composites films, SEM analysis was performed. SEM micrographs of the fracture surface of PLA and PLA adding 3 wt.% of TiO_2, SiO_2, $Ti_{70}Si_{30}$, $Ti_{50}Si_{50}$, $Ti_{40}Si_{60}$, and TiO_2SiO_2, and 5 wt.% of $Ti_{70}Si_{30}$ are shown in Figure 5a, and the surface of the PLA and PLA composites films after the tensile test is shown in Figure 5b.

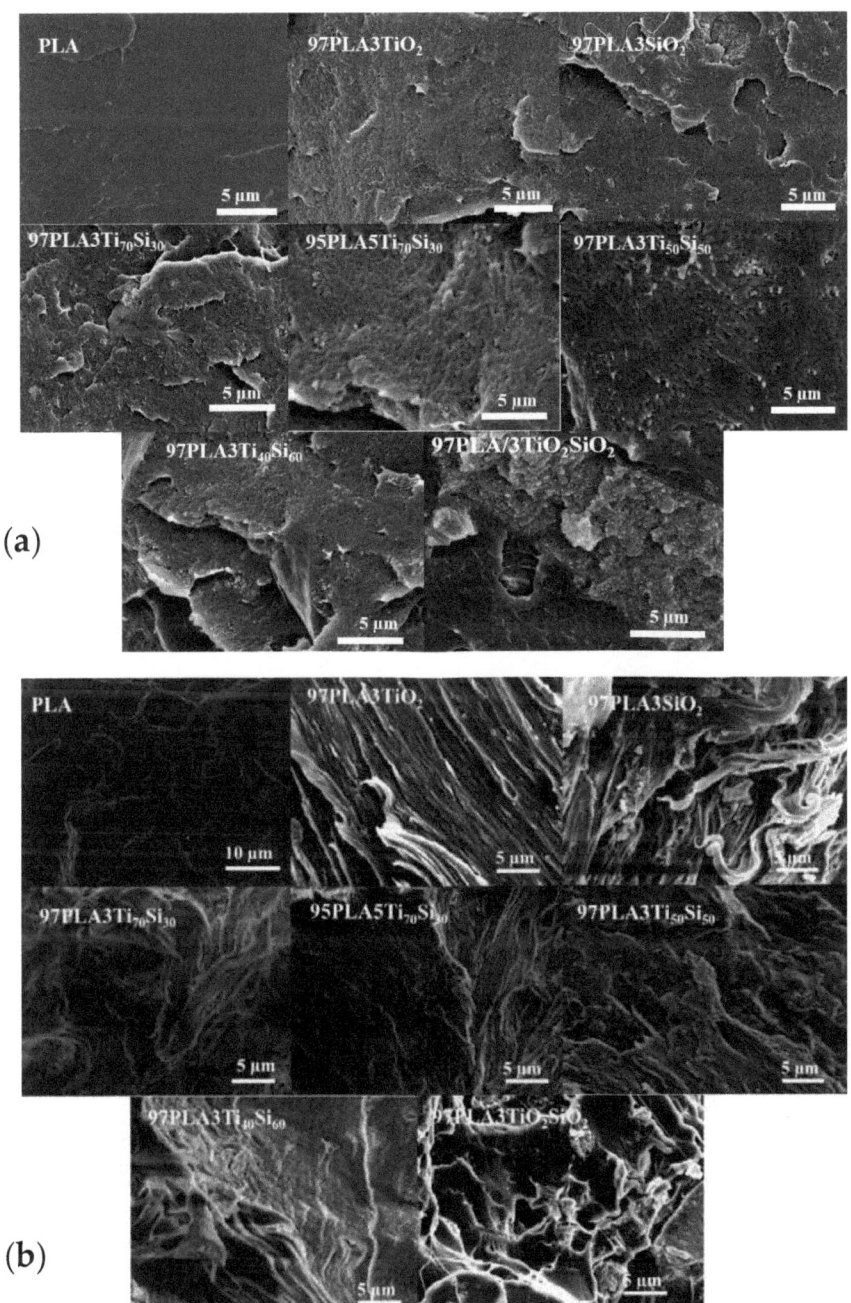

Figure 5. SEM micrographs (×2.5 k, WD = 13–15 mm, acceleration voltage 9–12 kV) of (**a**) the fracture surface and (**b**) after tensile testing of PLA, PLA/TiO$_2$, PLA/SiO$_2$, PLA/Ti$_x$Si$_y$, and PLA/TiO$_2$SiO$_2$ composites.

The SEM results showed that a relatively brittle and comparatively flat surface without holes and air bubbles was found on the fracture surface of the pure PLA films. Meanwhile, SEM images of all PLA composites exhibited roughness caused by adding 3 wt.% of TiO_2, SiO_2, $Ti_{70}Si_{30}$, $Ti_{50}Si_{50}$, $Ti_{40}Si_{60}$, and TiO_2SiO_2 nanoparticles, particularly at 3 wt.% of TiO_2SiO_2 (Figure 5a). The enhancement of the mechanical properties depended on the absence of voids, undamaged position of fillers, interfacial bonding between the fillers and matrix, and the absence of an agglomerate of fillers [49]. However, the white spots in the PLA composites micrographs illustrates the agglomerates of TiO_2, SiO_2, $Ti_{70}Si_{30}$, $Ti_{50}Si_{50}$, $Ti_{40}Si_{60}$, and TiO_2SiO_2 in the PLA matrix, leading to poor mechanical properties. In this work, although some agglomerations could be observed in all PLA composite films, 3wt.% of $Ti_{70}Si_{30}$ and $Ti_{50}Si_{50}$ was still kept intact within the PLA matrix (Figure 5b). As $Ti_{70}Si_{30}$-loading was increased to 5 wt.%, the position of $Ti_{70}Si_{30}$ in PLA was displaced, leading to the formation of a gap between the filler surface and PLA matrix. Therefore, it is an indication of poor interfacial adhesion between $Ti_{70}Si_{30}$ and PLA at high loading [50].

The EDX elemental mapping results (Figure 6b–e) suggested the existence of Ti_xSi_y mixed oxides in the PLA composites. Furthermore, EDX elemental analysis results (Figure 6f) of the selected area also confirmed the spatial distribution of the Si, Ti, and O elements of the Ti_xSi_y mixed oxide in the PLA composite. The distribution of Si and Ti in the particles was relatively uniform in the case of the PLA/$3Ti_{70}Si_{30}$ composite.

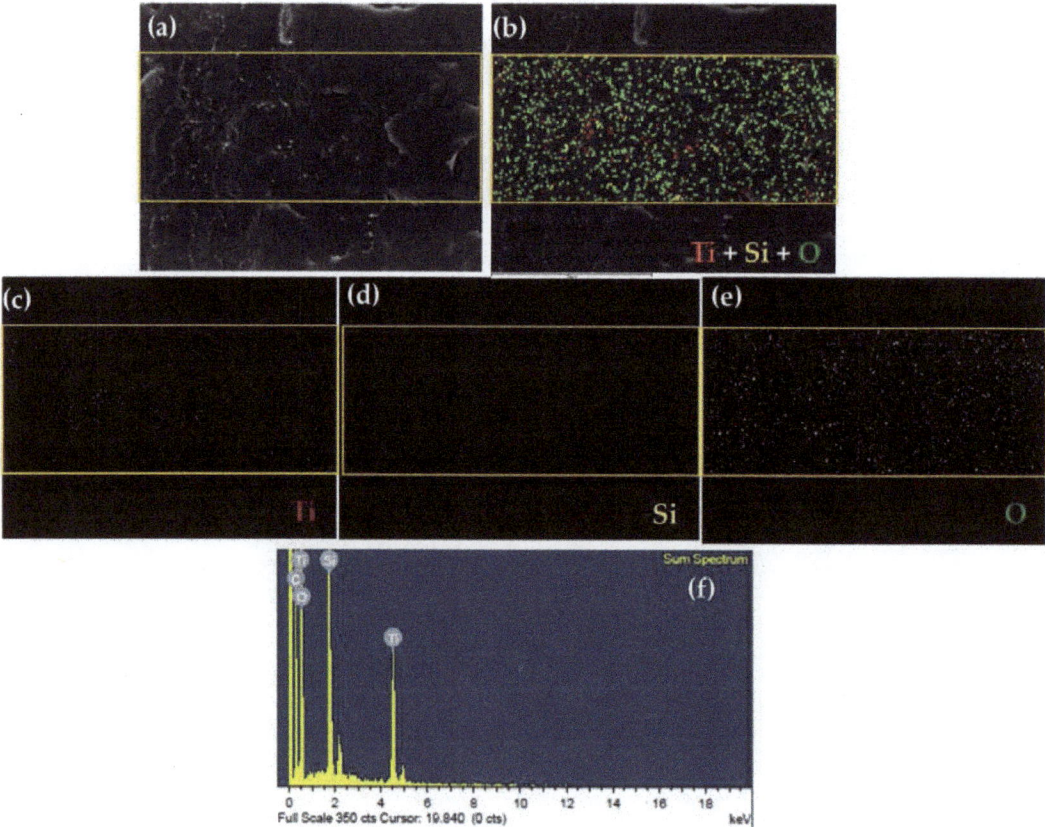

Figure 6. (a) SEM images (b) over all elements in the PLA/$Ti_{70}Si_{30}$ composite and the corresponding elemental mapping analysis of (c) Ti, (d) Si, (e) O, and (f) EDX spectra.

3.4. Water Vapor Transmission Rate (WVTR)

One of the most important properties of bio-based composites films is the ability to evaluate the moisture transfer from the environment to the product. The WVTR of the PLA and PLA composite films is shown in Figure 7. The WVTR of the PLA films was 0.316 g m^{-2} day^{-1} which was lower than the PLA films incorporated with 3wt.% of SiO$_2$, Ti$_{70}$Si$_{30}$, and Ti$_{50}$Si$_{50}$, which were 1.000, 1.023, and 0.523 g m^{-2} day^{-1}. In addition, the WVTR of the PLA/Ti$_{70}$Si$_{30}$ composite film increased with increasing Ti$_{70}$Si$_{30}$ content to 5 wt.%. It is common that, for a solid polymer, the water vapor transmission follows a simple mechanism including adsorbing at the entering face, dissolving, and rapidly creating equilibrium, diffusing through the film, and desorbing at the exit face [51]. The smaller the particle diameter of the nanoparticles is, the more the indirect pathway reducing the diffusion coefficient is produced [52,53]. In other words, the particle diameter is indirectly proportional to the diffusion coefficient. Consequently, the hydrophilicity of the PLA composite incorporating SiO$_2$, Ti$_{70}$Si$_{30}$, and Ti$_{50}$Si$_{50}$ was improved.

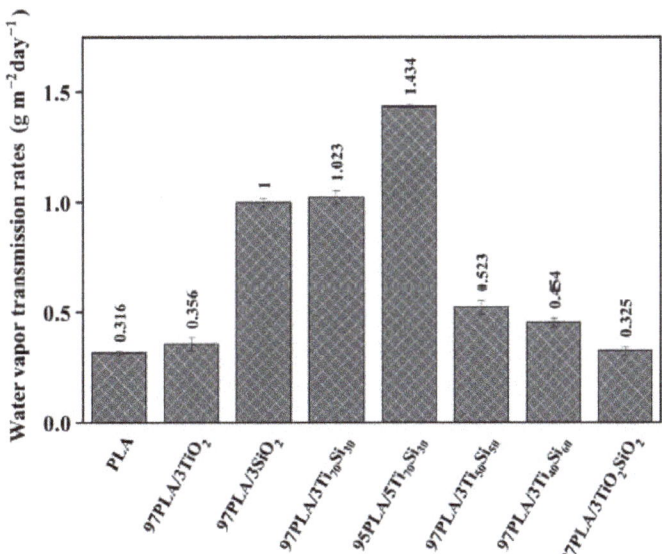

Figure 7. Water vapor transmission rate (WVTR) of PLA, PLA/TiO$_2$, PLA/SiO$_2$, PLA/Ti$_x$Si$_y$, and PLA/TiO$_2$SiO$_2$ composites.

3.5. Photocatalytic Degradation of Methylene Blue (MB)

In this study, the photocatalytic activity of PLA and PLA composite films were investigated by degrading methylene blue (MB). The decomposition of MB might be caused by UV irrigation without the presence of any photocatalyst.

Figure 8 shows changes in the concentration of MB in an aqueous solution under UV irrigation, which was a result of MB decomposition. The presence of 3 wt.% TiO$_2$, SiO$_2$, Ti$_{70}$Si$_{30}$, Ti$_{50}$Si$_{50}$, Ti$_{40}$Si$_{60}$, and TiO$_2$SiO$_2$ in the PLA film matrix exhibited MB degradation more efficiently than using photocatalysis solely. The efficiency to degrade MB was TiO$_2$ > Ti$_{70}$Si$_{30}$ > TiO$_2$SiO$_2$ > Ti$_{50}$Si$_{50}$ > Ti$_{40}$Si$_{60}$ > SiO$_2$, respectively. It was also found that an increase in Ti$_{70}$Si$_{30}$-loading to 5wt.% improved the efficiency of the photocatalytic activity of PLA. The photo-activity of the mixed oxide was evidently increased because the high content of the mixed oxide's increasing surface area of the filler effectively concentrated MB around the nanoparticle and produced high concentrations of organic compounds for the photocatalysis, which consequently improved the photocatalytic activity of PLA. It is known that photocatalytic activity occurs at the surface of the photocatalyst. Therefore, the surface area of PLA composites film, which in turn depends on the size of the nanoparticles, film morphology, and thickness, has an effect on photocatalytic reactivity [54]. The PLA composite film containing TiO_2 can degrade MB more effectively than that containing only photocatalysis. This may be due to two reasons. Firstly, MB was degraded directly by UVC. Secondly, TiO_2 received light energy more than band-gap energy and then the electron in the valence band (VB) was excited to the conduction band (CB), resulting in a generated hole (h^+) (Equation (8)). This hole could oxidize MB (Equation (9)) or oxidized H_2O to produce OH (Equation (10)). The e^- in CB could reduce O_2 at the surface of TiO_2 to generate O_2^- (Equation (11)). The appearance of radical (OH, O_2^-) and h^+ reacted with MB to generate a peroxide derivative and hydroxylate or degrade completely to CO_2 and H_2O [54]. The photodegradation mechanism can be summarized by Equations (8)–(11).

$$TiO_2 + UVC \rightarrow e^- + h^+ \quad (8)$$

$$h^+ + MB \rightarrow CO_2 + H_2O \quad (9)$$

$$h^+ + H_2O \rightarrow H^+ + OH \quad (10)$$

$$e^- + O_2 \rightarrow O_2^- \quad (11)$$

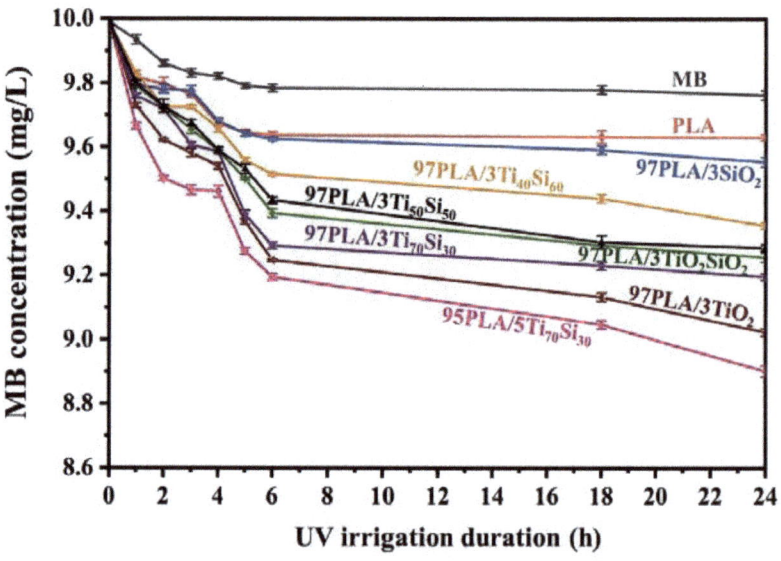

Figure 8. Concentration of methylene blue (MB) due to absorption of PLA, PLA/TiO_2, PLA/SiO_2, PLA/Ti_xSi_y, and PLA/TiO_2SiO_2 composite films under UV irrigation.

The presence of filler in the PLA film matrix shows more efficiency to degrading MB than using only photocatalysis. The enhanced photocatalytic properties of the PLA composites could be mainly attributed to promoted surface adsorption and mass transfer/diffusion, increased light absorption and utilization efficiency, and especially, the higher charge transfer and separation rate by the filler, increased light-harvesting ability, and promoting photoexcitation charge separation, which were the main reasons for improving photocatalytic activity [55,56]. In addition, the photocatalytic activity is influenced by the crystal structure, particle size, specific surface area, and porosity of nanoparticles. So, ultrafine powders of mixed oxide show good catalytic activity. However, agglomeration often takes place, resulting in the reduction or even complete loss of photocatalytic activity.

Due to its photocatalytic activity, $Ti_{70}Si_{30}$ nanoparticles with a high specific surface area (569 m^2 g^{-1}) [33] can degrade MB, making it a suitable material for photocatalytic application.

3.6. Light Transmittance and Opacity Measurements

UV light can create free radicals in products by a photochemical reaction, leading to a negative effect for food. Some of the unfriendly effects include the deterioration of antioxidants, destruction to vitamins and proteins, and a change in color. UV radiation is classified into UV-A (wavelength 320–400 nm), UVB (280–320 nm), and UV-C (200–280 nm) [57,58]. For good optical properties, the optical transmittance should exceed 90% in the visible range (measured from 400 to 800 nm). So, the optical transmittance was measured using a light source with a 600 nm wavelength, which is the central wavelength of the visible range. So, a lot of research uses this wavelength to evaluate opacity [41,59,60]. The addition of TiO_2, SiO_2, $Ti_{70}Si_{30}$, $Ti_{50}Si_{50}$, $Ti_{40}Si_{60}$, and TiO_2SiO_2 into the PLA matrix caused a significant decrease of transmittance in all UV regions (Table 4). The results show that the addition of filler into the PLA matrix caused a significantly decrease of transmittance in all UV regions. The presence of 3wt.% $Ti_{70}Si_{30}$ in the PLA film matrix succeeded in blocking more than 99.6% of the 240, 300, and 360 nm wavelengths, as representatives of UV-C, UV-B, and UV-A radiation, respectively, with a low opacity for the composite film. Moreover, the increase of $Ti_{70}Si_{30}$ oxide-loading up to 5wt.% into PLA improved the UV-blocking efficiency.

Table 4. Transmittance (%) and opacity values of PLA, PLA/TiO_2, PLA/SiO_2, PLA/Ti_xSi_y, and PLA/TiO_2SiO_2 composite films in the visible, UV-A, UV-B, and UV-C regions.

Sample	Transmittance, %				Opacity (AU·nm·mm^{-1})
	UV-C (240 nm)	UV-B (300 nm)	UV-A (360 nm)	Visible (600 nm)	
PLA	1.24	24.47	34.77	51.39	1.16
97PLA/3TiO_2	0.00	0.12	0.41	1.98	6.81
97 PLA/3SiO_2	0.04	9.48	18.03	38.37	2.31
97 PLA/3Ti70Si30	0.00	0.28	0.75	14.00	3.42
95 PLA/5Ti70Si30	0.00	0.00	0.36	4.06	6.33
97 PLA/3Ti50Si50	0.00	0.32	2.37	15.65	3.22
97 PLA/3Ti40Si60	0.00	0.03	0.81	10.48	3.92
97 PLA/3TiO_2SiO_2	0.00	0.19	1.43	7.30	4.74

The PLA films were transparent and colorless, and the addition of SiO_2 to PLA remained transparent, while other PLA composite films showed higher opacity than the pure PLA film. However, the transparency changes related to the increasing $Ti_{70}Si_{30}$ oxide contents from 3 up to 5 wt.% provided totally opaque films by more than two orders of magnitude in the opacity films, but this was still lower than the composite film of PLA with TiO_2. Similarly, the addition of TiO_2 and $Ti_{70}Si_{30}$ made the PLA composites' color appear whiter because of the characteristic whiteness of the TiO_2 and $Ti_{70}Si_{30}$ nanoparticles. Photographs of PLA, PLA/TiO_2, PLA/SiO_2, PLA/Ti_xSi_y, and PLA/TiO_2SiO_2 composites films are shown in Figure 9. These results suggest that the PLA composite produced with $Ti_{70}Si_{30}$ oxide was suitably applied to transparency packaging with good UV-blocking efficiency.

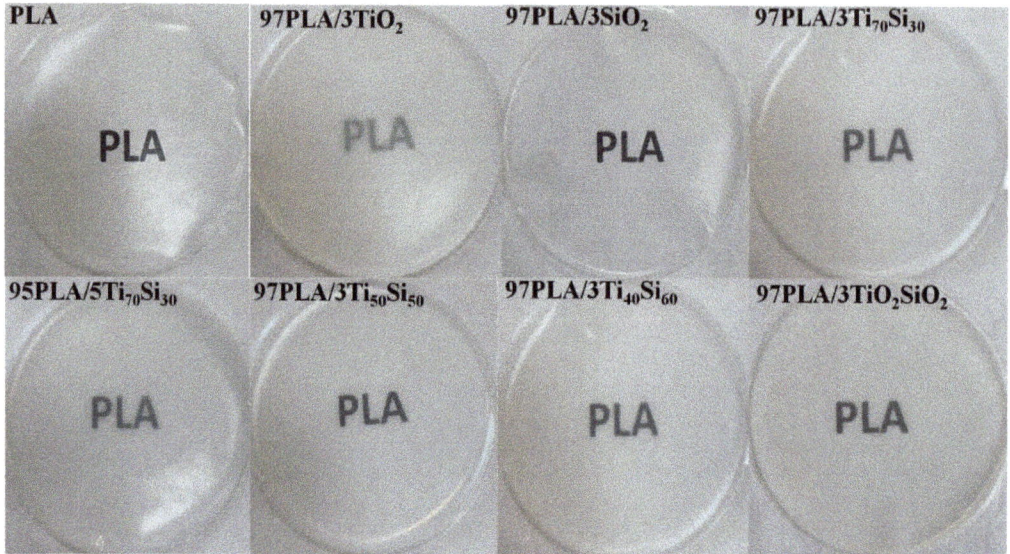

Figure 9. Photographs of films prepared from PLA, PLA/TiO$_2$, PLA/SiO$_2$, PLA/Ti$_x$Si$_y$, and PLA/TiO$_2$SiO$_2$ composites (250 ± 4.68 μm thickness).

3.7. Hydrolytic Degradation

Figure 10 shows the percentage weight loss of PLA and PLA composite films as a function of hydrolytic degradation time. Complete degradation of PLA was achieved at about 1200 min, while all of the PLA composite films were hydrolyzed faster than neat PLA. Interestingly, the incorporation of 3 wt.% of TiO$_2$, SiO$_2$, Ti$_{70}$Si$_{30}$, Ti$_{50}$Si$_{50}$, Ti$_{40}$Si$_{60}$, and TiO$_2$SiO$_2$ exhibited a much higher weight loss as a function of time than neat PLA. The presence of filler induced a much more apparent change of weight loss of hydrolytic degradation, which indicates the enhancement of a hydrolytic degradation ability of the PLA matrix. This was attributed by the addition of nanoparticles, which helped accelerate the hydrolytic degradation of the PLA matrix. Furthermore, 97PLA/3TiO$_2$, 97PLA/3SiO$_2$, 97PLA/3Ti$_{70}$Si$_{30}$, 97PLA/3Ti$_{50}$Si$_{50}$, 97PLA/3Ti$_{40}$Si$_{60}$, and 97PLA/3TiO$_2$SiO$_2$ were fully degraded at 840, 300, 420, 420, and 560 min, respectively. Moreover, the PLA composite containing 5 wt.% of Ti$_{70}$Si$_{30}$ degraded faster than all the composites and it was fully degraded in approximately 240 min. Consequently, it could be concluded that the rate of the hydrolytic degradation of PLA composite films can be controlled by the filler content. This result is in agreement with Buzarovka and Grozdanov (2012) [12].

3.8. In Vitro Degradation

The degradation of PLA in PLA composites involves several processes such as water uptake, ester cleavage and formation if there are oligomer fragments, the dissolution of the oligomer fragment, etc., [61]; as a result, factors affecting the hydrolysis tendency of PLA would control the degradation of PLA. The long-term hydrolytic degradation of PLA and PLA composite films in a phosphate buffered saline (PBS) (pH = 7.4 ± 0.2) solution at 37 °C was evaluated by mass loss in 56 days. Figure 11 illustrates the mass loss of the PLA and PLA composite with the degradation time. From 0 to 14 days, all of the samples exhibited a dramatic increase in mass loss with increasing immersion time. After this period, the mass loss of all samples accelerated gradually. PLA incorporating with 3 wt.% of TiO$_2$, SiO$_2$, Ti$_{70}$Si$_{30}$, Ti$_{50}$Si$_{50}$, Ti$_{40}$Si$_{60}$, and TiO$_2$SiO$_2$ exhibited higher weight loss as a function of immersion time than neat PLA. In this case, TiO$_2$, SiO$_2$, Ti$_{70}$Si$_{30}$, Ti$_{50}$Si$_{50}$, Ti$_{40}$Si$_{60}$, and TiO$_2$SiO$_2$ dispersed in the PLA matrix, the water molecules penetrated easier within the samples to generate the degradation process and might have be absorbed into the gap between the conglomeration of the nanoparticles due to the agglomeration of the nanofiller. Consequently, a long time is spent on diffusion into the PLA matrix. Therefore, the degradation rate increased in the first period and reached its maximum [62]. In addition, the mass loss of the PLA composite was also found to increase with an inclining amount of $Ti_{70}Si_{30}$ to 5 wt.%. Consequently, it could be concluded that the rate of the long-term degradation of the PLA composite films depended upon the content of the mixed oxide loading. This result was connected to the hydrophilicity of TiO_2, SiO_2, $Ti_{70}Si_{30}$, $Ti_{50}Si_{50}$, $Ti_{40}Si_{60}$, and TiO_2SiO_2, as well as the high-water absorption of the composites [63]. Regarding changes in the tensile strength, elongation at the break and Young's modulus of the PLA and PLA composite films are shown in Table 5. This table shows that the tensile strength and elongation at the break of the PLA and all of the PLA composite films decreased significantly after 28 days of in vitro degradation. The result suggests that the PLA and PLA composite films were mechanically stable during 28 days of in vitro degradation. The tensile strengths of the PLA and PLA composites decreased after 28 days of degradation with microcracks appearing on part of their surfaces. It was supposed that the PLA composites would lose their mechanical strengths quickly after the microcracks developed over the whole area of the fibers [64].

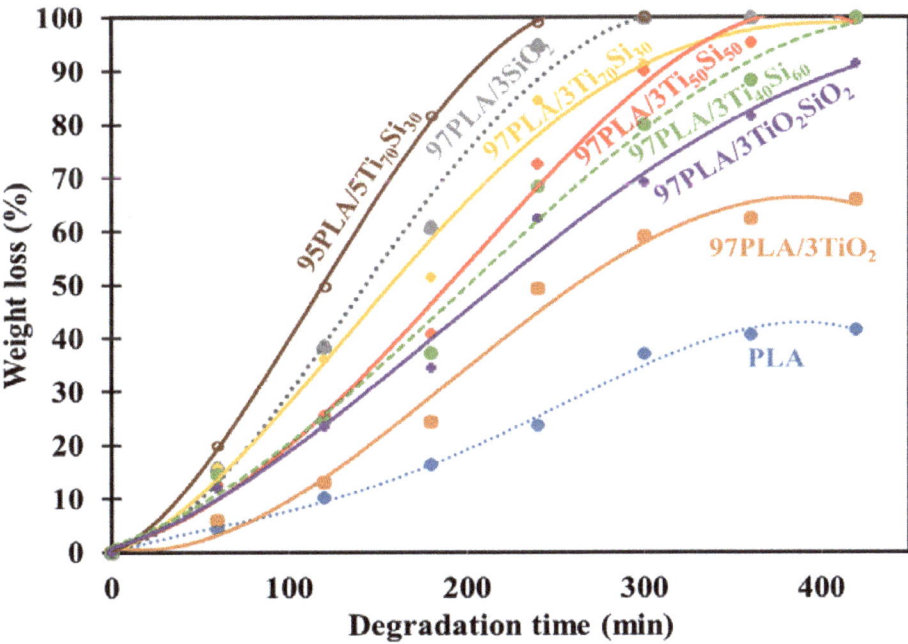

Figure 10. Weight loss of hydrolytic degradation of PLA, PLA/TiO_2, PLA/SiO_2, and PLA/Ti_xSi_y composite films as functions of degradation time.

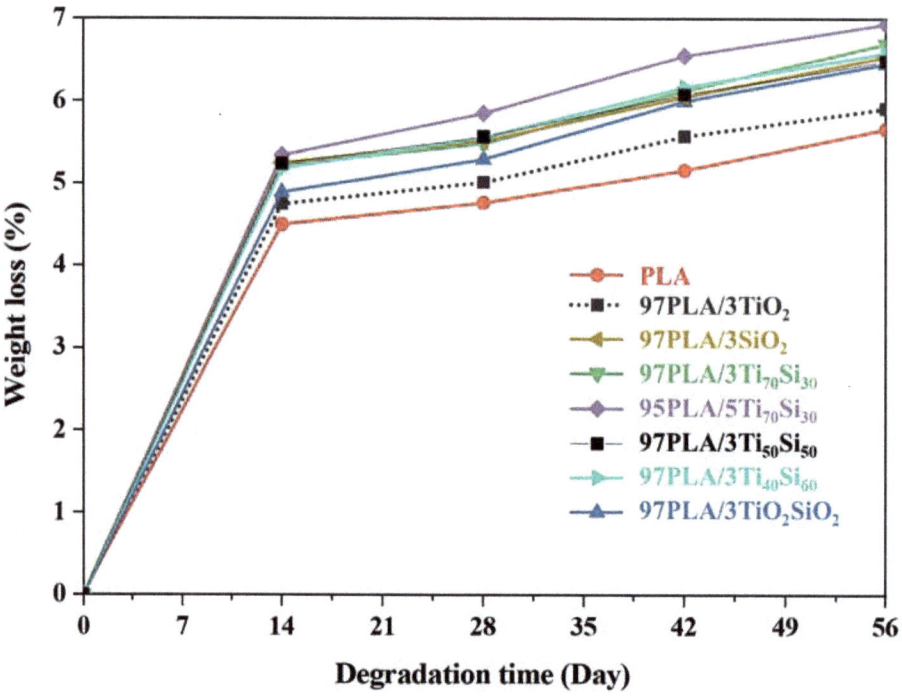

Figure 11. Weight loss of PLA, PLA/TiO$_2$, PLA/SiO$_2$, PLA/Ti$_x$Si$_y$, and PLA/TiO$_2$SiO$_2$ composites films after different periods of in vitro degradation.

Table 5. Tensile properties of properties of PLA, PLA/TiO$_2$, PLA/SiO$_2$, and PLA/Ti$_x$Si$_y$ composite films after different period of in vitro degradation.

Time (Day)	Sample	Tensile Strength (MPa)	Elongation at Break (%)	Young's Modulus (GPa)
0	PLA	33.94 ± 1.38	19.62 ± 8.64	0.98 ± 0.23
	97PLA/3TiO$_2$	28.22 ± 1.47	18.1 ± 4.24	1.01 ± 0.04
	97PLA/3SiO$_2$	29.75 ± 1.38	11.8 ± 4.96	1.07 ± 0.10
	97PLA/3Ti$_{70}$Si$_{30}$	35.64 ± 3.21	11.12 ± 1.81	1.18 ± 0.06
	95PLA/5Ti$_{70}$Si$_{30}$	32.06 ± 2.29	9.88 ± 2.81	1.01 ± 0.10
	97PLA/3Ti$_{50}$Si$_{50}$	35.31 ± 2.01	9.61 ± 2.57	1.29 ± 0.12
	97PLA/3Ti$_{40}$Si$_{60}$	30.67 ± 1.74	22.30 ± 2.53	1.02 ± 0.05
	97PLA/3TiO$_2$SiO$_2$	29.00 ± 1.42	58.37 ± 5.00	0.96 ± 0.07
14	PLA	29.43 ± 3.02	3.81 ± 1.61	1.51 ± 0.14
	97PLA/3TiO$_2$	23.05 ± 2.87	2.58 ± 0.54	1.46 ± 0.31
	97PLA/3SiO$_2$	23.53 ± 2.07	2.78 ± 0.27	1.39 ± 0.40
	97PLA/3Ti$_{70}$Si$_{30}$	33.56 ± 3.45	3.69 ± 0.61	1.74 ± 0.14
	95PLA/5Ti$_{70}$Si$_{30}$	31.75 ± 2.31	3.55 ± 0.70	1.49 ± 0.03
	97PLA/3Ti$_{50}$Si$_{50}$	30.75 ± 2.51	3.45 ± 0.80	1.29 ± 0.04
	97PLA/3Ti$_{40}$Si$_{60}$	27.99 ± 1.18	4.46 ± 0.95	1.21 ± 0.44
	97PLA/3TiO$_2$SiO$_2$	24.83 ± 2.35	3.25 ± 0.38	1.44 ± 0.56

Table 5. Cont.

Time (Day)	Sample	Tensile Strength (MPa)	Elongation at Break (%)	Young's Modulus (GPa)
28	PLA	19.95 ± 2.45	2.06 ± 0.38	1.75 ± 0.25
	97PLA/3TiO$_2$	18.69 ± 2.02	1.98 ± 0.23	2.02 ± 0.00
	97PLA/3SiO$_2$	n/a *	n/a	n/a
	97PLA/3Ti$_{70}$Si$_{30}$	29.35 ± 2.07	2.56 ± 0.12	1.91 ± 0.39
	95PLA/5Ti$_{70}$Si$_{30}$	10.03 ± 2.40	1.11 ± 0.33	n/a
	97PLA/3Ti$_{50}$Si$_{50}$	18.03 ± 2.80	2.11 ± 0.23	n/a
	97PLA/3Ti$_{40}$Si$_{60}$	17.56 ± 2.01	2.08 ± 0.13	n/a
	97PLA/3TiO$_2$SiO$_2$	17.39 ± 2.46	1.79 ± 0.35	n/a
42	PLA	6.95 ± 2.49	1.33 ± 0.19	n/a
	97PLA/3TiO$_2$	0.26 ± 0.11	0.81 ± 0.10	n/a
	97PLA/3SiO$_2$	n/a	n/a	n/a
	97PLA/3Ti$_{70}$Si$_{30}$	15.97 ± 2.62	1.52 ± 0.11	n/a
	95PLA/5Ti$_{70}$Si$_{30}$	n/a	n/a	n/a
	97PLA/3Ti$_{50}$Si$_{50}$	5.97 ± 2.63	0.62 ± 0.52	n/a
	97PLA/3Ti$_{40}$Si$_{60}$	n/a	n/a	n/a
	97PLA/3TiO$_2$SiO$_2$	3.24 ± 1.79	1.12 ± 0.64	n/a

* n/a = not available.

3.9. Antimicrobial Activity

Metal oxides hold greater antibacterial efficiency, and their reinforcement in polymer composites expressively expands the antimicrobial properties of the film, which is desired in biomedical and food packaging applications. Bacteria are generally characterized by the cell membrane, which is composed mostly of a homogeneous peptidoglycan layer (which consists of amino acids and sugar). Gram-positive bacteria such as *Staphylococcus aureus* have one cytoplasm membrane with multilayers of the peptidoglycan polymer and a thicker cell wall (20–80 nm) [65], whereas in Gram-negative bacteria such as *Escherichia coli*, the bacteria wall is composed of two cell membranes, and an outer membrane and a plasma membrane with a thin layer of peptidoglycan with a thickness of 7–8 nm [65].

TiO$_2$ nanoparticles are known for their antibacterial activity, and recent studies have confirmed their efficiency as antibacterial agents [65,66]. As a result, the antibacterial activity of PLA incorporated with 3 wt.% of Ti$_{70}$Si$_{30}$, Ti$_{50}$Si$_{50}$, and TiO$_2$SiO$_2$, and 5 wt.% of Ti$_{70}$Si$_{30}$, to form composites was compared to the antibacterial activity of PLA adding 3 wt.% of TiO$_2$. The results of the antimicrobial activity of the Gram-negative bacteria (*Escherichia coli* or *E. coli*) and Gram-positive bacteria (*Staphylococcus aureus* or *S. aureus*) of the PLA and PLA composites are shown in Tables 6 and 7, respectively. The number of bacteria *Escherichia coli* and bacteria *Staphylococcus aureus* on the PLA and PLA composite films at time 0 h (at dilution 10^{-3}) and 24 h (at dilution 10^0) are shown in Figure 12.

Table 6. Antimicrobial activity of Gram-negative bacteria (*Escherichia coli*) of PLA, PLA/TiO$_2$, PLA/Ti$_x$Si$_y$, and PLA/TiO$_2$SiO$_2$ composites.

Samples	Blank (Ut) (t = 24 h) Log CFU/mL	Sample (At) (t = 24 h) Log CFU/mL	Antimicrobial Activity [a] (R)
PLA	5.96 ± 0.01	5.87 ± 0.02	0.09
97PLA/3TiO$_2$	5.96 ± 0.01	0.00 ± 0.00	5.96
97PLA/3Ti$_{70}$Si$_{30}$	5.96 ± 0.01	0.00 ± 0.00	5.96
95PLA/5Ti$_{70}$Si$_{30}$	5.96 ± 0.01	0.00 ± 0.00	5.96
97PLA/3Ti$_{50}$Si$_{50}$	5.96 ± 0.01	3.21 ± 0.04	2.75
97PLA/3TiO$_2$SiO$_2$	5.96 ± 0.01	2.53 ± 0.04	3.43

[a] Antibacterial activity (R) $\geq$ 2 = antimicrobial effectiveness.

Table 7. Antimicrobial activity of Gram-positive bacteria (*Staphylococcus aureus*) of PLA, PLA/TiO$_2$, PLA/Ti$_x$Si$_y$, and PLA/TiO$_2$SiO$_2$ composites.

Samples	Blank (Ut) (t = 24 h) Log CFU/mL	Sample (At) (t = 24 h) Log CFU/mL	Antimicrobial Activity [a] (R)
PLA	4.35 ± 0.04	4.35 ± 0.04	0
PLA/3TiO$_2$	4.35 ± 0.04	0.00 ± 0.00	4.35
PLA/3Ti$_{70}$Si$_{30}$	4.35 ± 0.04	2.96 ± 0.01	2.04
PLA/5Ti$_{70}$Si$_{30}$	4.35 ± 0.04	0.00 ± 0.00	4.35
PLA/3Ti$_{50}$Si$_{50}$	4.35 ± 0.04	3.17 ± 0.08	1.83
PLA/3TiO$_2$SiO$_2$	4.35 ± 0.04	3.21 ± 0.01	1.79

[a] Antibacterial activity (R) $\geq$ 2 = antimicrobial effectiveness.

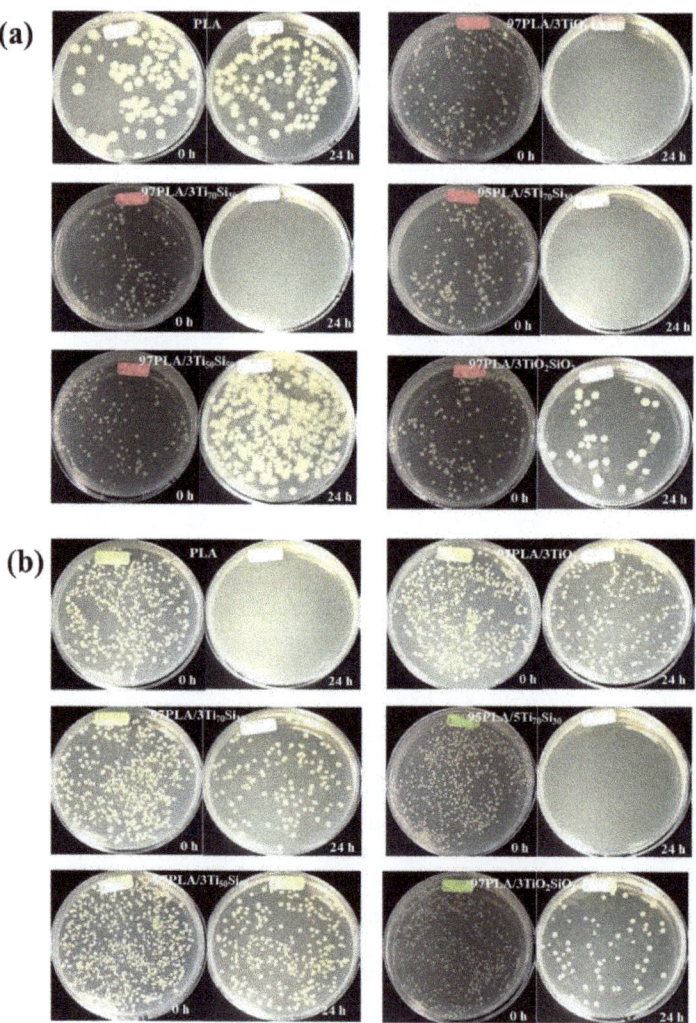

Figure 12. The number of (**a**) bacteria *Escherichia coli* and (**b**) bacteria *Staphylococcus aureus* on PLA and PLA composite films at time 0 h (at dilution 10^{-3}) and 24 h (at dilution 10^0).

The antimicrobial activity (R) values of Neat PLA film against *E. coli* and *S. aureus* were 0.09 and 0, respectively. This result shows that PLA has no significant antibacterial effects on *E. coli* and *S. aureus*. However, the 97PLA/3TiO$_2$ composite film exhibited the highest antimicrobial activity (R = 5.96 against *E. coli* and R = 4.35 against *S. aureus*). As shown in Table 6, all of PLA composites exhibited an antimicrobial activity agent against *E. coli* (if R ≥ 2 antimicrobial effectiveness). Moreover, the results confirm that PLA incorporated with 3wt.% of Ti$_{70}$Si$_{30}$ has sufficient antimicrobial effectiveness. Likewise, PLA with the addition of 3 wt.% of TiO$_2$ and Ti$_{70}$Si$_{30}$ exhibited an antibacterial effect on *S. aureus*. This was due to TiO$_2$ and Ti$_x$Si$_y$ oxide utilizing a similar mechanism against bacterial growth by directly damaging the bacterial surface. Adding TiO$_2$ and Ti$_x$Si$_y$ oxide into PLA could lead to reduced soluble protein expression by suppressing the synthesis of nucleic acids. Thus, TiO$_2$ and Ti$_x$Si$_y$ oxide antibacterial action against *S. aureus* was probably through inhibiting the synthesis of nucleic acid, thereby reducing protein synthesis against bacterial growth [67]. The mechanism referred to as the antimicrobial action of TiO$_2$ is commonly associated with reactive oxygen species (ROS) with high oxidative potentials produced under a band-gap irradiation photo-induced charge in the presence of O$_2$ [68]. Pleskova et al. investigated the bactericidal activity of the TiO$_2$ film and discovered that *S. aureus* is swelled by TiO$_2$ through damaging the cell membrane [69]. In addition, the increasement of Ti$_{70}$Si$_{30}$ loading to 5 wt.% improved the antimicrobial effectiveness of the PLA composites. However, 97PLA/3Ti$_{50}$Si$_{50}$ and 97PLA/3TiO$_2$SiO$_2$ exhibited low antimicrobial activity (R = 2.75, 3.43 against *E. coli* and R = 1.83, 1.79 against *S. aureus*). This was due to a higher SiO$_2$ content in Ti$_{50}$Si$_{50}$ and TiO$_2$SiO$_2$, resulting in a lower efficiency of antimicrobial activity. In addition, the variation in the microorganism structure between the Gram-negative *(E. coli)* and Gram-positive (*S. aureus*) bacteria may explain the difference in the antibacterial effect of samples against *E. coli* and *S. aureus*. Both bacteria have similar internal, but different external structures. The peptidoglycan layer of Gram-positive bacteria is thick and includes teichoic and lipoteichoic acids. A Gram-negative bacterium has a thin peptidoglycan layer and an outer membrane made up of proteins, phospholipids, and lipopolysaccharides. Therefore, *S. aureus* needs longer contact time or higher catalyst concentrations to achieve the same effect as *E. coli* [70].

4. Conclusions

The aim of this study was to examine the influence of 3wt.% of TiO$_2$, SiO$_2$, Ti$_{70}$Si$_{30}$, Ti$_{50}$Si$_{50}$, Ti$_{40}$Si$_{60}$, and TiO$_2$SiO$_2$, and 5 wt.% of Ti$_{70}$Si$_{30}$ on the mechanical properties, thermal properties, morphological properties, degradation behavior, and antimicrobial activity of PLA. The PLA and PLA composites films were obtained by the solvent casting method. The addition of Ti$_{70}$Si$_{30}$ and Ti$_{50}$Si$_{50}$ into the PLA film slightly improved the tensile strength and Young's modulus of PLA. The incorporation of 5 wt.% of Ti$_{70}$Si$_{30}$ was found to decrease the cold crystallization temperature and increased the degree of crystallinity of PLA. It can be concluded that Ti$_{70}$Si$_{30}$ nanoparticles can act as a good nucleating agents for PLA. The thermal stability of PLA was enhanced with the incorporation of TiO$_2$ and SiO$_2$. The water vapor transmission rate (WVTR) of PLA was significantly increased by the incorporation of SiO$_2$, Ti$_{70}$Si$_{30}$, and Ti$_{50}$Si$_{50}$ nanoparticles. This is due to the hydrophilicity of the nanoparticles. In addition, efficiency of degrading MB is TiO$_2$ > Ti$_{70}$Si$_{30}$ > TiO$_2$SiO$_2$ > Ti$_{50}$Si$_{50}$ > Ti$_{40}$Si$_{60}$ > SiO$_2$, respectively. Moreover, the increase in Ti$_{70}$Si$_{30}$ loading to 5 wt.% improved the efficiency of the photocatalytic activity of PLA. All of the nanoparticles were able to remove UV light and, in particular, TiO$_2$ and Ti$_{70}$Si$_{30}$ enhanced a stronger higher UV-shielding potential. The hydrolytic degradation and in vitro degradation of PLA are important properties of the variety of application such as biomedical application and food packaging. PLA incorporated with 3wt.% of SiO$_2$, Ti$_{70}$Si$_{30}$, Ti$_{50}$Si$_{50}$, Ti$_{40}$Si$_{60}$, and TiO$_2$SiO$_2$ exhibited much higher weight loss as a function of time than neat PLA. The weight loss of the PLA composite was also found to increase with increasing Ti$_{70}$Si$_{30}$ to 5 wt.%. Furthermore, PLA with the addition of TiO$_2$ and Ti$_{70}$Si$_{30}$ exhibited an excellent antibacterial effect on Gram-negative bacteria (*E. coli*) and Gram-positive bacteria (*S. aureus*), indicating the improved antimicrobial effectiveness of PLA composites.

Supplementary Materials: The following supporting information can be downloaded at: https://www.mdpi.com/article/10.3390/polym14163310/s1, Figure S1: Curve of (a) TGA thermogram, (b) DTG Thermogram of PLA, PLA/TiO$_2$, PLA/SiO$_2$, PLA/Ti$_x$Si$_y$, and PLA/TiO$_2$SiO$_2$ composites.; Figure S2: The changes in tensile properties of PLA, PLA/TiO$_2$, PLA/SiO$_2$, and PLA/Ti$_x$Si$_y$ composite films after different periods of in vitro degradation.

Author Contributions: Conceptualization, Y.R.; methodology, A.T., C.R. and Y.R.; validation, A.T., C.R. and Y.R.; formal analysis, A.T.; investigation, A.T., C.R. and Y.R.; resources, A.T., C.R. and Y.R.; data curation, A.T.; writing—original draft preparation, A.T.; writing—review and editing, A.T., C.R. and Y.R.; visualization, C.R. and Y.R.; supervision, C.R. and Y.R.; project administration, C.R. and Y.R.; funding acquisition, C.R. and Y.R. All authors have read and agreed to the published version of the manuscript.

Funding: This research was funded by Thailand Science Research and Innovation (TSRI), grant number 42853.

Institutional Review Board Statement: Not applicable.

Informed Consent Statement: Not applicable.

Data Availability Statement: Not applicable.

Acknowledgments: The authors are grateful to Suranaree University of Technology (SUT); to the Center of Excellence on Petrochemical and Materials Technology (PETROMAT); to the Science, Research and Innovation Promotion Fund from Thailand Science Research and Innovation (TSRI); and to the Research Center for Biocomposite Materials for the Medical Industry and Agricultural and Food Industry for the financial support.

Conflicts of Interest: The authors declare no conflict of interest.

References

1. Cadar, O.; Paul, M.; Roman, C.; Miclean, M.; Majdik, C. Biodegradation behaviour of poly(lactic acid) and (lactic acid-ethylene glycol-malonic or succinic acid) copolymers under controlled composting conditions in a laboratory test system. *Polym. Degrad. Stab.* **2012**, *97*, 354–357. [CrossRef]
2. Lim, L.T.; Auras, R.; Rubino, M. Processing technologies for poly(lactic acid). *Prog. Polym. Sci.* **2008**, *33*, 820–852. [CrossRef]
3. Deroiné, M.; Le Duigou, A.; Corre, Y.M.; Le Gac, P.Y.; Davies, P.; César, G.; Bruzaud, S. Accelerated ageing of polylactide in aqueous environments: Comparative study between distilled water and seawater. *Polym. Degrad. Stab.* **2014**, *108*, 319–329. [CrossRef]
4. Wang, Y.; Kong, Y.; Zhao, Y.; Feng, Q.; Wu, Y.; Tang, X.; Gu, X.; Yang, Y. Electrospun, reinforcing network-containing, silk fibroin-based nerve guidance conduits for peripheral nerve repair. *J. Biomater. Tissue Eng.* **2016**, *6*, 53–60. [CrossRef]
5. Athanasiou, K.A.; Niederauer, G.G.; Agrawal, C.M. Sterilization, toxicity, biocompatibility and clinical applications of polylactic acid/polyglycolic acid copolymers. *Biomaterials* **1996**, *17*, 93–102. [CrossRef]
6. Cheung, H.Y.; Lau, K.T. Study on a silkworm silk fiber/biodegradable polymer biocomposite. In Proceedings of the ICCM International Conferences on Composite Materials, Kyoto, Japan, 8–13 July 2007.
7. Yeh, J.T.; Huang, C.Y.; Chai, W.L.; Chen, K.N. Plasticized properties of poly (lactic acid) and triacetine blends. *J. Appl. Polym. Sci.* **2009**, *112*, 2757–2763. [CrossRef]
8. Burg, K.J.L.; Holder, W.D.; Culberson, C.R.; Beiler, R.J.; Greene, K.G.; Loebsack, A.B.; Roland, W.D.; Mooney, D.J.; Halberstadt, C.R. Parameters affecting cellular adhesion to polylactide films. *J. Biomater. Sci. Polym. Ed.* **1999**, *10*, 147–161. [CrossRef]
9. Rasal, R.M.; Janorkar, A.V.; Hirt, D.E. Poly(lactic acid) modifications. *Prog. Polym. Sci.* **2010**, *35*, 338–356. [CrossRef]
10. Paul, M.A.; Delcourt, C.; Alexandre, M.; Degée, P.; Monteverde, F.; Dubois, P. Polylactide/montmorillonite nanocomposites: Study of the hydrolytic degradation. *Polym. Degrad. Stab.* **2005**, *87*, 535–542. [CrossRef]
11. Shogren, R.L.; Doane, W.M.; Garlotta, D.; Lawton, J.W.; Willett, J.L. Biodegradation of starch/polylactic acid/poly(hydroxyester-ether) composite bars in soil. *Polym. Degrad. Stab.* **2003**, *79*, 405–411. [CrossRef]
12. Buzarovska, A.; Grozdanov, A. Biodegradable poly(L-lactic acid)/TiO$_2$ nanocomposites: Thermal properties and degradation. *J. Appl. Polym. Sci.* **2012**, *123*, 2187–2193. [CrossRef]
13. Yang, K.K.; Wang, X.L.; Wang, Y.Z. Progress in nanocomposite of biodegradable polymer. *J. Ind. Eng. Chem.* **2007**, *13*, 485–500.
14. Huang, S.M.; Hwang, J.J.; Liu, H.J.; Lin, L.H. Crystallization behavior of poly(L-lactic acid)/montmorillonite nanocomposites. *J. Appl. Polym. Sci.* **2010**, *117*, 434–442. [CrossRef]

15. Tang, H.; Chen, J.B.; Wang, Y.; Xu, J.Z.; Hsiao, B.S.; Zhong, G.J.; Li, Z.M. Shear flow and carbon nanotubes synergistically Induced nonisothermal crystallization of poly(lactic acid) and its application in injection molding. *Biomacromolecules* **2012**, *13*, 3858–3867. [CrossRef]
16. Murariu, M.; Paint, Y.; Murariu, O.; Raquez, J.-M.; Bonnaud, L.; Dubois, P. Current progress in the production of PLA–ZnO nanocomposites: Beneficial effects of chain extender addition on key properties. *J. Appl. Polym. Sci.* **2015**, *132*, 42480. [CrossRef]
17. Hakim, R.H.; Cailloux, J.; Santana, O.O.; Bou, J.; Sánchez-Soto, M.; Odent, J.; Raquez, J.M.; Dubois, P.; Carrasco, F.; Maspoch, M.L. PLA/SiO$_2$ composites: Influence of the filler modifications on the morphology, crystallization behavior, and mechanical properties. *J. Appl. Polym. Sci.* **2017**, *134*, 45367. [CrossRef]
18. Wu, G.; Liu, S.; Jia, H.; Dai, J. Preparation and properties of heat resistant polylactic acid (PLA)/Nano-SiO$_2$ composite filament. *J. Wuhan Univ. Technol. Mater. Sci. Ed.* **2016**, *31*, 164–171. [CrossRef]
19. Xiu, H.; Qi, X.; Bai, H.; Zhang, Q.; Fu, Q. Simultaneously improving toughness and UV-resistance of polylactide/titanium dioxide nanocomposites by adding poly(ether)urethane. *Polym. Degrad. Stab.* **2017**, *143*, 136–144. [CrossRef]
20. Zapata, P.A.; Palza, H.; Cruz, L.S.; Lieberwirth, I.; Catalina, F.; Corrales, T.; Rabagliati, F.M. Polyethylene and poly(ethylene-co-1-octadecene) composites with TiO$_2$ based nanoparticles by metallocenic "in situ" polymerization. *Polymer* **2013**, *54*, 2690–2698. [CrossRef]
21. Fonseca, C.; Ochoa, A.; Ulloa, M.T.; Alvarez, E.; Canales, D.; Zapata, P.A. Poly(lactic acid)/TiO$_2$ nanocomposites as alternative biocidal and antifungal materials. *Mater. Sci. Eng. C* **2015**, *57*, 314–320. [CrossRef]
22. Wu, F.; Lan, X.; Ji, D.; Liu, Z.; Yang, W.; Yang, M. Grafting polymerization of polylactic acid on the surface of nano-SiO$_2$ and properties of PLA/PLA-grafted-SiO$_2$ nanocomposites. *J. Appl. Polym. Sci.* **2013**, *129*, 3019–3027. [CrossRef]
23. Serenko, O.A.; Muzafarov, A.M. Polymer composites with surface modified SiO$_2$ nanoparticles: Structures, properties, and promising applications. *Polym. Sci. Ser. C* **2016**, *58*, 93–101. [CrossRef]
24. Ha, S.W.; Weitzmann, M.N.; Beck, G.R., Jr. Applications of silica-based nanomaterials in dental and skeletal biology. In *Nanobiomaterials in Clinical Dentistry*; Karthikeyan, W., Ahmed, J.K., Eds.; Elsevier: Amsterdam, The Netherlands, 2013; pp. 69–91.
25. Galindo, I.R.; Viveros, T.; Chadwick, D. Synthesis and characterization of titania-based ternary and binary mixed oxides prepared by the sol–gel method and their activity in 2-propanol dehydration. *Ind. Eng. Chem. Res.* **2007**, *46*, 1138–1147. [CrossRef]
26. Liau, S.Y.; Read, D.C.; Pugh, W.J.; Furr, J.; Russell, A.D. Interaction of silver nitrate with readily identifiable groups: Relationship to the antibacterialaction of silver ions. *Lett. Appl. Microbiol.* **1997**, *25*, 279–283. [CrossRef]
27. Marra, A.; Silvestre, C.; Duraccio, D.; Cimmino, S. Polylactic acid/zinc oxide biocomposite films for food packaging application. *Int. J. Biol. Macromol.* **2016**, *88*, 254–262. [CrossRef]
28. Sirelkhatim, A.; Mahmud, S.; Seeni, A.; Kaus, N.H.M.; Ann, L.C.; Bakhori, S.K.M.; Hasan, H.; Mohamad, D. Review on zinc oxide nanoparticles: Antibacterial activity and toxicity mechanism. *Nano-Micro Lett.* **2015**, *7*, 219–242. [CrossRef]
29. Li, W.; Zhang, C.; Chi, H.; Li, L.; Lan, T.; Han, P.; Chen, H.; Qin, Y. Development of antimicrobial packaging film made from poly(lactic acid) incorporating titanium dioxide and silver nanoparticles. *Molecules* **2017**, *22*, 1170. [CrossRef]
30. Shebi, A.; Lisa, S. Evaluation of biocompatibility and bactericidal activity of hierarchically porous PLA-TiO$_2$ nanocomposite films fabricated by breath-figure method. *Mater. Chem. Phys.* **2019**, *230*, 308–318. [CrossRef]
31. Swaroop, C.; Shukla, M. Polylactic acid/magnesium oxide nanocomposite films for food packaging applications. In Proceedings of the 21st International Conference on Composite Materials, Xi'an, China, 20–25 August 2017.
32. Teamsinsungvon, A.; Sutapun, W.; Ruksakulpiwat, C.; Ruksakulpiwat, Y. Preparation of titanium-silica binary mixed oxide to use as a filler in poly (lactic acid). *Suranaree J. Sci. Technol.* **2019**, *26*, 31–36.
33. Teamsinsungvon, A.; Ruksakulpiwat, C.; Amonpattaratkit, P.; Ruksakulpiwat, Y. Structural Characterization of Titanium–Silica Oxide Using Synchrotron Radiation X-ray Absorption Spectroscopy. *Polymers* **2022**, *14*, 2729. [CrossRef]
34. Matijević, E.; Budnik, M.; Meites, L. Preparation and mechanism of formation of titanium dioxide hydrosols of narrow size distribution. *J. Colloid Interface Sci.* **1977**, *61*, 302–311. [CrossRef]
35. Vorkapic, D.; Matsoukas, T. Effect of Temperature and Alcohols in the Preparation of Titania Nanoparticles from Alkoxides. *J. Am. Ceram. Soc.* **1998**, *81*, 2815–2820. [CrossRef]
36. Cheung, H.Y.; Lau, K.T.; Tao, X.M.; Hui, D. A potential material for tissue engineering: Silkworm silk/PLA biocomposite. *Compos. Part B Eng.* **2008**, *39*, 1026–1033. [CrossRef]
37. Boonying, S.; Sutapun, W.; Suppakarn, N.; Ruksakulpiwat, Y. Crystallization Behavior of Vetiver Grass Fiber-Polylactic Acid Composite. *Adv. Mater. Res.* **2012**, *410*, 55–58. [CrossRef]
38. Tang, Z.; Zhang, C.; Liu, X.; Zhu, J. The crystallization behavior and mechanical properties of polylactic acid in the presence of a crystal nucleating agent. *J. Appl. Polym. Sci.* **2012**, *125*, 1108–1115. [CrossRef]
39. Elsner, J.; Shefy-Peleg, A.; Zilberman, M. Novel biodegradable composite wound dressings with controlled release of antibiotics: Microstructure, mechanical and physical properties. *J. Biomed. Mater. Res. Part B Appl. Biomater.* **2010**, *93*, 425–435. [CrossRef]
40. Teo, P.S.; Chow, W.S. Water vapour permeability of poly(lactic acid)/chitosan binary and ternary blends. *Appl. Sci. Eng. Prog.* **2014**, *7*, 23–27.
41. Feng, S.; Zhang, F.; Ahmed, S.; Liu, Y. Physico-mechanical and antibacterial properties of PLA/TiO$_2$ composite materials synthesized via electrospinning and solution casting processes. *Coatings* **2019**, *9*, 525. [CrossRef]

42. Felfel, R.M.; Leander, P.; Miquel, G.F.; Tobias, M.; Gerhard, H.; Ifty, A.; Colin, S.; Virginie, S.; David, M.G.; Klaus, L. In vitro degradation and mechanical properties of PLA-PCL copolymer unit cell scaffolds generated by two-photon polymerization. *Biomed. Mater.* **2016**, *11*, 015011. [CrossRef]
43. Racksanti, A.; Janhom, S.; Punyanitya, S.; Watanesk, R.; Watanesk, S. An approach for preparing an absorbable porous film of silk fibroin–rice starch modified with trisodium trimetaphosphate. *J. Appl. Polym. Sci.* **2015**, *132*, 41517. [CrossRef]
44. Sarasua, J.R.; Prud'homme, R.E.; Wisniewski, M.; Le Borgne, A.; Spassky, N. Crystallization and melting behavior of polylactides. *Macromolecules* **1998**, *31*, 3895–3905. [CrossRef]
45. Chen, P.; Zhou, H.; Liu, W.; Zhang, M.; Du, Z.; Wang, X. The synergistic effect of zinc oxide and phenylphosphonic acid zinc salt on the crystallization behavior of poly (lactic acid). *Polym. Degrad. Stab.* **2015**, *122*, 25–35. [CrossRef]
46. Chen, R.Y.; Zou, W.; Wu, C.R.; Jia, S.K.; Huang, Z.; Zhang, G.Z.; Yang, Z.T.; Qu, J.P. Poly(lactic acid)/poly(butylene succinate)/calcium sulfate whiskers biodegradable blends prepared by vane extruder: Analysis of mechanical properties, morphology, and crystallization behavior. *Polym. Test.* **2014**, *34*, 1–9. [CrossRef]
47. Zhang, H.; Huang, J.; Yang, L.; Chen, R.; Zou, W.; Lin, X.; Qu, J. Preparation, characterization and properties of PLA/TiO$_2$ nanocomposites based on a novel vane extruder. *RSC Adv.* **2015**, *5*, 4639–4647. [CrossRef]
48. Buzarovska, A. PLA nanocomposites with functionalized TiO$_2$ nanoparticles. *Polym. Plast. Technol. Eng.* **2013**, *52*, 280–286. [CrossRef]
49. Garlotta, D.; Doane, W.; Shogren, R.; Lawton, J.; Willett, J.L. Mechanical and thermal properties of starch-filled poly(D,L-lactic acid)/poly(hydroxy ester ether) biodegradable blends. *J. Appl. Polym. Sci.* **2003**, *88*, 1775–1786. [CrossRef]
50. Yew, G.H.; Mohd Yusof, A.M.; Mohd Ishak, Z.A.; Ishiaku, U.S. Water absorption and enzymatic degradation of poly(lactic acid)/rice starch composites. *Polym. Degrad. Stab.* **2005**, *90*, 488–500. [CrossRef]
51. Hu, Y.; Topolkaraev, V.; Hiltner, A.; Baer, E. Measurement of water vapor transmission rate in highly permeable films. *J. Appl. Polym. Sci.* **2001**, *81*, 1624–1633. [CrossRef]
52. Choudalakis, G.; Gotsis, A.D. Permeability of polymer/clay nanocomposites: A review. *Eur. Polym. J.* **2009**, *45*, 967–984. [CrossRef]
53. Tantekin-Ersolmaz, Ş.B.; Atalay-Oral, Ç.; Tatlıer, M.; Erdem-Şenatalar, A.; Schoeman, B.; Sterte, J. Effect of zeolite particle size on the performance of polymer–zeolite mixed matrix membranes. *J. Membr. Sci.* **2000**, *175*, 285–288. [CrossRef]
54. Chen, L.; Zheng, K.; Liu, Y. Geopolymer-supported photocatalytic TiO$_2$ film: Preparation and characterization. *Constr. Build. Mater.* **2017**, *151*, 63–70. [CrossRef]
55. Zhuang, J.; Zhang, B.; Wang, Q.; Guan, S.; Li, B. Construction of novel ZnTiO$_3$/g-C$_3$N$_4$ heterostructures with enhanced visible light photocatalytic activity for dye wastewater treatment. *J. Mater. Sci. Mater. Electron.* **2019**, *30*, 6322–6334. [CrossRef]
56. Wang, Q.; Zhang, L.; Guo, Y.; Shen, M.; Wang, M.; Li, B.; Shi, J. Multifunctional 2D porous g-C$_3$N$_4$ nanosheets hybridized with 3D hierarchical TiO$_2$ microflowers for selective dye adsorption, antibiotic degradation and CO$_2$ reduction. *Chem. Eng. J.* **2020**, *396*, 125347. [CrossRef]
57. Asmatulu, R.; Mahmud, G.A.; Hille, C.; Misak, H.E. Effects of UV degradation on surface hydrophobicity, crack, and thickness of MWCNT-based nanocomposite coatings. *Prog. Org. Coat.* **2011**, *72*, 553–561. [CrossRef]
58. Oleyaei, S.A.; Zahedi, Y.; Ghanbarzadeh, B.; Moayedi, A.A. Modification of physicochemical and thermal properties of starch films by incorporation of TiO$_2$ nanoparticles. *Int. J. Biol. Macromol.* **2016**, *89*, 256–264. [CrossRef] [PubMed]
59. Cui, R.; Jiang, K.; Yuan, M.; Cao, J.; Li, L.; Tang, Z.; Qin, Y. Antimicrobial film based on polylactic acid and carbon nanotube for controlled cinnamaldehyde release. *J. Mater. Res. Technol.* **2020**, *9*, 10130–10138. [CrossRef]
60. Chu, Z.; Zhao, T.; Li, L.; Fan, J.; Qin, Y. Characterization of Antimicrobial Poly (Lactic Acid)/Nano-Composite Films with Silver and Zinc Oxide Nanoparticles. *Materials* **2017**, *10*, 659. [CrossRef]
61. Sinha Ray, S.; Yamada, K.; Okamoto, M.; Ueda, K. Polylactide-layered silicate nanocomposite: a novel biodegradable material. *Nano Lett.* **2002**, *2*, 1093–1096. [CrossRef]
62. Luo, Y.B.; Wang, X.L.; Wang, Y.Z. Effect of TiO$_2$ nanoparticles on the long-term hydrolytic degradation behavior of PLA. *Polym. Degrad. Stab.* **2012**, *97*, 721–728. [CrossRef]
63. Kaseem, M.; Hamad, K.; Ur Rehman, Z. Review of recent advances in polylactic acid/TiO$_2$ Composites. *Materials* **2019**, *12*, 3659. [CrossRef]
64. Yuan, X.; Mak, A.F.T.; Yao, K. In vitro degradation of poly(L- lactic acid) fibers in phosphate buffered saline. *J. Appl. Polym. Sci.* **2002**, *85*, 936–943. [CrossRef]
65. Fu, G.; Vary, P.S.; Lin, C.T. Anatase TiO$_2$ nanocomposites for antimicrobial coatings. *J. Phys. Chem. B* **2005**, *109*, 8889–8898. [CrossRef]
66. Joost, U.; Juganson, K.; Visnapuu, M.; Mortimer, M.; Kahru, A.; Nõmmiste, E.; Joost, U.; Kisand, V.; Ivask, A. Photocatalytic antibacterial activity of nano-TiO$_2$ (anatase)-based thin films: Effects on *Escherichia coli* cells and fatty acids. *J. Photochem. Photobiol. B Biol.* **2015**, *142*, 178–185. [CrossRef]
67. Jiang, X.; Lv, B.; Wang, Y.; Shen, Q.; Wang, X. Bactericidal mechanisms and effector targets of TiO$_2$ and Ag-TiO$_2$ against Staphylococcus aureus. *J. Med. Microbiol.* **2017**, *66*, 440–446. [CrossRef] [PubMed]
68. Verdier, T.; Coutand, M.; Bertron, A.; Roques, C. Antibacterial Activity of TiO$_2$ Photocatalyst Alone or in Coatings on *E. coli*: The Influence of Methodological Aspects. *Coatings* **2014**, *4*, 670–686. [CrossRef]

69. Pleskova, S.N.; Golubeva, I.S.; Verevkin, Y.K. Bactericidal activity of titanium dioxide ultraviolet-induced films. *Mater. Sci. Eng. C* **2016**, *59*, 807–817. [CrossRef] [PubMed]
70. Talebian, N.; Zare, E. Structure and antibacterial property of nano-SiO$_2$ supported oxide ceramic. *Ceram. Int.* **2014**, *40*, 281–287. [CrossRef]

Article

Determination of the Influence of Multiple Closed Recycling Loops on the Property Profile of Different Polyolefins

Johanna Langwieser [1,2,*], Andrea Schweighuber [3], Alexander Felgel-Farnholz [3], Christian Marschik [1], Wolfgang Buchberger [3] and Joerg Fischer [2]

1. Competence Center CHASE GmbH, Altenberger Straße 69, 4040 Linz, Austria; christian.marschik@chasecenter.at
2. Institute of Polymeric Materials and Testing, Johannes Kepler University Linz, Altenberger Straße 69, 4040 Linz, Austria; joerg.fischer@jku.at
3. Institute of Analytical and General Chemistry, Johannes Kepler University Linz, Altenberger Straße 69, 4040 Linz, Austria; andrea.schweighuber@jku.at (A.S.); alexander.felgel-farnholz@jku.at (A.F.-F.); wolfgang.buchberger@jku.at (W.B.)
* Correspondence: johanna.langwieser@chasecenter.at; Tel.: +43-664-9693520

Abstract: In a circular economy, polymeric materials are used in multiple loops to manufacture products. Therefore, closed-loops are also envisaged for the mechanical recycling of plastics, in which plastic is used for products that are in turn collected and reprocessed again and again to make further products. However, this reprocessing involves degradation processes within the plastics, which become apparent through changes in the property profile of the material. In the present paper, the influence of multiple recycling loops on the material properties of four different polyolefins was analyzed. Two different closed-loop cycles with industrially sized processing machines were defined, and each polyolefin was processed and reprocessed within the predefined cycles. For the investigation of the effect of the respective loops, samples were taken after each loop. The samples were characterized by high-pressure liquid chromatography coupled to a quadru-pole time-of-flight MS, high-temperature gel permeation chromatography, melt flow rate measurements, infrared spectroscopy, differential thermal analysis, and tensile tests. With increasing number of processing loops, the tested polyolefins showed continuous material degradation, which resulted in significant changes in the property profiles.

Keywords: mechanical recycling; closed-loop; polyolefins; circular testing; polymer degradation

1. Introduction

Nowadays, 60% of plastic products have a use phase between one and 50 years, which is the lapse of time a product is in use until it becomes waste [1]. The life cycle of each product is different, which means that the consumption and production quantity does not match. In 2018, 29.1 million tons post-consumer plastic waste were collected in Europe. Of those, around 42% were used for energy recovery, 33% were recycled and 25% were brought to landfills [1,2]. In 2018, the highest recycling rate of plastic packaging waste was recorded in Lithuania (69.3%) followed by Slovenia (60.4%) and Bulgaria (59.2%) [3].

On the European market over 600 companies work in the plastics recycling industry with only five countries covering 67% of the total recycling capacity (France, Germany, Italy, Spain, UK). 80% of the total recycling capacity in Europe is covered by polyethylene terephthalate (PET) and polyolefins (polyethylene and polypropylene) [4]. In 2018, Europe had a processing capacity to recycle 1.7 million tons of rigid polyethylene high-density (PE-HD) and polypropylene (PP) (1.2 million tons for post-consumer rigid polyolefins, 0.5 million tons for pre-consumer material). The 1.2 million tons of post-consumer rigid polyolefins can yield 0.8 million tons of post-consumer recyclates (PCR) with a reject of around 25% [5].

Even though a lot of material is already recycled, around 75% of the used plastics in Europe are either thermally recovered or landfilled [1]. To aim for higher recycling rates, different approaches are defined all over the world. The 2030 Agenda for Sustainable Development, for example, was accepted by member states of the United Nations in 2015. 17 Sustainable Development Goals (SDGs) were settled on, in which the 12th Goal is "To ensure sustainable consumption and production patterns", one target is to reduce waste generation by prevention, reduction, recycling and reuse [6].

For a circular economy for plastics three common methods are used: (i) thermal, (ii) chemical and (iii) mechanical recycling. Thermal recycling (i) is the incineration of waste with energy recovery. The energy in the form of heat is used to produce steam and in turn to generate electricity. Chemical recycling (ii) is the process of converting the polymer chains to low molecular weight compounds or, if possible, into the original plastic monomer [7,8]. This paper focuses on the third option.

Mechanical recycling (iii) uses physical processes such as milling, washing, sorting, drying, re-granulating, and compounding to transform plastic waste into useable polymer feedstock. It is a well-established and an economical way to reprocess different materials. Mechanical recycling keeps the polymer molecule mostly intact. The material source can be divided into pre-consumer and post-consumer, whereas pre-consumer feedstock is usually clean and of a single type or at least of a known composition. Post-consumer waste is often strongly contaminated and needs additional treatment steps [9–11].

An objective to be pursued by mechanical recycling is closed-loop recycling where material is used for products which in turn will be collected and reprocessed for similar products repeatedly. Within this closed-loop the material is processed multiple times and each step has a significant influence on the material properties. Polyolefins degrade in a physical or chemical way triggered by light, heat, humidity, chemicals or bioactive substances. This breakdown becomes noticeable by a change in the material's mechanical, optical or electrical properties (e.g., crazes, cracks, erosion, discoloring). Within the material, degradation happens by chain scission, crosslinking, chemical transformations or by formation of new functional groups [12,13].

Numerous studies focused on the influence of multiple recycling cycles on the quality of polyolefins [14–23]. These analyses can be classified into two groups. On the one hand, studies investigated closed-loops using only one processing technique such as Pinheiro et al. [14] and Canevarolo [15] who looked at the mechanical and thermo-oxidative degradation of PE-HD and PP using a 30 mm twin-screw extruder with different processing conditions and screw profiles. The material was reprocessed five times and characterized by their residence time distribution which was measured in-line at the die exit but only during the first extrusion, by IR-spectra using thick hot-pressed films, by molecular weight distribution using gel permeation chromatography. The results coming from the analyses displayed degradation in the form of crosslinking for PE-HD and chain scission for PP. Bernardo et al. [16] investigated the properties of different PE-HD and PE-LD grades after ten injection molding cycles where chain scission and crosslinking was observed. In their study the reprocessed material was mixed with virgin material in every loop. With multiple processing the injection molded specimens changed color which reflects the formation of conjugated double bonds. Other works looked at specific applications for closed-loops such as films for greenhouses [17], discontinuous carbon fiber polypropylene composites [18] or in general polyethylene-like materials [19].

On the other hand, studies looked at loops with multiple degradation, but not multiple processing steps, such as Jansson et al. [20] who subjected two post-consumer PPs to repeated extrusion (using a 19 mm diameter extruder) and thermo-oxidative ageing of the extruded sheets separately. The materials showed a decrease of their oxidative induction temperature with every aging and extrusion step and showed changes in their strain-at-break behavior. Furthermore, other publications used smaller lab sized equipment such as Oblak et al. (L/D ratio of 16/40), Jin et al. [22] (L/D ratio of 16/40) and Schweighuber et al. [23] (L/D ratio of 5/14).

The objective of this study was the determination of the influence of multiple recycling cycles on the material properties of selected polyolefins, using industrially sized processing machines. Four virgin materials including a high-density polyethylene, a low-density polyethylene, and two polypropylenes with different melt flow rates were processed in two closed-loop cycles. The first cycle involved an extrusion and granulation step based on an underwater pelletizing system. The second cycle was designed to simulate an additional product phase, which was realized by injection molded multipurpose specimens. The material was extruded and granulated, injection molded, and lastly milled. Both closed-loop cycles were repeated six times. Properties of the samples were characterized by high-pressure liquid chromatography coupled to a quadrupole time-of-flight MS (HPLC-QTOF-MS), high-temperature gel permeation chromatography (HT-GPC), melt flow rate (MFR) measurements, infrared (IR) spectroscopy, differential thermal analysis, and tensile tests.

2. Materials and Methods

2.1. Materials

Four different materials were used: (i) a polyethylene high-density (PE-HD) with a melt flow rate (MFR) of 4 g/10 min (190 °C/2.16 kg), which is a typical injection and compression molding grade. The material is typically used in industrial food and transport packaging; (ii) a polyethylene low-density (PE-LD) with an MFR of 1.9 g/10 min (190 °C/2.16 kg), which is industrially applied in film extrusion for different packaging; (iii) a polypropylene injection molding type, this material has an MFR of 12 g/10 min (230 °C/2.16 kg) and is used for house ware and thin wall packaging; (iv) a polypropylene with an MFR of 4 g/10 min used for thermoforming packaging. An overview of the materials including their abbreviations used in this work is depicted in Table 1, the structural formulas of the materials can be seen in Figure 1.

Table 1. Material designation including melt flow rates (MFR), solid densities and application fields.

Characteristic	PE-I	PE-E	PP-I	PP-E
Type	Polyethylene high-density Injection molding	Polyethylene low-density Extrusion	Polypropylene homopolymer Injection molding	Polypropylene homopolymer Extrusion
MFR | g/10 min	4	1.9	12	4
Solid density | kg/m^3	954	924	905	n.s.
Application	Industrial food transport packaging	Bags, films, flexible packaging, food packaging, liners, pouches	House ware thin wall packaging	Thermoforming packaging

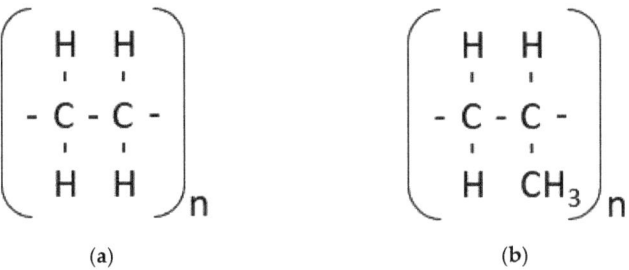

Figure 1. Structural formulas of the analyzed polyolefin grades for (**a**) polyethylene and (**b**) polypropylene.

2.2. Closed-Loop Recycling Cycles

To analyze the effect of multiple processing steps on the properties of our materials, two closed-loop recycling cycles were simulated. The first cycle (further referred to as

CL-E) used a 35 mm single-screw extruder (SML Maschinengesellschaft mbH, Lenzing, Austria) with a maximum mechanical power of 38.4 kW provided by the electrical drive unit and a maximum screw speed of 434 rpm. At the extruder head an EUP 50 under-water pelletizing system (Econ GmbH, Weisskirchen, Austria) was installed to granulate the extruded melt. To control the temperature of the extruder, it was equipped with six heating zones, as illustrated in Table 2, the feed opening (FO) and the heating zones one to three (Z1–Z3) are located directly on the barrel of the extruder, whereas adapter one (AD1) and adapter two (AD2) are responsible for heating the connections between the under-water pelletizing system and the extruder. The feed housing was water-cooled whereas the remaining barrel sections were cooled by forced air to avoid excessive heat generation. The materials were processed with a screw speed of 100 rpm which resulted in an output of approximately 25 kg/h. The melt temperatures for the PE- and PP-types were 200 °C and 230 °C, respectively. The bypass and the die plate were set at a temperature of 180 °C, the cooling water temperature was held beneath 30 °C. The granulation speed was around 1500 rpm, the dryer speed was around 800 rpm. The material was extruded and granulated six times and samples were taken before each extrusion step (marked with a red box in Figure 2).

Table 2. Temperature program in °C for the heating zones of the single-screw extruder.

Material	FO	Z1	Z2	Z3	AD1	AD2
PE-I	45 °C	190 °C	200 °C	200 °C	200 °C	200 °C
PE-E	45 °C	190 °C	200 °C	200 °C	200 °C	200 °C
PP-I	45 °C	190 °C	220 °C	220 °C	220 °C	220 °C
PP-E	45 °C	190 °C	220 °C	220 °C	220 °C	220 °C

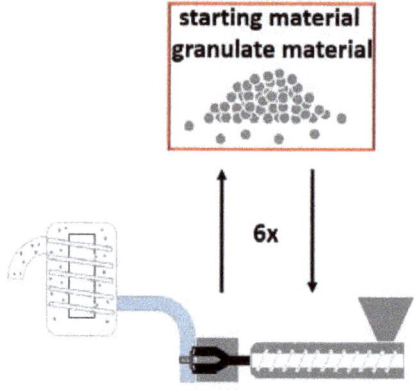

Figure 2. Schematic representation of the extrusion cycle (CL-E) using a 35/34D single screw extruder with an EUP 50 under-water pelletizing system.

The second cycle combined multiple processing steps by adding an injection molding step after granulation (further referred to as CL-EI). The main purpose of this modification was to additionally simulate the product processing phase of the material. In addition to the extrusion and the granulating step with the same processing parameters, the material within this cycle was injection molded to multipurpose specimens (ISO 17855-2 [24]) using an injection molding machine victory 60 (Engel Austria GmbH, Schwertberg, Austria) and milled with a granulator S-Max 2 Plus (Wittmann Battenfeld GmbH, Kottingbrunn,

Austria) at 27 rpm. The temperature program for the injection molding machine is shown in Table 3 whereas Z1 to Z4 are abbreviations for the heating zones on the plasticizing unit. Additional processing parameters for the injection molding machine can be seen in Table 4. In total the CL-EI was repeated six times, sample collection is marked with a red box in Figure 3.

Table 3. Temperature program for the injection molding machine.

Material	Die	Z1	Z2	Z3	Z4	Feed
PE-I	220 °C	220 °C	210 °C	205 °C	200 °C	50 °C
PE-E	220 °C	220 °C	210 °C	205 °C	200 °C	50 °C
PP-I	230 °C	230 °C	225 °C	220 °C	210 °C	50 °C
PP-E	230 °C	230 °C	225 °C	220 °C	210 °C	50 °C

Table 4. Additional processing parameters of the injection molding machine.

Parameter	Nominal-Values
Clamp force	kN	500
Back pressure	bar	450
Dosage volume	cm³	55
Cycle time	s	100
Cooling time	s	20
Back pressure time	s	15

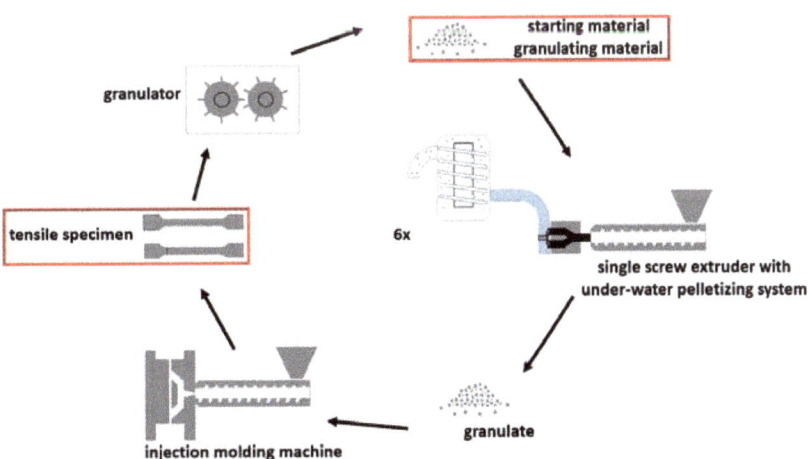

Figure 3. Schematic representation of the extrusion-injection molding cycle (CL-EI) using a 35/34D single screw extruder with an EUP 50 under-water pelletizing system, an injection molding machine victory 60, and granulator S-Max 2 Plus.

2.3. Characterization Methods

For both recycling cycles, selected properties were measured for all materials after each loop, including the consumption of stabilizers, the weight average molar masses and molar mass distributions, the melt flow rate, the carbonyl index, the degree of crystallinity and stress-strain diagrams.

2.3.1. High-Pressure Liquid Chromatography (HPLC)

To identify the consumption of the non-oxidized contents of the processing stabilizer, the secondary antioxidant Irgafos 168, and of the radical scavengers, the primary phenolic antioxidant Irganox 1010, an Agilent 1260 high-pressure liquid chromatograph coupled to an Agilent 6510 quadrupole time-of-flight MS (HPLC-QTOF-MS) with an electrospray ionization (ESI) source (Agilent Technologies Inc., Santa Clara, CA, USA) was used. For this purpose, the HPLC was equipped with a Kinetex C18 separation column (Phenomenex Inc., Aschaffenburg, Germany) and a C18 guard column (Phenomenex Inc., Aschaffenburg, Germany). Furthermore, tributyl phosphate was added to avoid stabilizer loss during sample preparation, and Cyanox 1790 was used as an internal standard.

2.3.2. Gel Permeation Chromatography (GPC)

A high-temperature gel permeation chromatograph (HT-GPC, PolymerChar S.A., Valencia, Spain) equipped with an IR 5 detector was used to obtain molar mass distributions (MMD) and weight average molar masses (M_w) of all collected samples from both loops. To dissolve the samples trichlorobenzene stabilized with butylated hydroxytoluene was used, and heptane was added as a flow marker.

2.3.3. Melt Flow Rate (MFR)

MFR measurements are single point measurements at low shear rates (depending on the measurement weight) which are often used for quality control of incoming materials. Since it is an easy and quick tool, it was used to compare each material sample of each cycle. For this purpose, the Aflow (ZwickRoell GmbH and Co. KG, Ulm, Germany) was used. The measurements were performed according to ISO 1133 [25] with a measurement weight of 2.16 kg and measuring temperatures of 190 °C for PE and 230 °C for PP. For each run about three grams of material were filled into the preheated cylinder and compacted two to three times. After five minutes of warm up time the measurement started automatically, and the piston pressed the molten material through the die. Five samples were cut from the middle strand and weighed. For each sample at least two measurements were carried out.

2.3.4. Infrared Spectroscopy (IR Spectroscopy)

The IR spectra were recorded using the Fourier transform infrared (FTIR) spectrometer Spectrum Two (Perkin Elmer Inc., Waltham, MA, USA). The measurements were conducted within the wavenumbers of 500 cm^{-1} to 4000 cm^{-1}. Each sample was measured at least twice. To evaluate the degradation within the materials, the carbonyl index (CI) was calculated by comparing the absorption peak for a carbonyl group with an existing reference peak of the investigated polymer. The carbonyl peak can be detected in the wavenumber range from 1850 to 1650 cm^{-1}. For the CI evaluation of PE, the CH$_2$ scissoring vibrations at approximately 1463 cm^{-1} and for PP, the CH$_2$ and CH$_3$ vibrations at around 1459 cm^{-1} were chosen. To calculate the CI, the area underneath the carbonyl peak was divided by the area underneath the reference peak as depicted in Equation (1) [26].

$$CI = \frac{A_{1850-1650 \text{ cm}^{-1}}}{A_{reference}}, \qquad (1)$$

2.3.5. Differential Thermal Analysis (DTA)

The heat of fusion ΔH_{melt} and the specific heat capacity were measured using a DSC 4000 (Perkin Elmer Inc., Waltham, MA, USA). Heating and cooling rate was 10 K/min from 30 °C to 230 °C. The mass of the specimens was 5.0 ± 0.5 mg. All measurements were performed in an air atmosphere. The degree of crystallinity was calculated based on Equation (2):

$$\alpha = \frac{\Delta H_{melt}}{\Delta H_{100\%}} \times 100\%, \qquad (2)$$

where $\Delta H_{100\%}$ refers to the heat of fusion of 100% crystalline polymer. For polyethylene, it is 293 J/g and for polypropylene it is 207 J/g [27]. For the PE-I, ΔH_{melt} was evaluated between 70 and 150 °C, for PE-E, in the range of 60 and 125 °C, and for both PP grades between 120 and 180 °C.

2.3.6. Tensile Test

The stress-strain diagrams were recorded using the universal testing machine Z020 (ZwickRoell GmbH and Co. KG, Ulm, Germany). Multipurpose specimens were tested according to ISO 527-2 [28]. At the beginning of the measurement an initial load of 0.1 MPa was reached. All specimens were measured with a testing speed of 1 mm/min in the strain region of 0.05% to 0.25% and then tested with a testing speed of 50 mm/min until failure via break occurred. For each sample at least ten measurements were conducted.

3. Results

This chapter discusses the results of our experimental study. To differentiate the two loops, grey lines for the extrusion-cycle (CL-E) and red lines for the extrusion-injection molding cycle (CL-EI) were used. In this paper the single point data is presented, the mentioned multi point data can be found in the Supplementary Materials.

3.1. Consumption of Stabilizers

According to the data sheets [29,30], Irgafos 168 (IF168) is a hydrolytically stable secondary antioxidant, which reacts during processing with hydroperoxides formed by autoxidation and thereby prevents polymer degradation induced by processing, protects the primary antioxidants, and thus extends the performance of primary antioxidants. Irganox 1010 (IX1010) is a sterically hindered primary phenolic antioxidant which protects organic substrates against thermo-oxidative degradation. Both stabilizers can be used in polyolefins.

Figure 4a depicts the amount of non-oxidized stabilizers left inside PE-I over the reprocessing loops. Compared to PE-I with about 0.27 w% of IF168 and about 0.02 w% of IX1010 within the virgin samples, no such stabilizers (IF168 and IX1010) were found in PE-E. For PE-I, only for the CL-EI a substantial and continuous reduction by increased processing of IF168 was obtained. IF168 was depleted after the fifth loop. While for the CL-E the IX1010 values stayed constant, for the CL-EI, an initial drop after the first loop and constant values afterwards were detected.

In Figure 4b,c, the stabilizer content in both PP grades after each loop is illustrated. PP-E possessed with approximately 0.35 w% for IF168 and 0.16 w% for IX1010 in the initial virgin state roughly double the stabilizer content than PP-I. The stabilizer contents of both materials showed a significant drop in both cycles. The remaining intact amounts of IF168 and IX1010 were in general lower after CL-EI than after CL-E. For both PPs, IF168 was sooner depleted than the IX1010, depicted in full IF168 loss for PP-E in the CL-EI after the fourth loop, for PP-I in the CL-E after the third loop, and for PP-I in the CL-EI after the first loop.

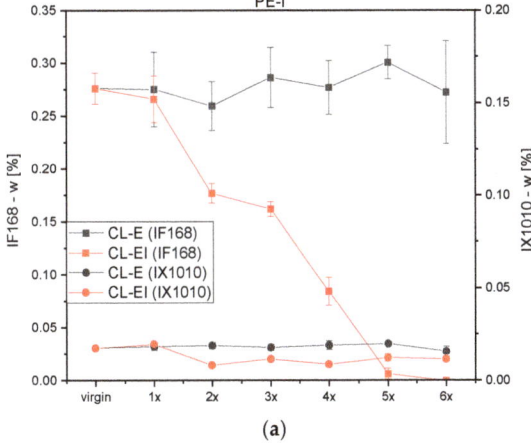

(a)

Figure 4. *Cont.*

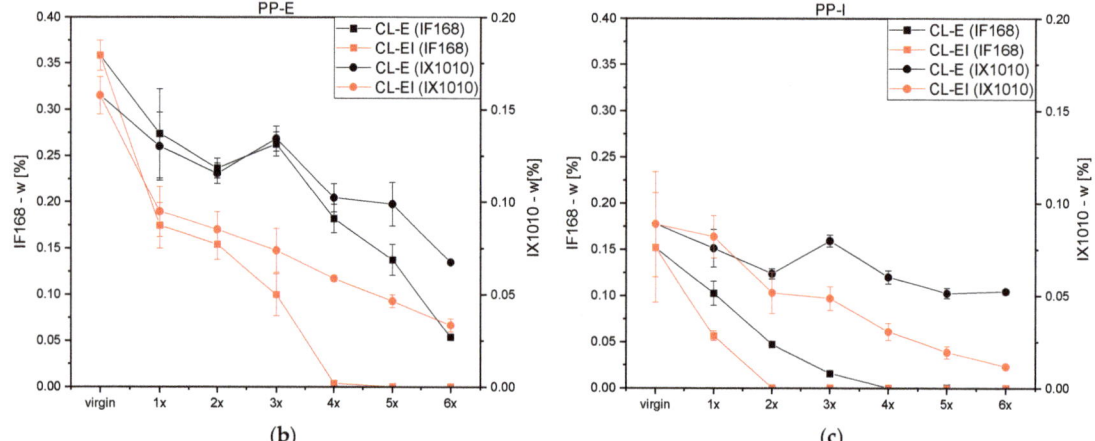

Figure 4. Content of stabilizers IF168 and IX1010 in (**a**) PE-I and both PP grades (**b**) PP-E and (**c**) PP-I after each loop for both recycling cycles.

3.2. Molar Mass Distributions (MMD) and Weight Average Molar Mass (M_w)

In Figure S1, the changes in molar mass distribution (MMD) curves and in Figure 5a the weight average molar mass (M_w) of PE-E after each loop of both processing cycles, CL-E and CL-EI, are illustrated. Overall, for the CL-E no significant changes in the distribution were observed. For the CL-EI the shape of the distribution curve changed to a flatter and wider one with increased processing. Just as Jin et al. [22] showed, also in our investigations of the CL-EI, a shift in the peak of the MMD to smaller values and a shape change at higher molecular weights was observed. These changes in the distribution curves suggest that the molecular weight increased due to crosslinking.

To acquire a better picture of the overall changes measured via GPC, M_w was evaluated. It stayed rather constant after the CL-E with values around 91,200 g/mol but significantly increased in course of the CL-EI up to ca. 111,000 g/mol. Degradation in polyethylene can lead to crosslinks between the polymer chains which stems from the reaction of freshly cut chains due to shear and temperature with the terminal vinyl unsaturation [14]. This results in an increase in molecular mass by branching and crosslinking.

Figure S2 depicts the MMD and Figure 5b M_w of PE-I. No significant changes in the shape of the MMD could be observed for either the CL-E or the CL-EI. The weight average molar mass dropped for both cycles during the first processing step and remained approximately constant between 90,000 and 95,000 g/mol afterwards.

In Figure S3 and Figure 5c, with increasing loops a shape change of the MMD in both processing cycles of PP-E was illustrated. Just as mentioned above, also for PP-E, the curves shifted to lower values. However, in contrast to PE-E, increasing peaks were detected, which was also discovered by Canevarolo et al. [15]. This change led to the assumption, that due to the thermo-mechanical degradation the longer chains break, and the resulting shorter chains are added to the middle part of the MMD curve. With increasing processing, the amount of the shorter chains got bigger, and the curve shifted upwards.

A significant decrease in M_w starting with about 350,000 g/mol in the virgin material was observed whereas weight average molar masses of the CL-EI were lower with the last and lowest value at 213,000 g/mol comparted to 90,000 g/mol for the CL-E. This behavior over all is an indicator for chain scission in a material, which is the preferred degradation behavior of PP. The chains continuously broke due to thermal and mechanical influences and thereby decreased their molecular mass. This decrease led to a narrowing and a peak increase of the MMD curve.

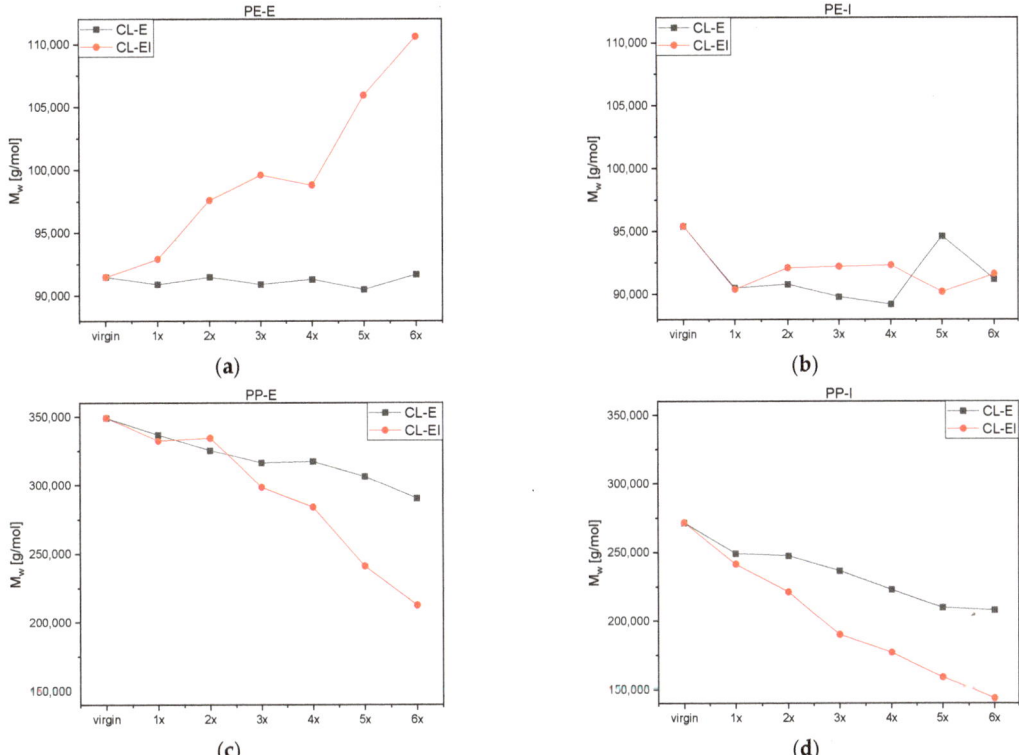

Figure 5. M_w after each loop for both processing cycles: (**a**) PE-E, (**b**) PE-I, (**c**) PP-E and (**d**) PP-I.

Figure S4 shows the MMD and Figure 5d the M_w of PP-I. Again, the typical degradation behavior of PP can be observed, and the molecular masses decreased continuously in both cycles, which led to an increase, a shift to lower values and a narrowing of the MMD. The M_w for the CL-EI was continuously lower than for the CL-E with the lowest value of 144,000 g/mol after six loops of CL-EI and of 210,000 g/mol after six loops of CL-E.

3.3. Melt Flow Rate (MFR)

The MFR is a single point measurement and a measure for the fluidity of a material at low shear rates [31]. The increase in molecular mass illustrated above means longer and more connected chains which results in hindered movement of the polymer and thus leads to a higher viscosity and lower MFR. The review by Yin et al. [10] states that based on the MFR, PP grades have good thermal stability and degrade later than PE-HD or PE-LD types, which show imminent MFR reductions and thus have poor thermal stability. In our investigations the opposite behavior was obtained.

The MFR results of PE-E and PE-I can be seen in Figure 6a,b. The CL-E hardly influenced the viscosity of the PE types while the CL-EI led to a decrease of MFR values for both, which may have been caused by aforementioned degradation effects. According to these measurements, the increased processing using the extrusion-injection molding cycle influenced the viscosity of the PE types more than just processing by extrusion, which was expected and was also detected by the methods discussed before.

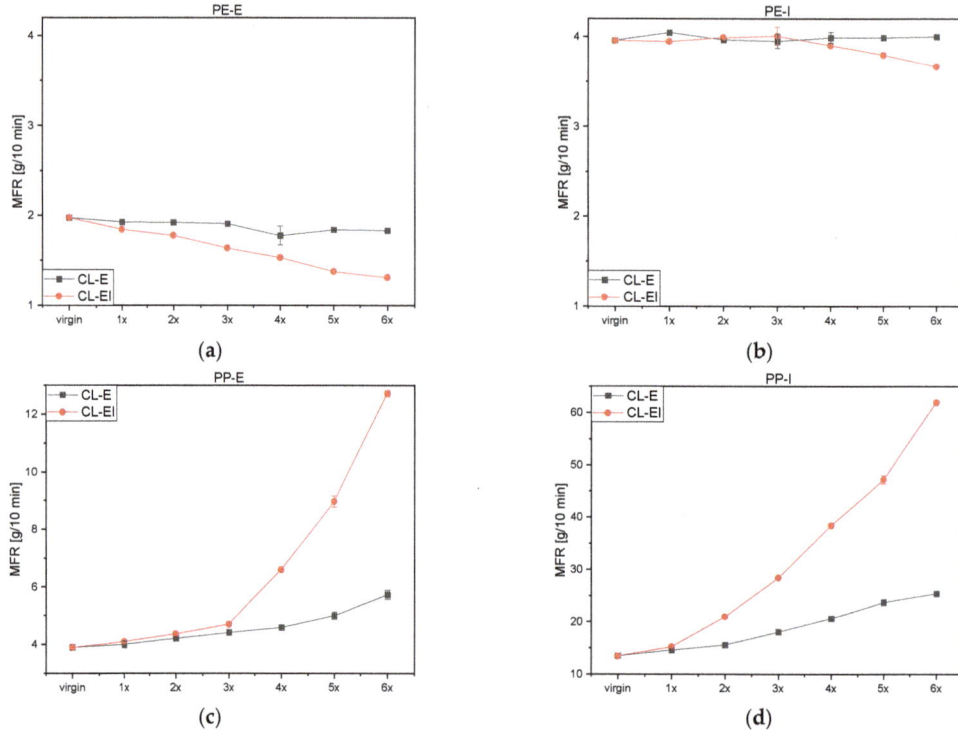

Figure 6. MFR values for both cycles (**a**) PE-E, (**b**) PE-I, (**c**) PP-E and (**d**) PP-I.

Both PP grades showed a significant MFR increase after both loops (see Figure 6c,d). PP commonly shows degradation in the form of chain scission as mentioned above. Due to shorter chains, the molecular mass decreased (see Figure 5) and the molecules were entangled to a lesser extent. This elicits better movement in the fluid state and the chains glide apart. Thus, in the same time a higher amount of material can flow through the measurement die, which resulted in an increased MFR. PP-E illustrated only a slight increase up until the third loop, where the MFR values started to increase in a non-linear way. PP-I shows a decrease of viscosity starting from the first loop.

3.4. Infrared Spectra (IR Spectra)

Figure S5 depicts the evaluated IR data of the PE-E used in this study. In Figures S5 and S6 characteristic peaks of PE can be distinguished [32]:

- 3000 and 2840 cm^{-1}: symmetric and antisymmetric stretching vibrations of CH$_2$ groups and
- 1463 cm^{-1}: scissoring vibrations of CH$_2$ groups.

In Figure S5a additional peaks at 3500–3000 cm^{-1} and 1700–1500 cm^{-1} were detected. These peaks may stem from the lubricant (Erucamide according to the data sheet) used. Due to the increased processing, the lubricant seemed to migrate out of the material onto its surface. Owing to the carbonyl group in the used lubricant, the CI was distorted which led to higher and more stable values until the lubricant wore off (see grey curve in Figure 7a).

In Figure S5b the additional peaks caused by the lubricant disappear. This indicates that due to the excessive processing in CL-EI, the lubricant migrated completely to the surface and was removed. With further processing the carbonyl peaks increased, which suggested that the material oxidized with further processing, as shown in Figure 7a in red.

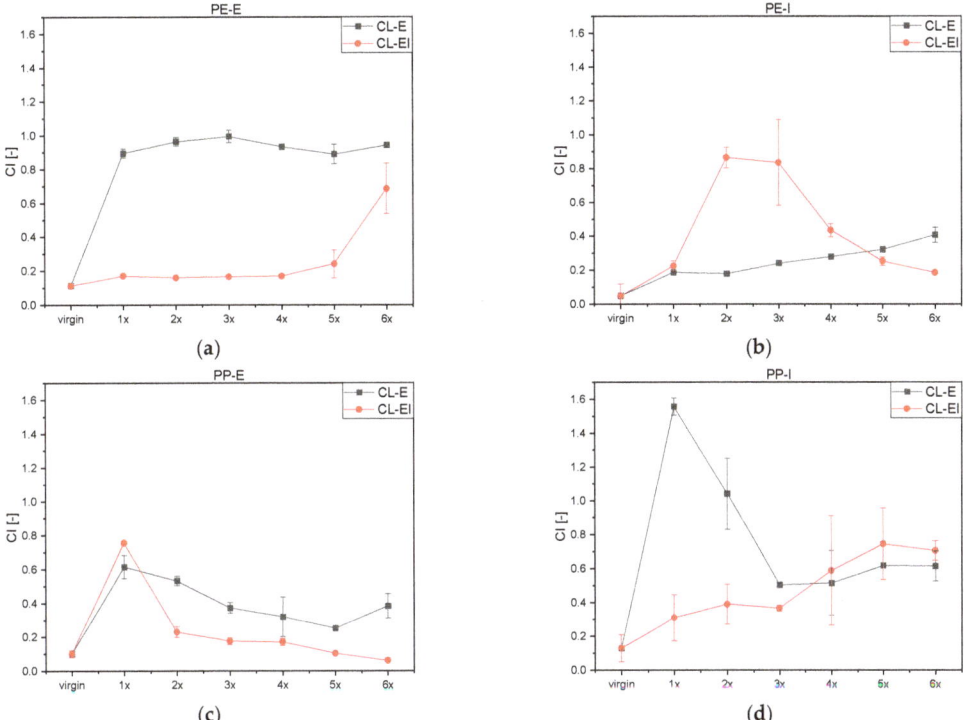

Figure 7. Calculated CI values for (**a**) PE-E, (**b**) PE-I, (**c**) PP-E and (**d**) PP-I.

PE-I showed next to the characteristic peaks' significant changes in the wavenumber range related to the absorption of the carbonyl groups. Since no further peaks can be observed in the spectrum, the increase of the carbonyl groups presumably stemmed from oxidation due to degradation. This behavior can be seen in Figure 7b. Processing according to the CL-E led to a continuous increase of the CI up to the sixth loop. The CL-EI influenced the material by a sudden increase after the second loop and a further continuous decrease.

The IR spectra of the PP grades can be seen in Figures S7 and S8. PP exhibited the following characteristic peaks [32]:

- 3000 to 2840 cm^{-1}: symmetric and antisymmetric stretching vibrations of CH_2 and CH_3 groups and
- 1459 cm^{-1} and 1376 cm^{-1}: bending vibrations of CH_2 and CH_3 groups.

For PP-E (Figure 7c), the CI reached its highest value for both cycles after the first loop and decreased furthermore. After the fifth loop of the CL-E, the CI of the material started to increase again.

For PP-I, distinct CI values for the CL-E could not be distinguished due to the appearance of additional peaks which may have been present due to a lubricant used, such as estimated for PE-E before. For the CL-EI a continuous increase of the CI could be seen in Figure 7d.

3.5. Melting Behavior and Degree of Crystallinity

Figures S9 and S10 illustrate the melting peaks of both polyethylene grades with no significant trends. Additionally, the degrees of crystallinity are shown in Figure 8a,b. According to literature [13], typical crystallinity values for PE-LD range between 40 and 50%, whereas PE-HD lies between 60 and 80%, which is in agreement for the tested materials PE-E and PE-I, respectively. For both PE grades, the CL-EI values were lower than the values of the CL-E. Furthermore, the standard deviations of PE-I were bigger than the PE-E's. Jin et al. [22] and Oblak. et al. [21] discovered rather constant degrees of crystallinities up until the 40th loop and 20th loop, respectively. This is in agreement with our measurements concerning PE, where no significant trend was obtained.

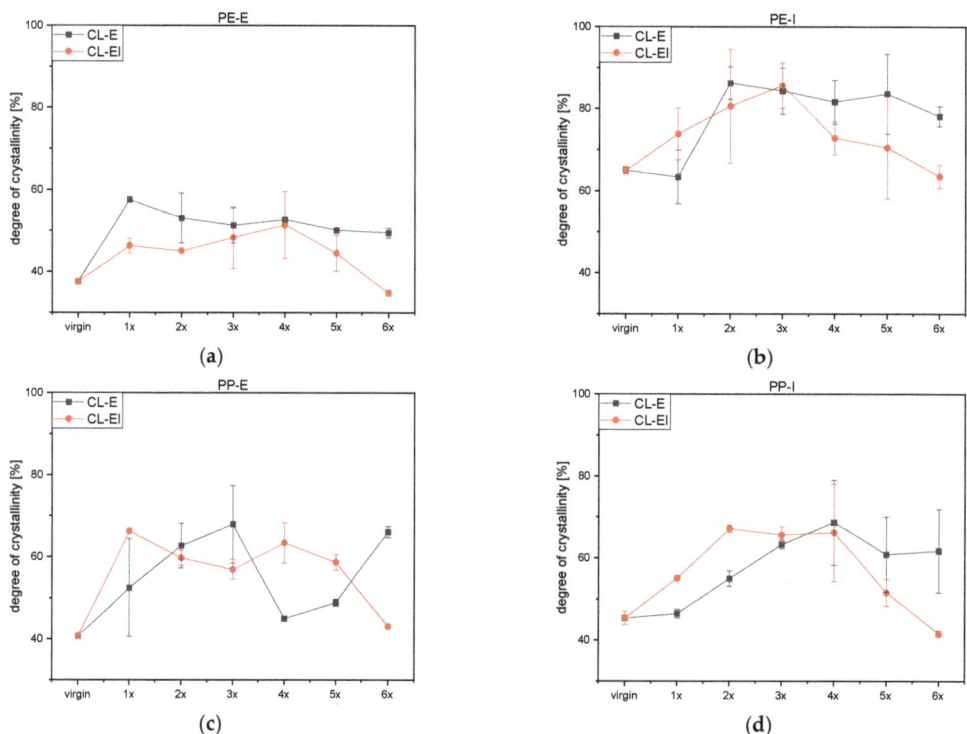

Figure 8. Calculated degrees of crystallinity after each loop for both cycles: (**a**) PE-E, (**b**) PE-I, (**c**) PP-E and (**d**) PP-I.

Figures S11 and S12 show the melting peaks of both PP grades, again with no significant trend. Furthermore, the degrees of crystallinity are depicted. Typical values of the crystallinity range between 30 and 60% [13]. Hence, within our investigations some results exceeded the typical range. In Figure 8c the PP-E shows increase and decrease in a random manner for both cycles even though the CL-E depicts stronger fluctuations than the CL-EI.

PP-I in Figure 8d on the contrary shows similar behavior to PE-I: For both cycles, the crystallinity increases and peaks at the fourth loop. However, with further processing the degree of crystallinity decreased again and was at its lowest at the last step. The CL-EI shows a steeper incline and a faster decline than the CL-E.

3.6. Mechanical Properties of the CL-EI

Due to the availability of multipurpose specimens, only specimens of the processing cycle CL-EI were characterized. As depicted in Figures S13 and S14, all four polyolefin grades point to ductile behavior after the first loop with a continuous increase of stress and strain until the specimens yield, followed by a decrease in stress, at least in three of four materials, and by an increasing strain up until failure.

Based on the negligible changes of the PE grades with continuous processing, the results of the second to fifth loop are not presented in this paper. PE-E only shows small changes after the sixth loop due to the CL-EI. Both curves of the PE-I show strain hardening behavior at high strains shortly before failure. After the sixth loop, the materials still exhibit ductile behavior, which leads to the conclusion that the CL-EI has a minimum influence on mechanical properties of the PE grades.

Both PP grades embrittle with continuous processing, which leads to decreasing strain at beak values depicted in Figure 9. This decline is presumably related to chain scission that was also concluded in previous chapters. For both materials, the yield stress after the first to sixth loop stays roughly the same.

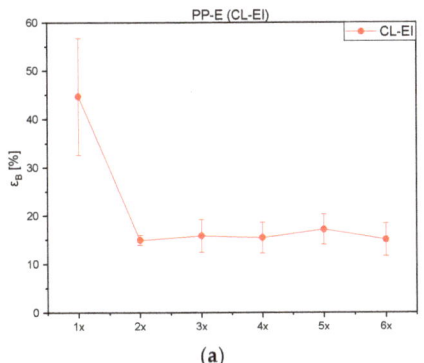 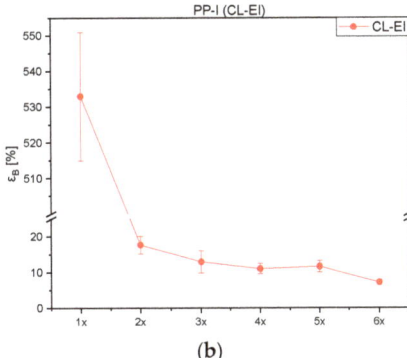

(a) (b)

Figure 9. Strain at break values over the course of the processing loops of (**a**) PP-E, and (**b**) PP-I.

4. Discussion

Other publications showed that degradation in PE happens by attacking the longer chains and forming crosslinks and additional branches in an uncontrolled manner [14,16]. This formation of longer chains and more interconnected chains leads to a flatter and wider MMD [22], lower values in MFR, an increase in CI due to the formation of carbonyl groups instead of crosslinks or branches [26], lower degrees of crystallinity due to hindered movement of the chains [21,22] and decreasing mechanical properties [19].

In this study, for both PE grades, no significant changes can be seen in Figure 10 for the CL-E with either characterization method (i.e., HPLC, GPC, and MFR). The stabilizers in PE-E showed already in the virgin material very low values. Due to the already low values of the stabilizers and the increased processing steps of the CL-EI, an increase in weight average molar mass started to happen. This increase can be caused by the formation of crosslinks and branches and leads to a decrease of melt flow rate as depicted in the Figure 10 below. Due to longer and connected chains, the movement of the whole melt is restricted, and less melt can flow through the measurement die.

The stabilizers in the PE-I continuously decreased over the course of the loops. However, the existence of stabilizers may be the reason for rather stable weight average molar mass values because the polymer chains are protected for a longer time. Thus, constant melt flow rates were expected and mostly derived.

The results from the IR measurements were not considered in this discussion for either PE type due to falsified values stemming from the migration of lubricants. Even though in this study the results do not add any valuable information, other publications support the assumption that the CI values are a good indicator of degradation [33]. Furthermore, due to negligible effects of the cycles on the mechanical properties of both PEs the results were not added either.

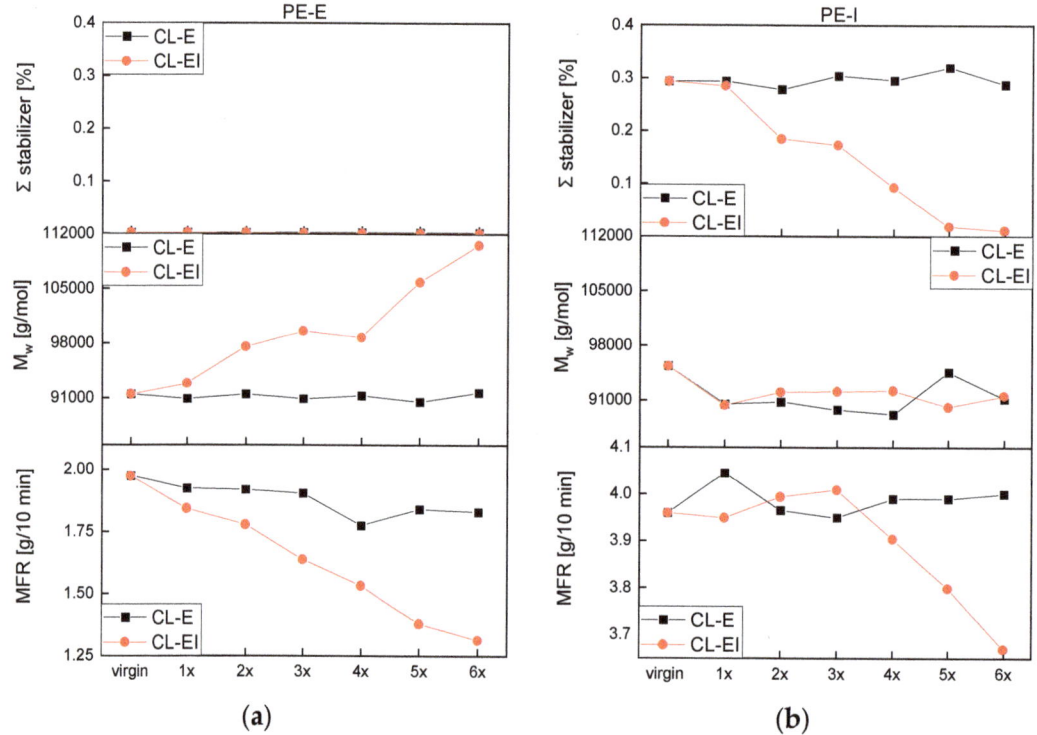

Figure 10. The sum of antioxidants (IF168 and IX1010), the weight average molar mass (M_w), and the melt flow rate (MFR) as a function of recycling steps for (**a**) PE-E, and (**b**) PE-I.

PP has been looked at in multiple publications before this study whereas the typical degradation behavior is chain scission, this usually leads to a higher and slimmer MMD and a decrease in M_w [15], therefore to an increase in MFR due to easier movement of the chains and thus unstable melt flow can be obtained. Similar to within the PE grades, the CI should increase by forming carbonyl groups [26]. In some cases, due to degradation the degree of crystallinity can increase which leads to embrittlement and a decrease in mechanical strength [20]. Figure 11 displays the same tendencies for both PP grades in both cycles. The stabilizers decreased with increased processing in both cycles. According to literature [34], the chains of the polymers are protected by the stabilizers; thus, the molar mass should start to be affected after full stabilizer loss. However, as discovered in our work and by Fischer et al. [34], PP starts to degrade immediately. The degradation via chain scission leads to lower weight average molar masses. The polymer chains become shorter and allow for better movement of the melt and with this a lower viscosity which means higher MFR values are to be expected. The PP-I showed the most severe changes in MFR with a value of over 60 g/10 min after the sixth loop. In general, for PP, a good correlation between the results of HPLC, GPC, and MFR was determined.

Similar to the above, distinct values of the CI could not be distinguished for either PP, for this reason no additional diagrams were added. Furthermore, due to non-significant yield stresses and already presented strain values in the chapter above, the mechanical properties are not repeated in this chapter.

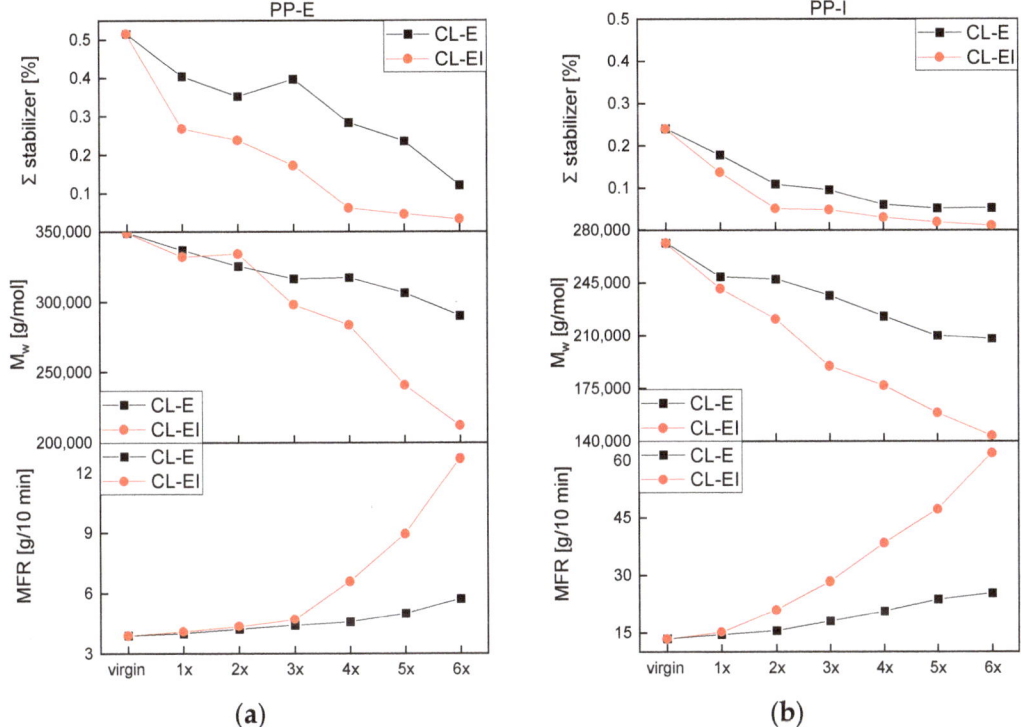

Figure 11. The sum of antioxidants (IF168 and IX1010), the weight average molar mass (M_w), and the melt flow rate (MFR) as a function of recycling steps for (**a**) PP-E, and (**b**) PP-I.

5. Conclusions

In this study, the degradation behavior of four different polyolefins was investigated. Polymers were chosen based on their chemical composition and on the suggested processing method in their data sheet. Two polyethylenes (PE), a polyethylene high-density (PE-HD) for injection molded products and a polyethylene low-density (PE-LD) for extruded films, and two polypropylenes (PP), one injection molding grade and one extrusion grade, were selected for our research. To investigate the effect of multiple processing loops on the material behavior, the polyolefins were processed with industrially sized machines in two predefined closed-loop cycles. The first cycle comprised of an extrusion and a granulating step, and the second cycle involved additional steps of injection molding and milling. Both cycles were repeated six times. After each of these closed-loops, samples were taken and the respective samples were characterized by high-pressure liquid chromatography, high-temperature gel permeation chromatography, melt flow rate measurements, infrared spectroscopy, differential thermal analysis, and tensile tests.

Investigations with the PE grades showed continuous aging and material degradation was indicated by increasing weight average molar masses and decreasing melt flow rates. Furthermore, the multiple processing loops of the PE types led to a decrease in stabilizer content. The most stable material was the PE-HD injection molding grade. While a significant stabilizer decrease was found in the closed-loop cycle comprising the higher number of processing steps, no significant changes were determined in the other properties especially the CI values, which were distorted, the degrees of crystallinity, with no significant trends, and the mechanical properties, where no relevant changes could be seen. Overall, the results of these investigations point to chain branching and crosslinking within the PE grades.

The degradation behavior of the PP grades was more pronounced than within the PE grades. Both PP materials showed a significant decrease in stabilizer content, a strong decrease in weight average molar masses, an increase in melt flow rate, and embrittlement by a decrease of mechanical strength with increasing cycles. Similar to with the PE grades, the carbonyl index results were falsified, and the degrees of crystallinity did not indicate degradation. Good correlations between the stabilizer contents, weight average molar masses and melt flow rates were found leading to the conclusion that chain scissions dominated for PP in both closed-loop cycles.

Supplementary Materials: The following supporting information can be downloaded at: https://www.mdpi.com/article/10.3390/polym14122429/s1. Figure S1. MMD of PE-E: (a) MMD after each loop of the CL-E and (b) MMD after each loop of the CL-EI. Figure S2. MMD of PE-I: (a) MMD after each loop of the CL-E and (b) MMD after each loop of the CL-EI. Figure S3. MMD of PP-E: (a) MMD after each loop of the CL-E and (b) MMD after each loop of the CL-EI. Figure S4. MMD of PP-I: (a) MMD after each loop of the CL-E and (b) MMD after each loop of the CL-EI. Figure S5. PE-E: IR absorbance spectra and enlarged spectra in the wavenumber range from 1750 to 1500 cm^{-1} of the (a) CL-E and (b) CL-EI. Figure S6. PE-I: IR absorbance spectra and enlarged spectra in the wavenumber range from 1750 to 1500 cm^{-1} of the (a) CL-E and (b) CL-EI. Figure S7. PP-E: IR absorbance spectra and enlarged spectra in the wavenumber range from 1750 to 1500 cm^{-1} of the (a) CL-E and (b) CL-EI. Figure S8. PP-I: IR absorbance spectra and enlarged spectra in the wavenumber range from 1750 to 1500 cm^{-1} of the (a) CL-E and (b) CL-EI. Figure S9. Melting peaks of the PE-E after each loop for (a) CL-E and (b) CL-EI. Figure S10. Melting peaks of the PE-I after each loop for (a) CL-E and (b) CL-EI. Figure S11. Melting peaks of the PP-E after each loop for (a) CL-E and (b) CL-EI. Figure S12. Melting peaks of the PP-I after each loop for (a) CL-E and (b) CL-EI. Figure S13. Melting peaks of the PP-I after each loop for (a) CL-E and (b) CL-EI. Figure S14. Melting peaks of the PP-I after each loop for (a) CL-E and (b) CL-EI.

Author Contributions: Conceptualization, J.L., C.M. and J.F.; methodology, J.L. and J.F; validation, J.L.; formal analysis, J.L., A.S. and A.F.-F.; investigation, J.L., A.S. and A.F.-F.; resources, C.M., W.B. and J.F.; data curation, J.L.; writing—original draft preparation, J.L.; writing—review and editing, C.M. and J.F.; visualization, J.L.; supervision, W.B. and J.F.; project administration, J.F.; funding acquisition, C.M. and J.F.; All authors have read and agreed to the published version of the manuscript.

Funding: The authors acknowledge financial support through the COMET Centre CHASE, funded within the COMET—Competence Centers for Excellent Technologies programme by the BMK, the BMDW and the Federal Provinces of Upper Austria and Vienna. The COMET programme is managed by the Austrian Research Promotion Agency (FFG).

Institutional Review Board Statement: Not applicable.

Informed Consent Statement: Not applicable.

Data Availability Statement: The data presented in this study are available on request from the corresponding author.

Acknowledgments: This work was supported by the Institute of Polymer Processing and Digital Transformation (JKU, Linz, Austria). Supported by Johannes Kepler Open Access Publishing Fund.

Conflicts of Interest: The authors declare no conflict of interest.

References

1. Plastics Europe. *Plastics-the Facts 2020*; Plastics Europe: Brussels, Belgium, 2020.
2. Statista Research Department. Recycling Rate of Plastic Packaging Waste in the European Union (EU-28) from 2010 to 2018. 5 July 2021. Available online: https://www.statista.com/statistics/881967/plastic-packaging-waste-recycling-eu/ (accessed on 15 November 2021).
3. Eurostat. More than 40% of EU Plastic Packaging Waste Recycled. 13 January 2021. Available online: https://ec.europa.eu/eurostat/web/products-eurostat-news/-/ddn-20210113-1 (accessed on 15 November 2021).
4. Plastics Recyclers Europe. *Report on Platics Recycling Statistics 2020*; Plastics Recyclers Europe: Brussels, Belgium, 2020.
5. Plastics Recyclers Europe. *HDPE & PP Market in Europe—State of Play*; Plastics Recyclers Europe: Brussels, Belgium, 2020.

6. United Nations—Department of Economic and Social Affairs. 20 October 2021. Available online: https://sdgs.un.org/#goal_section (accessed on 8 May 2022).
7. Rudolph, N.; Kiesel, R.; Aumnate, C. *Understanding Plastics Recycling*; Carl Hanser Verlag: Munich, Germany, 2017.
8. Schyns, Z.; Shaver, M. Mechanical recycling of packaging plastics: A review. *Macromelcular Rapid Commun.* **2021**, *42*, 2000415. [CrossRef] [PubMed]
9. Hamad, K.; Kaseem, M.; Deri, F. Recycling of waste from polymer materials: An overview of the recent works. *Polym. Degrad. Stabil.* **2013**, *98*, 2801–2812. [CrossRef]
10. Yin, S.; Tuladhar, R.; Shi, F.; Shanks, R.A.; Combe, M.; Collister, T. Mechanical reprocessing of polyolefin waste: A review. *Polym. Eng. Sci.* **2015**, *55*, 2899–2909. [CrossRef]
11. Soto, J.M.; Blazquez, G.; Calero, M.; Quesada, L.; Godoy, V.; Martin-Lara, M.A. A real case study of mechanical recycling as an alternative for managing of polyethylene plastic film presented in mixed municipal solid waste. *J. Clean. Prod.* **2018**, *203*, 777–787. [CrossRef]
12. Shah, A.A.; Hasan, F.; Hameed, A.; Ahmed, S. Biological degradation of plastics: A comprehensive review. *Biotechnol. Adv.* **2008**, *26*, 246–265. [CrossRef] [PubMed]
13. Domininghaus, H.; Eyerer, P.; Elsner, P.; Hirth, T. *Die Kunststoffe und ihre Eigenschaften*; Springer: Berlin/Heidelberg, Germany, 2005.
14. Pinheiro, L.A.; Chinelatto, M.A.; Canevarolo, S.V. The role of chain scission and chain branching in high density polyehtylene dring thermo-mechanical degradation. *Polym. Degrad. Stabil.* **2004**, *86*, 445–453. [CrossRef]
15. Canevarolo, S.V. Chain scission distribution function for polypropylene degradation during multiple extrusions. *Polym. Degrad. Stabil.* **2000**, *70*, 71–76. [CrossRef]
16. Bernardo, C.A.; Cunha, A.M.; Mendes, A.A. Study of the degradation mechanisms of polyethylene during reprocessing. *Polym. Degrad. Stabil.* **2011**, *98*, 1125–1133.
17. La Mantia, F.P. Closed-loop recycling. A case study of films for greenhouses. *Polym. Degrad. Stabil.* **2010**, *95*, 285–288. [CrossRef]
18. Tapper, R.J.; Longana, M.L.; Yu, H.; Hamerton, I.; Potter, K.D. Development of a closed-loop recycling process for discontinuous carbon fibre polypropylene composites. *Composites B* **2018**, *146*, 222–231. [CrossRef]
19. Häußler, M.; Eck, M.; Rothauer, D.; Mecking, S. Closed-loop recycling of polyethylene-like materials. *Nature* **2021**, *590*, 423–427. [CrossRef] [PubMed]
20. Jansson, A.; Möller, K.; Gevert, T. Degradation of post-consumer polypropylene materials exposed to simulated recycling—Mechanical properties. *Polym. Degrad. Stabil.* **2003**, *82*, 37–46. [CrossRef]
21. Oblak, P.; Gonzalez-Gutierrez, J.; Zupancic, B.; Aulova, A.; Emri, I. Mechanical properties of extensively recycled high density polyehtylene (HDPE). *Mater. Today Proc.* **2016**, *3*, 1097–1102. [CrossRef]
22. Jin, H.; Gonzalez-Gutierrez, J.; Oblak, P.; Zupancic, B.; Emri, I. The effect of extensive mechanical recycling on the properties of ow density polyethylene. *Polym. Degrad. Stabil.* **2012**, *97*, 2262–2272. [CrossRef]
23. Schweighuber, A.; Felgel-Farnholz, A.; Bögl, T.; Fischer, J.; Buchberger, W. Investigations on the influence of multiple extrusion on the degradation of polyolefins. *Polym. Degrad. Stabil.* **2021**, *192*, 109689. [CrossRef]
24. *ISO 17855-2:2016*; ISO/TC 61/SC 9 Thermoplastic Materials. Plastics—Polyethylene (PE) Moulding and Extrusion Materials—Part 2: Preparation of Test Specimens and Determination of Properties. ISO: Geneva, Switzerland, 2016.
25. *ISO 1133-1:2011*; ISO/TC 61/SC 5 Physical-Chemical. Plastics—Determination of The Melt Mass-Flow Rate (MFR) and Melt Volume-Flow Rate (MVR) of Thermoplastics—Part 1: Standard Method. ISO: Geneva, Switzerland, 2011.
26. Almond, J.; Sugumaar, P.; Wenzel, M.; Hill, G.; Wallis, C. Determination of the carbonyl index of polyehtylene and polypropylene using specified area under band methodology with ATR-FTIR spectroscopy. *E-Polymers* **2020**, *20*, 369–381. [CrossRef]
27. Ehrenstein, G.W.; Riedel, G.; Trawiel, P. *Thermal Analysis of Plastics*; Carl Hanser Verlag: Munich, Germany, 2004.
28. *ISO 527-2:2012*; ISO/TC 61/SC 2 Mechanical Behavior. Plastics—Determination of Tensile Properties—Part 2: Test Conditions for Moulding and Extrusion Plastics. ISO: Geneva, Switzerland, 2012.
29. BASF Schweiz AG. *Technical Information—Irgafos 168*; BASF Schweiz AG: Basel, Switzerland, 2010.
30. BASF Corporation. *Technical Data Sheet—Irganox 1010*; BASF Corporation: Charlotte, NC, USA, 2019.
31. Baur, E.; Harsch, G.; Moneke, M. *Werkstoff-Führer Kunststoffe*; Carl Hanser Verlag: München, Germany, 2019.
32. Verleye, G.A.L.; Roeges, N.P.G.; De Moor, M.O. *Easy Identification of Plastics and Rubbers*; Rapra Technology Limited: Shropshire, UK, 2001.
33. Zweifel, H.; Maier, R.D.; Schiller, M. *Plastics Additives Handbook*; Carl Hanser Verlag: Munich, Germany, 2009.
34. Fischer, J.; Lang, R.W.; Bradler, P.R.; Freudenthaler, P.J.; Buchberger, W.; Mantell, S.C. Global and local aging and differently stabilized polypropylenes exposed to hot chlorinated water without and with superimposed mechanical environmental loads. *Polymers* **2019**, *11*, 1165. [CrossRef] [PubMed]

Article

Evaluation and Characterization of Ultrathin Poly(3-hydroxybutyrate) Fibers Loaded with Tetraphenylporphyrin and Its Complexes with Fe(III) and Sn(IV)

Svetlana G. Karpova [1], Natalia A. Chumakova [2,3], Anton V. Lobanov [1,3,4], Anatoly A. Olkhov [1,4], Alexandre A. Vetcher [5,6,*] and Alexey L. Iordanskii [3,*]

[1] N.M. Emanuel Institute of Biochemical Physics, Russian Academy of Sciences, 4 Kosygin St., 119991 Moscow, Russia; karpova@sky.chph.ras.ru (S.G.K.); avlobanov@mail.ru (A.V.L.); aolkhov72@yandex.ru (A.A.O.)
[2] Department of Chemistry, Lomonosov Moscow State University, 1 Kolmogorov St., 119991 Moscow, Russia; harmonic2011@yandex.ru
[3] N.N. Semenov Federal Research Center for Chemical Physics, Russian Academy of Sciences, 4 Kosygin St., 119334 Moscow, Russia
[4] Academic Department of Innovational Materials and Technologies Chemistry, Plekhanov Russian University of Economics, 36 Stremyanny Ln, 117997 Moscow, Russia
[5] Institute of Biochemical Technology and Nanotechnology (IBTN), Peoples' Friendship University of Russia (RUDN), 6 Miklukho-Maklaya St., 117198 Moscow, Russia
[6] Complementary and Integrative Health Clinic of Dr. Shishonin, 5 Yasnogorskaya St., 117588 Moscow, Russia
* Correspondence: avetcher@gmail.com (A.A.V.); aljordan08@gmail.com (A.L.I.)

Abstract: The effect of small additions (1–5 wt.%) of tetraphenylporphyrin (TPP) and its complexes with Fe (III) and Sn (IV) on the structure and properties of ultrathin fibers based on poly(3-hydroxybutyrate) (PHB) has been studied. A comprehensive study of biopolymer compositions included X-ray diffraction (XRD), differential scanning calorimetry (DSC), spin probe electron paramagnetic resonance method (EPR), and scanning electron microscopy (SEM). It was demonstrated that the addition of these dopants to the PHB fibers modifies their morphology, crystallinity and segmental dynamics in the amorphous regions. The annealing at 140 °C affects crystallinity and molecular mobility in the amorphous regions of the fibers, however the observed changes exhibit multidirectional behavior, depending on the type of porphyrin and its concentration in the fiber. Fibers exposure to an aqueous medium at 70 °C causes a nonlinear change in the enthalpy of melting and challenging nature of a change of the molecular dynamics.

Keywords: poly(3-hydroxybutyrate); biodegradable polyester; ultrafine electrospun fibers; tetraphenylporphyrin; metalloporphyrin complexes; Fe(III); Sn(IV); X-ray diffraction; DSC; spin probe EPR method; SEM

1. Introduction

The widespread employment of nanotechnology in modern medicine, the transition from traditional macro- and micro- to submicron and nanoscale medication forms (MF), as well as the implementation of ultrafine implants and diagnostic systems, cause the scientific community to pay close attention to bio-based polymer materials that are completely decomposed in the living systems without the formation of toxic products. Poly(3-hydroxybutyrate) (PHB) as a basic representative of polyhydroxyalkanoates' family (PHA) is such bacterial biodegradable and biocompatible polymer with great commercial perspectives and high sustainability [1–3]. PHB micro/nano fibers have been fabricated by electrospinning (ES), which is well-developed for creating nanoscale polymer carriers with adjustable morphology and properties [4,5] The ES technique provides the design of nonwoven fibrous membranes (mats) with a large inherent surface-to-volume ratio [6] that is extremely important for biomedical, packaging, and environmental applications.

The crucial characteristics of the biodegradable MF are the kinetic parameters of drug release in vitro or in vivo which are determined by the combination of drug diffusion and biopolymer decomposition rate in the limited space of fibrous micro- nanocarrier. Since the drug diffusional transport depends strongly on the inherent structure of the biopolymer, its morphology and crystallinity, all of these characteristics can significantly affect the kinetic profile of drug release, and, eventually, the effectiveness of the MF implementation. Structural changes in fibrillar mats during storage and can be caused by multiple factors, such as water absorption, heating, oxidation, ozonolysis, action of UV radiation, as well as the degradation effect of microorganisms. The listed factors can act simultaneously or consequently, depending on the operating conditions of the MF and the environment.

The effective method of direct impact on the structure of polymer materials is to dope it with the modifiers of organic and inorganic natures. In our preliminary reports, we demonstrated the influence of a series of additives on the structure of PHB-based fibrous materials. As the modifiers we used dipyridamole [7], chitosan [8], TiO_2 and silicon nanoparticles [9], Fe-chloroporphyrin [10], Zn-porphyrin [11], Mg-chloroporphyrin [12], etc., that could be used as the therapeutic agents. The effect of low-molecular substances on the structure of the crystalline and amorphous phases of PHB fibers was shown in the above-mentioned publications. All these substances interact with the polyester groups of PHB. As a result of such interactions, both deceleration and acceleration of crystallinity, orientation, and relaxation of macromolecules have occurred. Obviously, for the formation of matrices with desired properties, it is necessary to establish the relationship between the structure of bioactive dopants and their impact on the structural and dynamic parameters of the fibrous material.

The challenges related to the variation of free volume [13,14] and structural organization [15,16] of the micro- and nanofibers stimulate molecular/segmental mobility investigations. The free volumes in polar fibrous polymers were found to be localized mainly at the chain ends [13], contributing to total free volume. However, for the biopolymers with poor polarity such as a highly crystalline PHB, the free volume should locate predominantly in intercrystalline amorphous areas where the transitive macromolecules are situated. The comprehensive exploration of drug delivery therapeutic systems on the base of polysorbate-80 and the cyclodextrin derivative has been performed to disclose the macro- and microstructure characterization in combination with free volume distribution study [15]. The innovative method of ortho-positronium annihilation was implemented there to display the free volume variation as response on intrinsic characteristics of drug delivery system [15]. The correlation among drug release, free volume concentration and segmental dynamics is a key factor of diffusion processes that control the drug release profiles in planar and fibrillar polymer systems [16,17].

Of particular interest is the incorporation of porphyrin metal complexes as the special dopants into the PHB polymer. It is well known from the numerous literature papers that the complexes of metals with TPPs have unique photocatalytic and antimicrobial properties [18,19]. Metalloporphyrin complexes could promote the formation of the reactive oxygen species, such as superoxide radical anion, peroxide and hydroxyl radicals, and hydrogen peroxide, the cytostatic activity of these substances is well known. These radical and radical-ion particles cause oxidative destructive reactions in cells, in other words, antimicrobial effect [20,21].

Along with the bactericide activity, due to the specific geometry and electronic structure, the complexes of metals with porphyrins have a significant effect on the crystallization and segmental orientation of macromolecules [22]. The pristine porphyrin molecules as well as their metal complexes are amphiphilic [23] or even hydrophobic [24] that promote their aggregation in the form of nanoparticles [23,25–27]. Depending on the nature of the metal included in metalloporphyrins, they exhibit a variable tendency to aggregation [28] that determines their ability to act as nucleating agents during polymer crystallization. In addition, porphyrins and their metal complexes have several binding sites, which facilitate the appearance of coordination interactions with the adjacent molecules [29]. Metal cations with chloride extra-ligands contained in the structure of porphyrin metal complexes can exchange ligands for polar fragments of the environment (for example, polymeric), such as oxygen-containing hydroxy- or carboxy-entities.

Complexes of TPP with Sn(IV) ($SnCl_2$-TPP) and Fe(III) (FeCl-TPP) are the most promising PHB dopants for creating polymeric materials for medical use. Sn-porphyrin metal-complexes are currently employed in the design of photocatalysts that promote the destruction of organic toxicants [29] and photosensitizers for medical diagnostics and therapy [30]. A significant advantage of the complexes of porphyrins with Sn (IV) is the presence of two extra ligands located on opposite sides of the plane of the porphyrin macrocycle (Figure 1). This structural peculiarity is responsible for the almost complete disability of $SnCl_2$-TPP to aggregate in comparison with the complexes of porphyrins and metals in the oxidation states +2 and +3. FeCl-TPP molecules could have a strong interaction with PHB macromolecules; however, the presence of only one Cl as a ligand on only one side of the cycle leads to the potential ability of molecular aggregation.

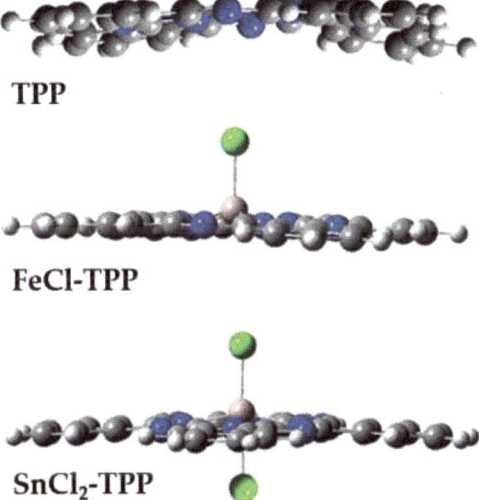

Figure 1. Side view on TPP, FeCl-TPP, and $SnCl_2$-TPP.

Summarizing the above-mentioned topics, the goal of this study is the comparative analysis of the effect of TPP and its complexes with Sn (IV) and Fe (III) on the structural and dynamic parameters of a fibrous material based on PHB, as well as the detection of external factors' impacts (ozonolysis, hydrolysis, annealing at 140 °C, etc.) on the structural and dynamic characteristics of the studied biodegradable polymer-porphyrin compositions.

2. Materials and Methods

2.1. Materials

For ES we used microbiologically synthesized PHB Series 16F Biomer (Schwalbach am Taunus, Germany) with a viscosity-average molecular mass (M_v) 2.06×10^5 D, density $\rho = 1.248$ g/cm^3, melting point Tm = 177 °C, initial degree of crystallinity ~63%.

TPP, $FeCl_3$-TPP complex and $SnCl_2$-TPP complex were obtained according to the procedure described earlier [31,32]. Their structural formulas are exhibited in Figure 1. The TPP complexes of Fe and Sn are readily soluble in organochlorine solvents, such as chloroform ($CHCl_3$), but insoluble in water. To obtain the solution prepared for ES formation, the porphyrin dopants were dissolved in CHCl$_3$ and then added to solutions of PHB in CHCl$_3$. Thus, the both components have the same cosolvent.

The ES-formation solutions of PHB, PHB/TPP, PHB/FeCl$_3$-TPP, and PHB/SnCl$_2$-TPP in CHCl$_3$ were prepared at a temperature of 60 °C by stirring on an automatic magnetic stirrer to complete uniformity. The concentration of PHB in the solution was 7 wt.%, the content of TPP, FeCl$_3$-TPP, SnCl2-TPP was equal to 1, 3, and 5 wt.% relatively to the mass of PHB.

Fibers were obtained by ES on a single capillary laboratory setup with the following parameters: capillary diameter—0.1 mm, voltage—12 kV, distance between electrodes—18 cm, solution conductivity—10 µS/cm as described in [32] The fibers in the form of the mats were annealed at 140 °C ± 1 °C in the vacuum oven within the time interval 30–240 min. Right after then fibers were rapidly cooled to the room temperature.

2.2. Methods

X-ray structural analysis (XRD) of the samples was carried out by transmission recording. High-resolution two-dimensional scattering patterns were obtained using the S3-Micropix small- and wide-angle X-ray scattering system (CuKα radiation, λ = 1.542 Å) (Hecus X-ray Systems GmbH (Graz, Austria)). A Pilatus 100K (DECTRIS Ltd. (Baden, Switzerland)) detector was used, as well as a PSD 50M linear devise for the argon flow, a high voltage of 50 kV and a current of 1 mA on the Xenocs Genix (Xenocs SAS (Grenoble, France)) source tube. Fox 3D X-ray optics (Xenocs SAS (Grenoble, France)) were used to form the X-ray beam; the diameters of the forming slits in the collimator were 0.1 and 0.2 mm. To eliminate X-ray scattering in air, the X-ray mirror unit and the camera during signal accumulation were placed under the vacuum of $(2–3) \times 10^{-2}$ torr. The signal accumulation time was varied in the range 600–5000 s.

X-band electron paramagnetic resonance (EPR) spectra were recorded on an EPR-V automatic spectrometer (N.N. Semenov Federal Research Center for Chemical Physics, Russian Academy of Sciences, Moscow, Russia). To avoid saturation effects, the microwave power did not exceed 1 mW. The modulation amplitude was always significantly less than the resonance line width and did not exceed 0.5 G. A stable nitroxide radical TEMPO was used as a spin probe. The radical was incorporated into the fibers from the gas phase at a temperature of 50 °C for an hour. The concentration of the radical in the polymer was determined by double integration of the EPR spectra. As the reference was vacuum degassed TEMPO solution in CCl$_4$ with a radical concentration of around 1×10^{-3} mol/L.

Simulation of EPR spectra was performed using the computer program described in This program is the modified version of the program described in [33]. The simulation was performed using nonlinear least-square algorithm. The values of the parameters under consideration were selected in such a way as to minimize the sum of squared deviations between the calculated spectra and the experimental ones. The initial program [33] allows calculating ESR spectra in the framework of Brownian rotational diffusion. The modified program [34] also allows taking into account the lognormal distribution of the rotational correlation times of paramagnetic molecules and calculating the spectra under the assumption of simultaneous rotation and quasi-librations (high-frequency low-amplitude vibrations of molecules near the equilibrium position [35,36]) of the radicals. The spectra were simulated using the following principal values of the g-tensor and the tensor of hyperfine interaction of an unpaired electron with a ^{14}N nucleus: g_{xx} = 2.0093, g_{yy} = 2.0063, g_{zz} = 2.0022, A_{xx} = 7.0 G, A_{yy} = 5.0 G, A_{zz} = 35.0 G. The value of A_{zz} was determined experimentally from the EPR spectrum of TEMPO in PHB recorded at the temperature of liquid nitrogen. The obtained value is close to the value of A_{zz} given in [37].

Hydrolysis of samples in aqueous medium was studied at 70 °C ± 1 °C. Before the introduction of the radical, the samples exposed to water were dried in the vacuum oven to constant weight for 100 hr.

The differential scanning calorimetry (DSC) study was accomplished on a NETZSCH DSC 204 F1 instrument (NETZSCH-Geratebau GmBH (Selb, Germany) in an Ar atmosphere with a heating rate of 10 K/min. The average statistical error in the measurement of thermal effects was ±3%. The enthalpy of melting was calculated using the NETZSCH Proteus program. Thermal analysis was conducted according to the standard procedure [38]. Peak separation was performed using the NETZSCH Peak Separation 2006.01 software.

The scanning electron microscopy (SEM) to obtain geometric parameters of the fibrous materials was performed by TM-3000 scanning electron microscope (Hitachi, Ltd. (Tokyo, Japan)) at an accelerating voltage of 20 kV. Au layer of 10–20 nm was deposited on the surface of a nonwoven fibrous material sample. The computation and measurements of PHB fibrillar diameters were achieved using the software tool Java-based ImageJ version 1.52a (National Institutes of Health, Bethesda, MD, USA) and OriginPro 2018 (Origin Lab Corporation, Northampton, MA, USA). The histograms of diameter distribution for the loaded fibers of PHB have been built on the base of statistic estimation for 100 fibrillar elements patterned at SEM microphotographs.

3. Results and Discussion

3.1. Effect of Porphyrin and Metal-Porphyrines Concentrations on PHB Fiber Morphology

The pristine PHB fibers demonstrate the large concentration of ellipsoid structures as the anomalous bead-like fragments existing in the fibrous mats along with conventional cylindrical form being typical for electrospun fibers, see Figure 2A. As it was stated in the literature and a series of our works [39–42], the reason for the ellipsoids' formation is low electrical conductivity (<1 μS/cm) and low surface tension of ES polymer solutions. The size of the ellipsoid beads in the transverse direction is 7.5 ± 2.5 μm, and in the longitudinal direction is ~17 ± 7 μm. The average diameter of the cylindrical fragments of the fiber is located in the range 1.5–4.5 μm. Comparative data on fiber diameter distribution is presented on inserts of Figure 2.

When the metal complexes of porphyrins were embedded into PHB fibers, the polymer morphology changed dramatically. At the $SnCl_2$-TPP complex content 1–5 wt.%, the ellipsoid elements have completely disappeared; and for 1% of the complex, the fiber diameter distribution has the maximum at 1.75 ± 0.25 μm. The distribution asymmetry (insert in Figure 2B) manifested in the predominance of thin fibers' fraction < 1.75 μm is likely a consequence of the jet splitting. With an increase in the concentration of FeCl-TPP from 3 to 5%, the fibers with a diameter of 3 μm prevail; see the corresponding histograms in Figure 2C,D.

The above observations can be explained as follows. When the metal-TPP complexes were added to the polymer electrospinning solution, ionic conductivity arises, which leads to the leveling of the surface tension of the solution drop and to the stability of the electrospinning process. The relative decrease in the fiber diameter is likely to be in correlation with the value of the electrical conductivity of the solution used for the polymer fibers formation. The bed-like entities disturbed the normal statistic distribution of the fibrils which caused to anomalous histogram. In this regard, we reconstructed the Figure 2A under condition that the anomalous beds should be excluded from statistical evaluation.

It is necessary to note that polydispersity with different locations of the distribution maximum in fiber diameters was observed for all compositions under study (see the corresponding histograms in Figure 2). Naturally, the polydispersity is due to the splitting effect in the primary polymer jet during electrospinning. The splitting effect can be associated with both an increase in the electrical conductivity of solution and a decrease in the surface tension at the polymer-air interface. Presumably, both effects could arise as a result of the metalloporphyrin introduction.

Thus, the introduction of TPP complexes leads to a change in the geometry and morphology of the fibers in the nonwoven fibrous material. The changes are undoubtedly associated with structural features of the PHB filament, which we will describe in the further sections by the methods of XRD, DSC, and EPR of the paramagnetic probe.

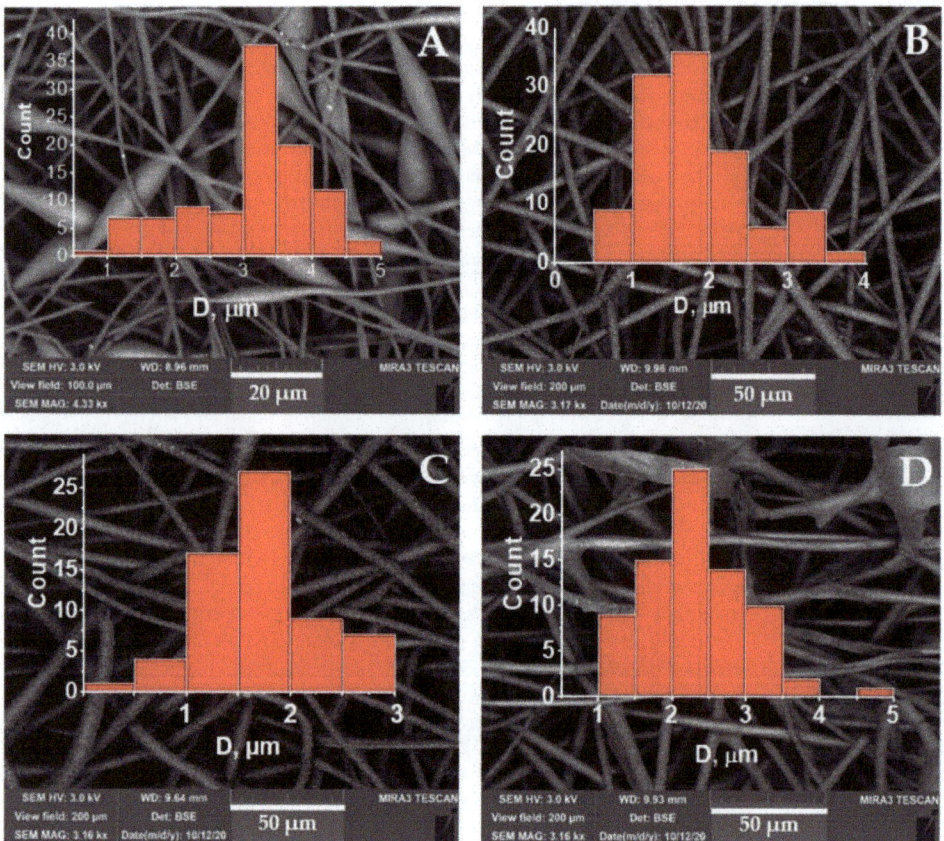

Figure 2. SEM images of the PHB fibrillar structures loaded by the metal-porhyrine complex of SnCl$_2$. The concentration of SnCl$_2$—PPT: (**A**) 0%, (**B**) 1%, (**C**) 3%, (**D**) 5%.

3.2. XRD Patterns of PHB Fibers Comprising the Porphyrines' Complexes

The intermolecular interaction among the metal—porphyrin complex particles, as well as among these particles and PHB molecules, significantly depends on the nature of the complexes. During ES, the porphyrin molecules are attracted each other and, as a result, much larger aggregates are observed. The interaction of these particles with PHB macromolecules could be extremely small. In contrast with the pristine porphirin the FeCl-TPP complex has a strong interaction with PHB macromolecules. The presence of an extra-ligand chloride in the metal complex leads to intermolecular repulsion, i.e., the molecules FeCl-TPP are aggregated into smaller particles. In contrast, SnCl$_2$-TPP complexes are statistically distributed in the polymer matrix due to the repulsive forces between a pair of Cl$^-$ groups (see Figure 1). The interaction of SnCl$_2$-TPP with PHB macromolecules is weaker than that of FeCl-TPP. This picture of intermolecular interactions determines the structural and dynamic characteristics of ultrathin fibers.

The study of initial PHB fibers and fibers from PHB loaded with TPP, SnCl$_2$-TPP, and FeCl-TPP of various compositions just after ES and after annealing at 140 °C, was carried out by X-ray diffraction at wide and small angles. Figure 3 demonstrates the typical high-angle diffractograms of PHB/FeCl-TPP fibers.

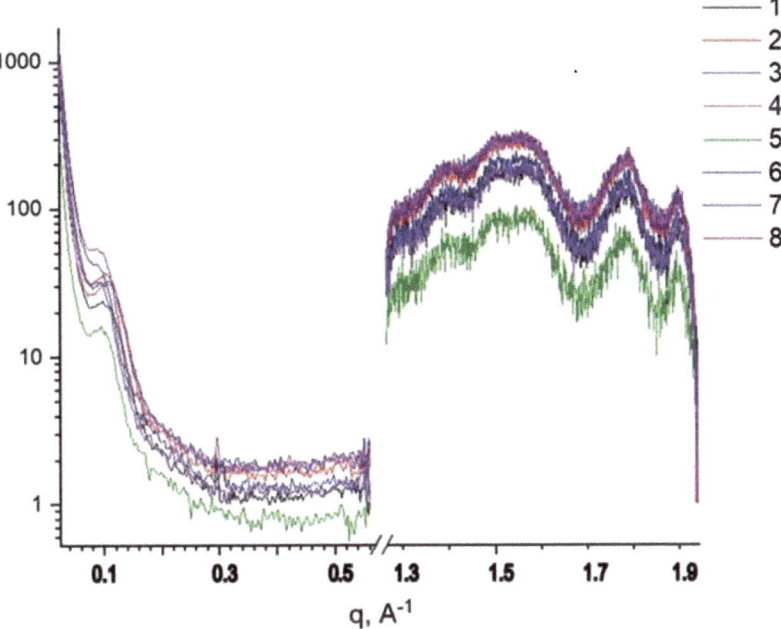

Figure 3. Diffractograms of PHB fibers containing metal-porphyrin complex (FeCl-TPP) before (1–4) and after annealing (5–8). PHB. nonannealed (1) and annealed (5); PHB-TPP-FeCl (1%): nonannealed (2) and annealed (6); PHB-TPP-FeCl (3%): nonannealed (3) and annealed (7); PHB-TPP-FeCl (5%): nonannealed (4) and annealed (8).

The average effective crystallite size L_{hkl} in the crystallographic hkl direction was determined from the integral half-width of the line of the corresponding X-ray reflection using the Selyakov-Scherrer formula,

$$\Delta_{hkl}(2\theta) = \lambda / L_{hkl} \cos\theta_m \quad (1)$$

The value of the large period was calculated by the formula,

$$d = n\lambda / 2\theta_m \quad (2)$$

where d—long period, λ = 1.542 Å wavelength of CuKα—radiation, θ_m—diffraction angle, and n—the order of reflection.

Figure 4a exhibits the dependences of the degree of crystallinity and the relative longitudinal size of crystallites on the concentration of the dopants. It is remarkably visible, that both parameters increase with the introduction of FeCl-TPP (the size of the crystallites increased by 30%). In PHB/TPP and PHB/(SnCl$_2$-TPP) fibers, the changes in the degree of crystallinity and the longitudinal size of crystallites are within the experimental errors. The significant effect of FeCl-TPP on the crystal structure of the polymer is due to strong interaction of the additive molecules with the polymer macromolecules and the formation of dopant particles that can serve as the nuclei of crystallization. It should be noted that with an increase in the FeCl-TPP concentration from 3% to 5%, only a slight decrease in the degree of crystallinity and the longitudinal size of crystallites is observed. It takes place due to the aggregation of complexes into larger particles, which does not lead to a further increase in the proportion of crystallites during ES.

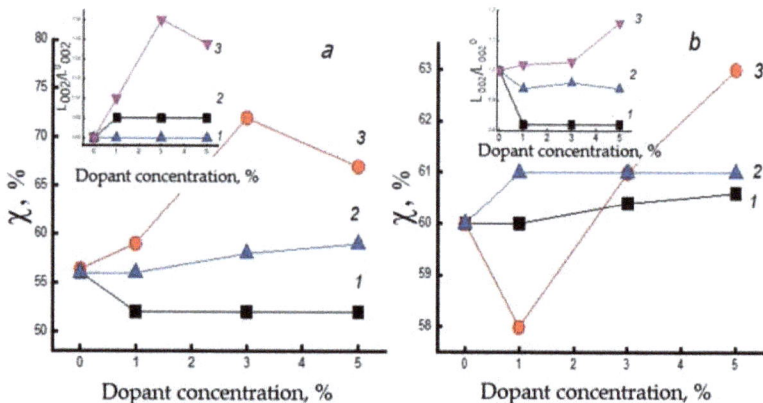

Figure 4. Crystallinity degree (χ) dependence and relative longitudinal size (L_{002}/L^0_{00}) evolution (the inserts) as functions of the dopant concentration. (**a**) initial fibers, (**b**) the same fibrous samples after annealing at 140 °C. 1—TPP, 2—SnCl$_2$-TPP, 3—FeCl-TPP.

Thermal sterilization of polymeric materials remains the most common method in clinical practice. The recommended hot air annealing temperature required for the death of all microorganisms is 160 °C, however, the mode allowed for processing materials based on PHB fibers is limited to 140 °C, which is due to the onset of polymer degradation. Figure 4b exhibits the dependences of the degree of crystallinity and the longitudinal size of crystallites of the fibers annealed at 140 °C for 2 h on the concentration of the dopants. One can see that when 1% of FeCl-TPP is added, a decrement in the degree of PHB crystallinity is observed, and with an increment in the concentration of the metal complex, the degree of crystallinity increases sharply. The observed effects can be explained on the basics of knowledge, that crystallites and amorphous regions in a polymer generally do not correspond to the minimum free energy. The tendency of an amorphous-crystalline system to a minimum of free energy is facilitated by annealing the polymer, when macromolecules acquire sufficient mobility, while the crystallites tend to increase the longitudinal size. It seems that an increment in the degree of crystallinity can be expected upon annealing the sample. However, at FeCl-TPP concentration of 1%, the degree of crystallinity decreases. We explain the observed phenomenon as follows. At 140 °C, linear structures, edged surfaces, and defected crystal regions are thawed, while FeCl-TPP particles diffuse into such systems and fix in them due to the strong intermolecular interaction of this metal complex with PHB molecules. As a result, such areas are decompressed and do not give a signal in XRD. Annealing of PHB/TPP and PHB/(SnCl$_2$-TPP) fibers at 140 °C leads to an insignificant increment in the degree of crystallinity.

3.3. Dynamic Characteristics of the Amorphous Phase of Ultrafine Fibers Loaded with Porphyrines

The structure of the amorphous regions is largely determined by the degree of crystallinity of the polymer. As a result, the addition of porphyrin metal complexes to the PHB fiber changes not only the degree of PHB crystallinity but also the morphology and molecular dynamics in the amorphous regions. The molecular mobility of the amorphous regions of the polymer was studied by the spin probe technique using the stable nitroxide radical TEMPO. The typical EPR spectra of the radical in the polymer matrixes under consideration are shown in Figure 5.

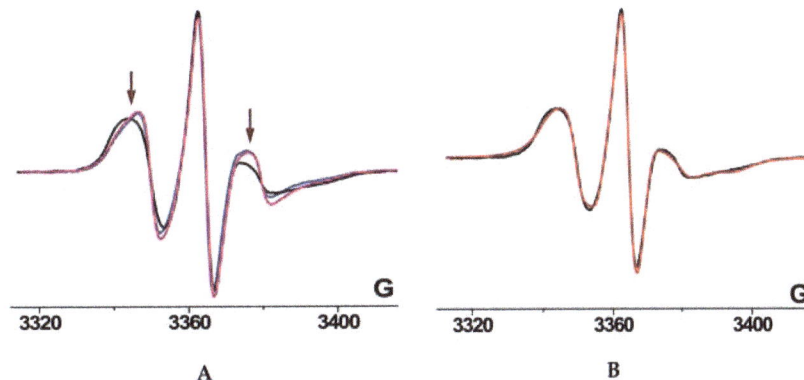

Figure 5. EPR spectra changes after annealing at 140 °C. (**A**) EPR spectra of TEMPO radical in PHB samples containing 5% TPP, before annealing (blue line) and after annealing for 90 (pink line) and 240 min (black line). The spectra are normalized to the intensity of the central component; the arrows indicate the regions of the greatest change in the line shape. (**B**) Spectrum simulation for the samples annealed for 240 min (black line—experimental spectrum, red line—simulation result).

The widely used method for determining the rotational parameters of spin probes in polymers is the computer simulation of the EPR spectra. It was found that for high-quality simulation of the spectra of the studied systems it is necessary to take into account the anisotropy of the rotational mobility of radicals, the continuous (lognormal) distribution of paramagnetic molecules over the rotational diffusion coefficients, as well as high-frequency low-amplitude vibrations of paramagnetic molecules near the equilibrium position—quasi-vibration [32–34,43]. As an illustration, Figure 5b shows the result of computer simulation of EPR spectrum of TEMPO in PHB containing 5% TPP after annealing at 140 °C for 240 min. Table 1 presents the parameters of the rotational mobility of the spin probes in the samples, the spectra of which are shown in Figure 5.

Table 1. Parameters of the rotational mobility of TEMPO radicals in PHB containing 5% TPP at different times of sample annealing. σ is the width of the lognormal distribution of paramagnetic molecules with respect to rotational mobility; L—amplitudes of quasi-vibrations.

	0 min	90 min	240 min
D_x (s^{-1})/$\tau_{c,x}$ (s)	<2 × 10^6/>8 × 10^{-8}	<2 × 10^6/>8 × 10^{-8}	<2 × 10^6/>8 × 10^{-8}
D_y (s^{-1}) (±0.1 × 10^7)/$\tau_{c,y}$ (s)	6.4 × 10^7/2.6 × 10^{-9}	6.0 × 10^7/2.8 × 10^{-9}	5.5 × 10^7/3.0 × 10^{-9}
D_z (s^{-1}) (±0.1 × 10^7)/$\tau_{c,z}$ (s)	5.8 × 10^7/2.9 × 10^{-9}	5.7 × 10^7/2.9 × 10^{-9}	2.7 × 10^7/6.2 × 10^{-9}
σ_y (±0.05)	0.61	0.76	0.55
σ_z*	2.50	2.50	2.50
L_y (±2°)	45°	46°	41°
L_z (±2°)	54°	55°	60°
τ_c*	2.4 × 10^{-9}	2.3 × 10^{-9}	5.6 × 10^{-9}

* The accuracy of determining the width of the distribution of radicals by rotational mobility around the molecular Z axis is low; in the resulting attempt at computer modeling, this value did not vary.

It was found that in all systems under study, paramagnetic molecules rotate anisotropically. The coefficients of the rotational diffusion of the radicals around X axis of g-tensor (N–O bond) do not exceed 2 × 10^6 c^{-1}, while the coefficients of rotational diffusion around Y and Z axes are in the range 10^7–10^8 c^{-1}. Such anisotropy may indicate the interaction of probe molecules with functional groups of polymer molecules due to p-orbitals of nitrogen and oxygen atoms. A wide distribution of rotational mobility of the radicals was also revealed. This is caused by inhomogeneity of the structure of the amorphous region of the polymers. Indeed, the amorphous phase is a set of structures characterized by different packing densities and different molecular dynamics of polymer chains. The EPR spectrum is a superposition of the spectra of radicals located in different regions of the amorphous phase and, therefore, having different mobility.

It should be noted that simulation of EPR spectra of low-molecular-weight dopants in polymers in the range of rotational correlation times of 10^{-7}–10^{-8} s is a very time-consuming procedure. In addition, the present work is aimed at the qualitative analysis of changes in the radicals' mobility as a result of treatment of the polymer matrix, and not at the interpretation of the exact values of the rotational parameters. In such a case it is useful to introduce a parameter that qualitatively characterizes the rotational mobility of radicals, can be determined without spectra simulation, and has sufficient sensitivity to small changes in the shape of the spectrum, such as the changes shown in Figure 5b. It can be seen that the greatest difference in the spectra is observed in the region of high-field and low-field components (indicated by arrows in the figure); therefore, as a parameter of the spectrum shape, we chose the value calculated as follows [44,45]:

$$\tau_c^* = \Delta H_{+1} \times [(I_{+1}/I_{-1})^{0.5} - 1] \times 6.65 \times 10^{-10} \, [s] \quad (3)$$

Here ΔH_{+1} is peak-to-peak width of the low-field spectral component; I_{+1} and I_{-1} are intensities of the low-field and high-field components correspondently.

This formula was proposed for the determination of the rotational correlation times of the nitroxide radicals of the piperidine series in the case of their isotropic rotation in the range of rotational correlation times of 5×10^{-11} s $\leq \tau_c^* \leq 1 \times 10^{-9}$ s. In this region, the EPR spectrum consists of three well-resolved components, the width of which is described within the framework of the Redfield theory [46]. In our case, this parameter qualitatively characterizes the shape of the EPR spectrum and can be considered as a characteristic correlation time. The values of τ_c^* for the three above systems are shown in Table 1. It can be seen that this value reflects the deceleration of the rotation of radicals in the PHB/TPP system (5%) upon sample annealing.

Figure 6a shows the change in the rotational correlation time of spin probes with an increase in the concentration of complexes. The highest growth of τ_c^* is in the PHB/FeCl-TPP system; in the PHB/TPP fibers, the correlation time increases not so significantly. In PHB/SnCl$_2$-TPP fibers, a decrease in τ_c^* is observed, although the degree of crystallinity determined using DSC increases with an increase in the additive content.

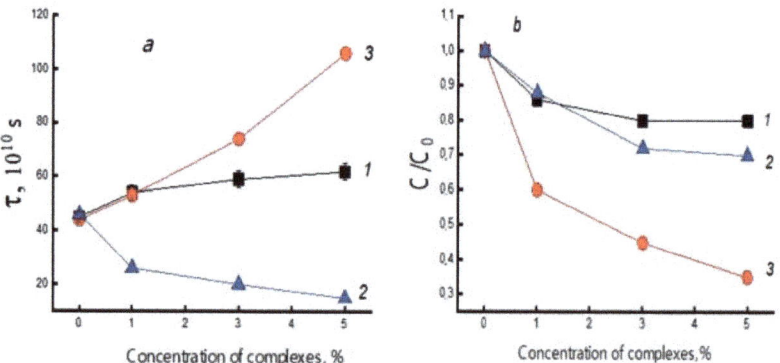

Figure 6. Dependences of probe correlation time τ (**a**) and relative probe concentration C/C_0 (**b**) on dopant content: TPP (1), SnCl$_2$-TPP (2), and FeCl-TPP (3).

As shown above, doping of PHB with PHB/FeCl-TPP causes an increase in the polymer crystallinity. This phenomenon is accompanied by increase in the correlation time, and, consequently, a slowdown in the molecular dynamics of macromolecules. In PHB/TPP, the effect of slowing mobility is weak. For the polymer doped with SnCl$_2$-TPP, a decrease in the rotational correlation time of the probe is observed, that is, an increase in the mobility of the polymer chains. This phenomenon is caused by the loosening effect of this metal complex, which is distributed in the system at the molecular level, and, therefore, the number of individual particles per unit volume of the polymer, in this case, is maximal.

Figure 6b demonstrates the dependences of the radical concentration on the polymer composition. It is seen that an increase in the PHB content leads to a decrease in the concentration of the radical. The strongest changes were observed in the polymer doped with $FeCl_3$-TPP due to the strongest change in the degree of crystallinity in these samples. It is known that impurity molecules do not diffuse into crystal structures; their concentration in dense amorphous regions is insignificant.

3.4. Effect of Thermal Treatment on the Structural and Dynamic Parameters of PHB Doped with Porphyrin and Metalloporphyrins

Of great scientific and practical interest is the study of the effect of sterilization on materials and products for medical use. The choice of optimal conditions for polymer processing, in view of the fact that the aggressive action of temperature, radiation, oxidizer (oxygen, ozone), along with the disinfection of the material and the destruction of pathogenic microorganisms, can lead to a significant deterioration in a structural hierarchy and hence in polymer properties.

3.4.1. Annealing of Samples at 140 °C

During ES performance, as a result of cooling and solidification, the polymer structure of ultrathin fibers can be far enough from the state of thermodynamic equilibrium. The imperfection in the crystalline phase and biopolymer morphology is manifested in the insufficient orientation of the segments in the fiber, as well as in an atypically low degree of crystallinity. To facilitate the transition of the polymer structure to equilibrium state the thermal annealing is used. The temperature impact allows intensifying segmental mobility and transferring the system to a more thermodynamically equilibrium state. The enhancing in the crystalline phase and especially additional crystallization occurs with the participation of transient polymer molecules located in intercrystalline area. Therefore, one should expect a change in the dynamics of spin probe that reflects the segmental mobility of PHB molecules in the intercrystalline fields. When comparing the mobility of the probe in the initial and annealed polymer samples contained different dopants, an opportunity appears to determine the effect of the additives on the polymer stability and dynamics of the intercrystalline structure.

Figure 7 shows the dependencies of the characteristic rotational correlation time of spin probes on the annealing duration of the samples at 140 °C. It is seen that τ^* decreases (rotational mobility of the radicals increases) significantly during annealing both for pristine PHB and PHB containing TPP, $SnCl_2$-TPP, and FeCl-TPP additives. The observed dependences can be explained as follows. During the fibers' fabrication, an essential proportion of the transitive macromolecules in transitive conformation are formed, but their segmental mobility is frozen as a result of the fiber jet solidification at room temperature. Annealing at 140 °C enhances macromolecular mobility that after gradual cooling leads to additional crystallization with the growth of crystallite sizes owing to the present of transitive straightened macromolecules. The spin probe molecules cannot penetrate into crystallites; therefore, their rotational mobility reflects the molecular mobility only in amorphous areas and in the areas of transitive macromolecules. After crystallization of the transitive areas the radicals are located predominantly in the amoprhous areas, so their average rotational mobility increases. Indeed, the absorption capacity of the fibers for the probe molecules decreases with an increase in the annealing time. For example, in the initial PHB fibers, the concentration of the radical was 0.75×10^{15} spin/cm^3, while after annealing for 90 min, it decreased to 0.53×10^{15} spin/cm^3, i.e., decreased 1.4 times. At the same conditions, the concentration of the radical in the PHB/$SnCl_2$-TPP composition decreases 2.1-fold, and in the system PHB/FeCl-TPP—10-fold after annealing the samples with 5% additive for 2 h. The greatest decrease in content of the probe for the system PHB/FeCl-TPP corresponds to the greatest decrease in the rotational correlational time of the radicals.

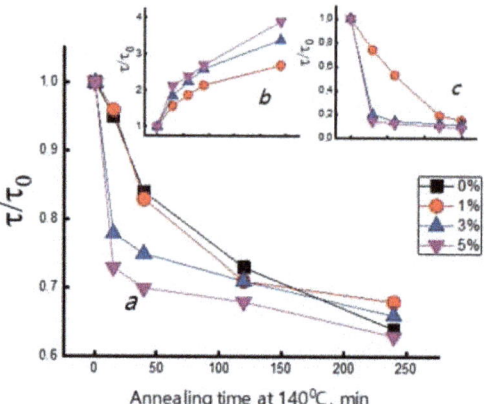

Figure 7. Dependence of the molecular dynamics expressed as relative correlation time (τ/τ_0) from annealing time at 140 °C for the PHB electrospun fibers with (**a**) TPP, (**b**) SnCl$_2$-TPP, (**c**) FeCl-TPP.

3.4.2. Exposure of Samples in Water at 70 °C

The structure and segmental dynamics of biodegradable materials for biomedicine significantly affect diffusion of drugs loaded into polymers and, consequently, the kinetic profiles of active component release. Since the polymer in the body functionalizes often in the aquatic environment, it is important to identify the changes in its structure and segmental mobility as a result of water action. Even a slight increase in the water content affects the diffusion processes of active agents in a polymer matrix. Previously, we studied the effect of an aqueous medium on the structure of PHB [47–52]. It was shown that under the influence of water, the PHB films degrade, the conformation of the polymeric chains altered, and the proportion of straightened segments decreases. It was necessary to establish the extent of these structural changes in the polymer compositions under investigation.

The exposure of fibrous materials to water medium in the present work was performed at 70° because at room temperature the saturation of polymer volume with water could last several days. When kept in water at 70 °C, the following processes occur simultaneously in the polymer:

(1) In accordance with the above description, heating the samples in water medium causes an increase in the degree of crystallinity and compaction of amorphous structures;
(2) Water molecules penetrating into the polymer matrix produce a plasticizing effect, as a result of which the degree of crystallinity also increases;
(3) When water molecules interact with PHB molecules and TPP or metalloporphyrin complexes, the hydrated complexes are formed, which loosen the structure of amorphous regions. After removal of boundless water, hydrated complexes could remain in the fiber;
(4) Water molecules penetrate into the accessible surfaces of crystallites and linear structures of amorphous areas, destroying these structures, as a result of which the degree of crystallinity decreases.

The EPR probe was introduced into the fibers after their heating in an aqueous medium and subsequent drying in vacuum during 2 days until constant weight. The test results are shown in Figure 8. It is seen that in the samples containing various porphyrin complexes in diverse concentrations and exposed in water for different times, the intensity of the above processes is essentially changed. In PHB fibers loaded by TPP-SnCl$_2$ (curves in Figure 8a), the mobility of spin probes increases (correlation time decreases) more than twice after maximal exposure (for 5 h). In the fibers with TPP completely different patterns of changes in molecular mobility are observed; at all concentrations of the additive and exposure time, the correlation time monotonically increases. In fibers containing FeCl-TPP, water exposure causes a complicated dependence of τ* on the additive concentration as the curves with maximum, see Figure 8b.

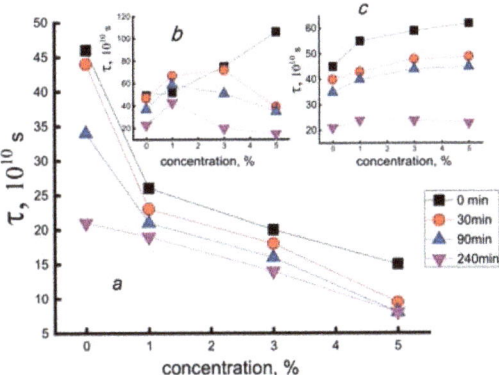

Figure 8. Dependence of time correlation of spin probe on the concentration of $SnCl_2$-TPP (**a**), FeCl-TPP (**b**), and TPP (**c**) at a different times of water exposure at 70 °C.

To compare the dopants impact on PHB structure the enthalpies of melting for crystalline fraction of PHB after water treatment are presented in Figure 9. In the case of $SnCl_2$-TPP dopant (curves 1–4) at all times of annealing, there are clearly observed the extreme dependences of ΔH on wt.%. After the minimum point at 1 wt.%, the further growth in the enthalpies is observed that means the increasing of the crystallinity of the polymer and is in good agreement with time correlation dependence in Figure 8a. In the case of TPP dopant the extremum in enthalpy is manifested not so clearly (curves 5–7). On the basis of comparing of Figures 8c and 9b we can suppose that treatment of PHB-TPP polymer in water at 70 °C causes an increase in the proportion of straightened polymer chains.

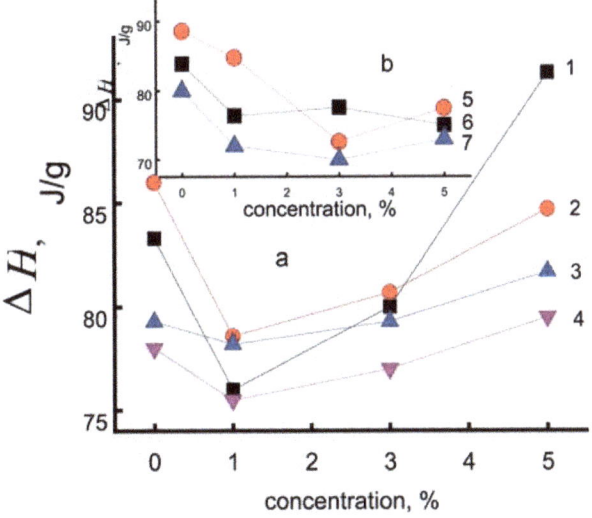

Figure 9. Dependence of melting enthalpy on the $SnCl_2$-TPP (**a**) and TPP (**b**) content. Durability of water treatment (in min) at 70 °C: 1—0, 2—30, 3—120, 4—240, 5—0, 6—120, 7—240.

In general, the results obtained indicate that the doping of PHB with TPP and its metallo-complexes has a significant and often multidirectional effect on structural changes in the polymer during its processing for the purpose of sterilization. This observation opens up the possibility of manufacturing materials from PHB with predetermined properties.

4. Conclusions

Generally, it can be concluded that doping of PHB with TPP and complexes TPP with Sn(IV) and Fe(III) leads to significant change in the geometry and morphology of the polymer fibers fabricated using ES method. According to the dynamic and structural exploring the annealing of ultrathin PHB fibers as well as their exposure in aqueous medium strongly affects the structure of the amorphous and crystalline regions of the polymer, moreover, the changing of the structure depends on the nature and amount of the additives. The data obtained should be taken into account when developing a sterilization regime for medical devices made of PHB.

Author Contributions: Conceptualization, A.L.I. and A.A.V.; methodology, S.G.K., N.A.C., A.V.L. and A.A.O.; software, A.A.O.; validation, S.G.K., N.A.C., A.V.L. and A.A.O.; formal analysis, S.G.K., N.A.C., A.V.L., A.A.O. and A.A.V.; investigation, S.G.K., N.A.C., A.V.L. and A.A.O.; resources, A.L.I.; data curation, A.A.O.; writing—original draft preparation, A.L.I. and A.A.V.; writing—review and editing, A.L.I. and A.A.V.; visualization, A.L.I. and A.A.O.; supervision, A.L.I. and A.A.V.; project administration A.L.I.; funding acquisition (see below). All authors have read and agreed to the published version of the manuscript.

Funding: This paper has been supported by the RUDN University Strategic Academic Leadership Program (recipient A.A.V.).

Institutional Review Board Statement: Not applicable.

Informed Consent Statement: Not applicable.

Data Availability Statement: Not applicable.

Acknowledgments: The authors are grateful to U.J. Haenggi and the company Biomer®, (Krailling, Germany) for the kind supply poly(3-hydroxybutyrate) and A.Kh. Vorobiev (Moscow State University, Department of Chemistry) for the generous providing of software for EPR spectra simulation. Special profound gratitude should be expressed to S.N. Chvalun (National Research Center «Kurchatov Institute») for the X-ray data presented and the fruitful discussion of the crystalline structure characterization. The authors would like to thank V.D. Bystrykh for her assistance with the edition of the submission's final version. A.A.V. is grateful to the RUDN University Strategic Academic Leadership Program for the support.

Conflicts of Interest: The authors declare no conflict of interest.

References

1. Anjana; Raturi, G.; Shree, S.; Sharma, A.; Panesar, P.S.; Goswami, S. Recent, approaches for enhanced production of microbial polyhydroxybutyrate: Preparation of biocomposites and applications. *Int. J. Biol. Macromol.* **2021**, *182*, 1650–1669. [CrossRef] [PubMed]
2. Briassoulis, D.; Athanasoulia, P.T.-G. Alternative optimization routes for improving the performance of poly(3-hydroxybutyrate) (PHB) based plastics. *J. Clean. Prod.* **2021**, *318*, 128555. [CrossRef]
3. Sirohi, R.; Pandey, J.P.; Gaur, V.K.; Gnansounou, E.; Sindhu, R. Critical overview of biomass feedstocks as sustainable substrates for the production of PHB. *Bioresour. Technol.* **2020**, *311*, 123536. [CrossRef] [PubMed]
4. Zhao, X.-H.; Niu, Y.-N.; Mi, C.-H.; Gong, H.-L.; Yang, X.-Y.; Cheng, J.-S.-Y.; Zhou, Z.-Q.; Liu, J.-X.; Peng, X.-L.; Wei, D.-X. Electrospinning nanofibers of microbial polyhydroxyalkanoates for applications in medical tissue engineering. *J. Polym. Sci.* **2021**, *59*, 1994–2013. [CrossRef]
5. Chou, S.-F.; Carson, D.; Woodrow, K.A. Current strategies for sustaining drug release from electrospun nanofibers. *J. Control. Release* **2015**, *220*, 584–591. [CrossRef] [PubMed]
6. Xie, J.; Shen, H.; Yuan, G.; Lin, K.; Su, J. The effects of alignment and diameter of electrospun fibers on the cellular behaviors and osteogenesis of BMSCs. *Mater. Sci. Eng.* **2021**, *120*, 111587. [CrossRef]
7. Karpova, S.G.; Ol'khov, A.A.; Shilkina, N.G.; Kucherenko, E.L.; Iordanskii, A.L. Influence of Drug on the Structure and Segmental Mobility of Poly(3-Hydroxybutyrate) Ultrafine Fibers. *Polym. Sci. Ser. A* **2017**, *59*, 58–66. [CrossRef]

8. Karpova, S.G.; Lomakin, S.M.; Popov, A.A.; Ol'khov, A.A.; Iordanskii, A.L.; Shilkina, N.S.; Gumargalieva, K.Z.; Berlin, A.A. Nonwoven blend composites based on poly(3-hydroxybutyrate)–chitosan ultrathin fibers prepared via electrospinning. *Polym. Sci. Ser. A* **2016**, *58*, 76–86. [CrossRef]
9. Olkhov, A.A.; Staroverova, O.V.; Bonartsev, A.P.; Zharkova, I.I.; Sklyanchuk, E.D.; Iordanskii, A.L.; Rogovina, S.Z.; Berlin, A.A.; Ishchenko, A.A. Structure and properties of ultrathin poly-(3-hydroxybutirate) fibers modified by silicon and titanium dioxide particles. *Polym. Sci. Ser. D* **2015**, *8*, 100–109. [CrossRef]
10. Karpova, S.G.; Olkhov, A.A.; Bakirov, A.V.; Chvalun, S.N.; Shilkina, N.G.; Popov, A.A. Poly(3-hydroxybutyrate) Matrices Modified with Iron(III) Complexes with Tetraphenylporphyrin. Analysis of the Structural Dynamic Parameters. *Russ. J. Phys. Chem. B* **2018**, *12*, 142–154. [CrossRef]
11. Karpova, S.G.; Ol'khov, A.A.; Krivandin, A.V.; Shatalova, O.V.; Lobanov, A.V.; Popov, A.A.; Iordanskii, A.L. Effect of Zinc–Porphyrin Complex on the Structure and Properties of Poly(3-hydroxybutyrate) Ultrathin Fibers. *Polym. Sci. Ser. A* **2019**, *61*, 70–84. [CrossRef]
12. Karpova, S.G.; Ol'khov, A.A.; Lobanov, A.V.; Popov, A.A.; Iordanskii, A.L. Biodegradable compositions of ultrathin poly-3-hydroxybutyrate fibers with MnCl–tetraphenylporphyrin complexes. Dynamics, structure and properties. *Nano-Technol. Russ.* **2019**, *14*, 132–143. [CrossRef]
13. Kotresh, T.M.; Ramani, R.; Jana, N.; Minu, S.; Shekar, R.I.; Ramachandran, R. Supermolecular Structure, Free Volume, and Glass Transition of Needleless Electrospun Polymer Nanofibers. *ACS Appl. Polym. Mater.* **2021**, *3*, 3989–4007. [CrossRef]
14. Sharma, S.K.; Pujari, P.K. Role of free volume characteristics of polymer matrix in bulk physical properties of polymer nanocomposites: A review of positron annihilation lifetime studies. *Prog. Polym. Sci.* **2017**, *75*, 31–47. [CrossRef]
15. Kazsoki, A.; Szabó, P.; Domján, A.; Balázs, A.; Bozó, T.; Kellermayer, M.; Farkas, A.; Balogh-Weiser, D.; Pinke, B.; Darcsi, A.; et al. Microstructural distinction of electrospun nanofibrous drug delivery systems formulated with different excipients. *Mol. Pharm.* **2018**, *15*, 4214–4225. [CrossRef] [PubMed]
16. Papp, J.; Szente, V.; Süvegh, K.; Zelkó, R. Correlation between the free volume and the metoprolol tartrate release of Metolose patches. *J. Pharm. Biomed. Anal.* **2010**, *51*, 244–247. [CrossRef]
17. Švajdlenková, H.; Šauša, O.; Adichtchev, S.V.; Surovtsev, N.V.; Novikov, V.N.; Bartoš, J. On the Mutual Relationships between Molecular Probe Mobility and Free Volume and Polymer Dynamics in Organic Glass Formers: Cis-1,4-poly(isoprene). *Polymers* **2021**, *13*, 294. [CrossRef]
18. Wei, M.; Wan, J.; Hu, Z.; Peng, Z.; Wang, B.; Wang, H. Preparation, characterization and visible-light-driven photocatalytic activity of a novel Fe(III) porphyrin-sensitized TiO_2 nanotube photocatalyst. *Appl. Surf. Sci. Part B* **2017**, *391*, 267–274. [CrossRef]
19. Bayat, F.; Karimi, A.R.; Adimi, T. Design of nanostructure chitosan hydrogels for carrying zinc phthalocyanine as a photosensitizer and difloxacin as an antibacterial agent. *Int. J. Biol. Macromol.* **2020**, *159*, 598–606. [CrossRef]
20. Sousa, J.F.M.; Pina, J.; Gomes, C.; Dias, L.D.; Pereira, M.M.; Murtinho, D.; Dias, P.; Azevedo, J.; Mendes, A.; de Melo, J.S.S.; et al. Transport and photophysical studies on porphyrin-containing sulfonated poly(etheretherketone) composite membranes. *Mater. Today Commun.* **2021**, *29*, 102781. [CrossRef]
21. Lobanov, A.V.; Kholujskaya, S.N.; Komissarov, G.G. The H_2O_2 as donor of electrons in catalytic reduction of inorganic carbon. *Russ. J. Phys. Chem. B* **2004**, *23*, 44–48.
22. Szuwarzynski, M.; Wolski, K.; Krukc, T.; Zapotoczny, S. Macromolecular strategies for transporting electrons and excitation energy in ordered polymer layers. *Prog. Polym. Sci.* **2021**, *121*, 101433. [CrossRef]
23. Gradova, M.A.; Zhdanova, K.A.; Bragina, N.A.; Lobanov, A.V.; Mel'nikov, M.Y. Aggregation state of amphiphilic cationic tetraphenylporphyrin derivatives in aqueous microheterogeneous systems. *Russ. Chem. Bull.* **2015**, *64*, 806–811. [CrossRef]
24. Rabiee, N.; Yaraki, M.T.; Garakani, S.M.; Ahmadi, S.; Lajevardi, A.; Bagherzadeh, M.; Rabiee, M.; Tayebi, L.; Tahriri, M.; Hamblin, M.R. Recent advances in porphyrin-based nanocomposites for effective targeted imaging and therapy. *Biomaterials* **2020**, *232*, 119707. [CrossRef]
25. Zheng, F.; Zhang, Y.; Han, Y.; Zhang, L.; Bouyssiere, B.; Shi, Q. Aggregation of petroporphyrins and fragmentation of porphyrin ions: Characterized by TIMS-TOF MS and FT-ICR MS. *Fuel* **2021**, *289*, 119889. [CrossRef]
26. Lobanov, A.V.; Gromova, G.A.; Gorbunova, Y.G.; Tsivadze, A.Y. Supramolecular associates of double-decker lanthanide phthalocyanines with macromolecular structures and nanoparticles as the basis of biosensor devices. *Prot. Met. Phys. Chem. Surf.* **2014**, *50*, 570–577. [CrossRef]
27. Zhang, X.; Wasson, M.C.; Shayan, M.; Berdichevsky, E.K.; Ricardo-Noordberg, J.; Singh, Z.; Papazyan, E.K.; Castro, A.J.; Marino, P.; Ajoyan, Z.; et al. A historical perspective on porphyrin-based metal–organic frameworks and their applications. *Coord. Chem. Rev.* **2021**, *429*, 213615. [CrossRef]
28. Faustova, M.; Mollaev, M.; Zhunina, O.; Nikolskaya, E.; Lobanov, A.; Shvets, V.; Yabbarov, N. Cytotoxic activity evaluation of metalloporphyrins in binary catalyst system. *FEBS Open Bio.* **2018**, *8*, 475–476.
29. Yoo, H.-Y.; Yan, S.; Ra, J.W.; Jeon, D.; Goh, B.; Kim, T.-Y.; Mackeyev, Y.; Ahn, Y.-Y.; Kim, H.-J.; Wilson, L.J.; et al. Tin porphyrin immobilization significantly enhances visible-light-photosensitized degradation of Microcystins: Mechanistic implications. *Appl. Catal. B Environ.* **2016**, *199*, 33–44. [CrossRef]
30. Arnold, D.P.; Blok, J. The coordination chemistry of tin porphyrin complexes. *Coord. Chem. Rev.* **2004**, *248*, 299–319. [CrossRef]
31. Nakagaki, S.; Machado, G.S.; Stival, J.F.; dos Santos, E.H.; Silva, G.M.; Wypych, F. Natural and synthetic layered hydroxide salts (LHS): Recent advances and application perspectives emphasizing catalysis. *Prog. Solid State Chem.* **2021**, *64*, 100335. [CrossRef]

32. Van, S.P.; Birrell, G.B.; Griffith, O.H. Rapid anisotropic motion of spin labels. models for motion averaging of the ESR parameters. *J. Magn. Reson.* **1974**, *15*, 444–459. [CrossRef]
33. Budil, D.E.; Sanghyuk, L.; Saxena, S.; Freed, J.H. Nonlinear-least-squares analysis of slow-motion EPR spectra in one and two dimensions using a modified Levenberg–Marquardt algorithm. *J. Magn. Reson. Ser. A* **1996**, *120*, 155–189. [CrossRef]
34. Chernova, D.A.; Vorobiev, A.K. Molecular mobility of nitroxide spin probes in glassy polymers: Models of the complex motion of spin probes. *J. Appl. Polym. Sci.* **2011**, *121*, 102–110. [CrossRef]
35. Isaev, N.P.; Kulik, L.V.; Kirilyuk, I.A.; Reznikov, V.A.; Grigor'ev, I.A.; Dzuba, S.A. Fast stochastic librations and slow small-angle rotations of molecules in glasses observed on nitroxide spin probes by stimulated electron spin echo spectroscopy. *J. Non-Cryst. Solids* **2010**, *356*, 1037–1104. [CrossRef]
36. Karpova, S.G.; Olkhov, A.A.; Popov, A.A.; Iordanskii, A.L.; Shilkina, N.G. Characteristics of the Parameters of Superfine Fibers of Poly(3-hydroxybutyrate) Modified with Tetraphenylporphyrin. *Inorg. Mater. Appl. Res.* **2021**, *12*, 44–54. [CrossRef]
37. Karpova, S.G.; Ol'khov, A.A.; Tyubaeva, P.M.; Shilkina, N.G.; Popov, A.A.; Iordanskii, A.L. Composite Ultrathin Fibers of Poly-3-hydroxybutyrate and a Zinc Porphyrin: Structure and Properties. *Russ. J. Phys. Chem. B* **2019**, *13*, 313–327. [CrossRef]
38. Castellón, E.; Günther, H.; Mehling., S.; Hiebler, L.; Cabeza, F. Determination of the enthalpy of PCM as a function of temperature using a heat-flux DSC—A study of different measurement procedures and their accuracy. *Energy Res.* **2008**, *32*, 1258–1265. [CrossRef]
39. Olkhov, A.A.; Staroverova, O.V.; Gol'dshtrakh, M.A.; Khvatov, A.V.; Gumargalieva, K.Z.; Iordanskii, A.L. Electrospinning of biodegradable poly-3-hydroxybutyrate. Effect of the characteristics of the polymer solution. *Russ. J. Phys. Chem. B* **2016**, *10*, 830–838. [CrossRef]
40. Filatov, Y.N.; Filatov, I.Y.; Smul'skaya, M.A. Role of macromolecular factor in polymer solution for electrospinning process. *Fibre Chem.* **2017**, *49*, 151–160. [CrossRef]
41. Shepa, I.; Mudra, E.; Dusza, J. Electrospinning through the prism of time. *Mater. Today Chem.* **2021**, *21*, 100543. [CrossRef]
42. Rodríguez-Tobías, H.; Morales, G.; Grande, D. Comprehensive review on electrospinning techniques as versatile approaches toward antimicrobial biopolymeric composite fibers. *Mater. Sci. Eng.* **2019**, *101*, 306–322. [CrossRef] [PubMed]
43. Saalmueller, J.W.; Long, H.W.; Volkmer, T.; Wiesner, U.; Maresch, G.G.; Spiess, H.W. Characterization of the motion of spin probes and spin labels in amorphous polymers with two-dimensional field-step ELDOR. *J. Polym. Sci. Part B Polym. Phys.* **1996**, *34*, 1093–1104. [CrossRef]
44. Buchachenko, A.L.; Wasserman, A.M. *Stable Radicals*; Chemistry: Moscow, Russia, 1973; 408p.
45. Karpova, S.G.; Ol'khov, A.A.; Chvalun, S.N.; Tyubaeva, P.M.; Popov, A.A.; Iordanskii, A.L. Comparative Structural Dynamic Analysis of Ultrathin Fibers of Poly-(3-hydroxybutyrate) Modified by Tetraphenyl–Porphyrin Complexes with the Metals Fe, Mn, and Zn. *Nanotechnol. Russ.* **2019**, *14*, 367–381. [CrossRef]
46. Redfield, A.G. The theory of relaxation processes. *Adv. Magn. Reson.* **1966**, *1*, 19–31.
47. Razumovskii, L.P.; Lordanskii, A.L.; Zaikov, G.E.; Zagreba, E.D.; McNeill, I.C. Sorption and diffusion of water and organic solvents in poly(l-hydroxybutyrate) films. *Polym. Degrad. Stab.* **1994**, *44*, 171–175. [CrossRef]
48. Iordanskii, A.L.; Kamaev, P.P.; Hänggi, U.J. Modification via preparation for poly(3-hydroxybutyrate) films: Water-transport phenomena and sorption. *J. Appl. Polym. Sci.* **2000**, *76*, 475–480. [CrossRef]
49. Kamaev, P.P.; Aliev, I.I.; Iordanskii, A.L.; Wasserman, A.M. Molecular dynamics of the spin probes in dry and wet poly(3-hydroxybutyrate) films with different morphology. *Polymer* **2001**, *42*, 515–520. [CrossRef]
50. Pankova, Y.N.; Shchegolikhin, A.N.; Iordanskii, A.L.; Zhulkina, A.L.; Olkhov, A.A.; Zaikov, G.E. The characterization of novel biodegradable blends based on polyhydroxybutyrate: The role of water transport. *J. Mol. Liq.* **2010**, *156*, 65–69. [CrossRef]
51. Ventura, H.; Claramunt, J.; Rodríguez-Perez, M.A.; Ardanuy, M. Effects of hydrothermal aging on the water uptake and tensile properties of PHB/flax fabric biocomposites. *Polym. Degrad. Stab.* **2017**, *142*, 129–138. [CrossRef]
52. Sangroniz, A.; Sarasua, J.R.; Iriarte, M.; Etxeberria, A. Survey on transport properties of vapours and liquids on biodegradable polymers. *Eur. Polym. J.* **2019**, *120*, 109232. [CrossRef]

Article

Progressing Ultragreen, Energy-Efficient Biobased Depolymerization of Poly(ethylene terephthalate) via Microwave-Assisted Green Deep Eutectic Solvent and Enzymatic Treatment

Olivia A. Attallah [1,2,†], Muhammad Azeem [1,*,†], Efstratios Nikolaivits [3], Evangelos Topakas [3] and Margaret Brennan Fournet [1]

1. Materials Research Institute, Technological University of the Shannon: Midlands Midwest, N37 HD68 Athlone, Ireland; oadly@ait.ie (O.A.A.); mfournet@ait.ie (M.B.F.)
2. Pharmaceutical Chemistry Department, Faculty of Pharmacy, Heliopolis University, Cairo-Belbeis Desert Road, El Salam, Cairo 11777, Egypt
3. Biotechnology Laboratory, Industrial Biotechnology & Biocatalysis Group, School of Chemical Engineering, National Technical University of Athens, 15780 Athens, Greece; stratosnikolai@gmail.com (E.N.); vtopakas@chemeng.ntua.gr (E.T.)
* Correspondence: m.azeem@research.ait.ie
† These authors contributed equally to this work.

Abstract: Effective interfacing of energy-efficient and biobased technologies presents an all-green route to achieving continuous circular production, utilization, and reproduction of plastics. Here, we show combined ultragreen chemical and biocatalytic depolymerization of polyethylene terephthalate (PET) using deep eutectic solvent (DES)-based low-energy microwave (MW) treatment followed by enzymatic hydrolysis. DESs are emerging as attractive sustainable catalysts due to their low toxicity, biodegradability, and unique biological compatibility. A green DES with triplet composition of choline chloride, glycerol, and urea was selected for PET depolymerization under MW irradiation without the use of additional depolymerization agents. Treatment conditions were studied using Box-Behnken design (BBD) with respect to MW irradiation time, MW power, and volume of DES. Under the optimized conditions of 20 mL DES volume, 260 W MW power, and 3 min MW time, a significant increase in the carbonyl index and PET percentage weight loss was observed. The combined MW-assisted DES depolymerization and enzymatic hydrolysis of the treated PET residue using LCC variant ICCG resulted in a total monomer conversion of ≈16% (w/w) in the form of terephthalic acid, mono-(2-hydroxyethyl) terephthalate, and bis-(2-hydroxyethyl) terephthalate. Such high monomer conversion in comparison to enzymatically hydrolyzed virgin PET (1.56% (w/w)) could be attributed to the recognized depolymerization effect of the selected DES MW treatment process. Hence, MW-assisted DES technology proved itself as an efficient process for boosting the biodepolymerization of PET in an ultrafast and eco-friendly manner.

Keywords: enzymatic hydrolysis; deep eutectic solvents; polyethylene terephthalate; Box-Behnken design; microwave depolymerization

1. Introduction

All-green routes to continuous circular material and commodity production, unmaking and remaking in a manner analogous to nature's many resource cycles, remain largely elusive for plastics [1]. Polyethylene terephthalate (PET) plastic value chain is a pertinent example of many current linear mine, use, and dispose economic processes. PET is highly recalcitrant and widely used in the manufacturing of packaging materials, beverage bottles, and synthetic fibers due to its high mechanical and thermal properties, nontoxicity, and excellent transparency [2]. The unabated increase in the demand for PET production is a grave environmental concern given the poor degradation rates of PET in soil and air [3]. Mechanical and chemical processing are the current mainstay approaches for PET recycling, with each having considerable limitations [4]. Loss of transparency of mechanically recycled PET and the presence of traces of reactive antimony catalyst restrict the application of recycled PET in food and beverage packaging [4,5]. On the other hand, chemical recycling, which comprises glycolysis, methanolysis, aminolysis, and hydrolysis to depolymerize PET into its monomers [6–10], requires long reaction times, large volumes of non-green solvents for reaction, and several product purification processes [11]. Recently, a number of alternative techniques are being explored for PET depolymerization, including the incorporation of efficient catalytic systems in depolymerization reactions [12,13], supercritical technology [14], and microwave-assisted methods [15,16]. However, despite the achievement of increased reaction rates, the need for harsh reaction conditions and use of non-green solvents remain a considerable challenge [17,18]. Recently, plastic biodepolymerization has been proposed as an environmentally friendly and promising technology for PET recycling [19]. Greener approaches for PET recycling, such as complete solubilization of PET in natural deep eutectic solvents and thin-layer film synthesis from PET polymer waste for nanofiltration, have also been employed recently as sustainable routes for PET recycling [18,20,21]. Herein, a novel multistep approach that echoes nature's sequential depolymerization steps for naturally occurring polymers, namely weathering, arthropodal digestion, and microbial and enzymatic degradation, is presented. A MW-assisted DES technique was combined with enzymatic hydrolysis to obtain enhanced PET depolymerization compared with enzymatic hydrolysis alone. Such a recycling methodology would be advantageous due to its low energy requirements and operation under mild conditions during plastic degradation/depolymerization [22]. A series of impeding factors, namely the need for low physical dimension preparations of the polymer as suitable substrates, slow catalytic activity, enzymatic thermal degradation at high processing temperatures [20], low interaction levels with the chemical structures of linear polymers [23], and high polymer crystallinity and hydrophobicity [24], serve to hinder the efficiency of biobased plastic recycling. Assisting techniques designed to overcome these barriers, which can render the plastic more amenable to biodepolymerization/biodegradation and augment the probability of depolymerization events using biobased agents, are required to progress towards sustainable plastic resource cycling. Recently, new combinations of physiochemical treatment techniques have been applied to overcome existing hindrances to biodepolymerization [25]. For instance, Falah et al. [26] proposed several sequential physiochemical treatments, including ultraviolet, high temperature, and nitric acid solvent treatment, prior to exposing PET for enzymatic degradation. The authors observed the development of cracks on the PET surface after treatment, which led to some enhancement in the enzymatic degradation of PET. Quartinello et al. [27] used a sequential chemoenzymatic treatment to facilitate depolymerization of PET from textile waste under mild conditions. The chemical treatment was performed under neutral conditions (pressure = 40 bar and temperature = 250 °C) to depolymerize PET into high-purity terephthalic acid (TPA) and small oligomers with a total monomer conversion of 85% within 90 min. Enzymatic hydrolysis was then performed using Humicola insolens cutinase to yield 97% pure TPA. Furthermore, Gong et al. [28] used a combination of alkaline hydrolysis and microbial strains (T = 37 °C, pH = 12, time = 48 h) and found enhanced conversion of PET into its functional monomers as a result of faster microbial growth and reduction in particle size of PET. Deep eutectic solvents (DESs) are a new class of ionic liquids that are becoming prominent for plastics depolymerization due their unique characteristics [1]. The use of DESs as catalysts in depolymerization reactions can make reaction conditions milder and decrease reaction times [29]. Recently, these solvents have been utilized as catalysts in microwave (MW)-assisted PET depolymerization reactions due to their strong MW heating characteristics and the synergic hydrogen bond formation of these solvents with PET polymer chains [8,15]. Different compositions of DESs have been employed and evaluated for the enhancement of PET depolymerization under mild conditions, as elaborated in

Table S1. To the best of our knowledge, the combination of MW-assisted DES technique and enzymatic hydrolysis to obtain enhanced PET depolymerization compared to enzymatic hydrolysis alone has not been previously explored and presents a strong progress towards the achievement of all-green permanent resource circularity in tandem with nature. Noticeably, treating PET in DESs under MW irradiation can increase PET chain flexibility and change the physicochemical properties of the polymer [6]. Thus, such MW-assisted DES treatment can provide an enhanced monomer conversion yield upon PET depolymerization.

In this study, an all-green, environmentally friendly sequential PET depolymerization approach was employed comprising treatment of PET using MW-assisted DES technique without the use of additional depolymerization agents followed by enzymatic hydrolysis using a variant of LCC cutinase. A green DES of ternary composition of choline chloride, glycerol, and urea was selected for PET treatment in the presence of MW irradiation. Recently, the Box-Behnken design (BBD) has been utilized in many studies for the optimization of reaction processes [30,31]. In this work, optimized MW treatment conditions were determined using BBD with respect to MW irradiation time, MW power, and volume of DES. The crystallinity index, carbonyl index, and weight loss of residual PET were used as the studied responses for BBD. Residual PET resulting from the optimized MW treatment process was further exposed to a four-day hydrolysis process by a thermostable polyesterase, and the total depolymerization efficiency was evaluated. The success of the combined techniques proposed in this study is expected to provide a green, environmentally friendly approach for plastic recycling.

2. Materials and Methods

2.1. Materials

PET granules were purchased from Alpek Polyester UK Ltd. (Lazenby, UK) and converted into micron sized fine powder using a centrifugal miller (Retsch Verder Scientific, Haan, Germany). Glycerol (99%), choline chloride (98%, ChCl), and urea (98%) were purchased from Sigma-Aldrich (Dorset, UK). All other chemicals were obtained from Aldrich (Darmstadt, Germany) and were of analytical grade and readily available to use without any purification.

2.2. Preparation of DES

The synthesis of the ternary DES was based on the method provided by [32]. Prior to the preparation of DES, ChCl was dried overnight at 65 °C in the oven. A DES of triplet composition based on urea/glycerol/ChCl was synthesized with 1:1:1 molar ratio by continuously mixing and heating at 80 °C until a homogeneous, clear, and transparent liquid was formed within 10 min.

2.3. MW Treatment Experiments

The MW treatment experiments were carried out by mixing 1 g of powdered PET in varied volumes of synthesized DES while stirring for 15 min. The prepared suspensions were then exposed to MW irradiation at specified MW power and time. After MW treatments, residual PET was filtered and washed three times with distilled water to obtain clean residual PET, and DES was regenerated. The PET residues were then dried in an oven at 70 °C overnight and kept in sealed containers for further analysis.

2.4. Experimental Design

A three-factor, three-level Box-Behnken design (BBD) (Design Expert Stat-Ease Inc., Minneapolis, MN, USA) was implemented for optimization of the proposed MW-assisted DES technique. A total of 15 runs with three center points were set up to study the following three factors: MW irradiation time (min) (X_1), microwave power (W) (X_2), and volume of DES (ml) (X_3). The responses were concluded as the weight loss (%), carbonyl index, and crystallinity index of residual PET. The chosen values for the studied factors were constructed using reported literature and preliminary experiments (Table 1).

Table 1. Variables and levels in Box-Behnken experimental design for PET pretreatment.

	Level			
Independent Variables	−1	0	1	Constrains
X_1: MW time (min)	1	2	3	In the range
X_2: MW power (W)	100	250	400	In the range
X_3: Volume of DES (mL)	20	35	50	In the range

2.5. Enzymatic Hydrolysis of PET Materials

For the enzymatic depolymerization, LCC variant ICCG (LCCv) was used [33]. The coding sequence of LCCv was codon optimized for expression in *E. coli* and cloned into pET26b(+) vector (GenScript Biotech B.V., Leiden, the Netherlands). The expression and purification of the recombinant protein was performed as described previously [34]. The purity of the resulting enzymatic preparation was checked on SDS-PAGE electrophoresis (12.5% (*w/v*)), and protein concentration was determined by measuring the absorbance at 280 nm based on the calculated molar extinction coefficient.

MW-treated PET samples used for enzymatic depolymerization were washed twice with ultrapure water in order to remove residual monomers that could potentially inhibit enzymatic action. Reactions took place in 10 mL of 100 mM potassium phosphate buffer, pH 8, containing 100 mg of PET residue and 4 μM of enzyme. Control reactions without the addition of enzyme were also realized. All reactions were incubated at 55 °C under shaking for 4 days. After that, 0.1% (*v/v*) of 6M HCl was added in each reaction and centrifuged at 4000× *g* at 10 °C. Supernatants were collected and analyzed by HPLC (Perkin Elmer, Boston, MA, USA) [34] in order to determine the concentration of the resulting water-soluble degradation products. The remaining material was washed 3 times with ultrapure water, freeze-dried, and weighed. Experiments were run in triplicates, and the standard deviation was estimated.

2.6. Instrumental Characterization

The PET samples before and after MW-assisted DES treatment were analyzed by FTIR spectroscopy (Perkin Elmer, Washington, MA, USA) at a spectral region of 4000–600 cm^{-1}. The carbonyl index was determined based on the obtained results using the baseline method. Ratios of ester carbonyl peak intensity at 1713 cm^{-1} to that of the normal C–H bonding mode at 1408 cm^{-1} in PET were calculated as follows [35]:

$$Carbonyl\ index = \frac{Absorption\ at\ 1713\ cm^{-1}}{Absorption\ at\ 1408\ cm^{-1}} \qquad (1)$$

The thermal behavior of the samples was evaluated by a DSC Perkin Elmer 4000 (Perkin Elmer Washington, MA, USA) with Pyris Software version 13.3.1 (Perkin Elmer Washington, MA, USA) under an inert nitrogen stream. About 10 mg of specimen was sealed in an aluminum pan. The DSC scans were recorded while heating from 30 to 275 °C at a heating rate of 10 °C min^{-1} and then cooled to 30 °C. The crystallinity index was calculated according to the following equation [36]:

$$Crystallinity\ index = (\Delta H_m / W \Delta H_{m0}) \times 100 \qquad (2)$$

where ΔH_m (Jg^{-1}) is the heat of fusion of the PET sample, ΔH_{m0} is the heat of fusion for completely crystalline PET (140 Jg^{-1}) [37], and W(g) is the weight fraction of residual PET in the samples.

The percentage weight loss of PET was determined at onset temperature of degradation (T_0) using a thermogravimetric analyzer Pyris TGA (Perkin Elmer, Washington, MA, USA). The polymer samples were placed in a standard aluminum pan and heated from 30 to 600 °C at the rate of 10 °C min^{-1} under nitrogen flow of 50 mL min^{-1}. The PET weight loss (%) was calculated as follows:

$$\text{PET weight loss (\%)} = (100 - \text{weight percent of PET at } T_0) \tag{3}$$

3. Results

3.1. Properties of DES

Ternary DES has been reported as a new type of DES to widen the range of DES applications owing to the extra functionality provided by their components. One pertinent area where ternary DESs were recently applied is CO$_2$ capture, which demonstrates that DESs have great flexibility in terms of synthesis, forms, and applications [32]. In the current study, we selected a green ternary DES with ChCl as a hydrogen bond acceptor and glycerol and urea as hydrogen bond donors to provide an initial depolymerization of PET when coupled with MW irradiation and to facilitate PET enzymatic hydrolysis. A schematic diagram of the proposed interactions of DES with PET is demonstrated in Figure 1. The advantage of this DES lies in its components being green, inexpensive, and largely available with the capacity to provide synergistic effects on PET depolymerization [7,15,27,33]. As shown in Figure 1, depolymerization of PET is postulated to involve a form of glycolysis reaction due to the presence of glycerol within the ternary DES [15]. It is also known that a quaternary ammonium compound, such as ChCl and DES itself, could act as a catalyst in mild glycolysis [7,8,15,26]. Simultaneously, the H-bond action between glycerol and urea is expected to change the charge density of the hydroxyl (OH) group in glycerol and increase the electronegativity of the oxygen atom in the glycerol OH group. Hence, the nucleophilicity of the oxygen becomes stronger, thereby supporting preferential attack to the carbon of the ester group in PET [38,39].

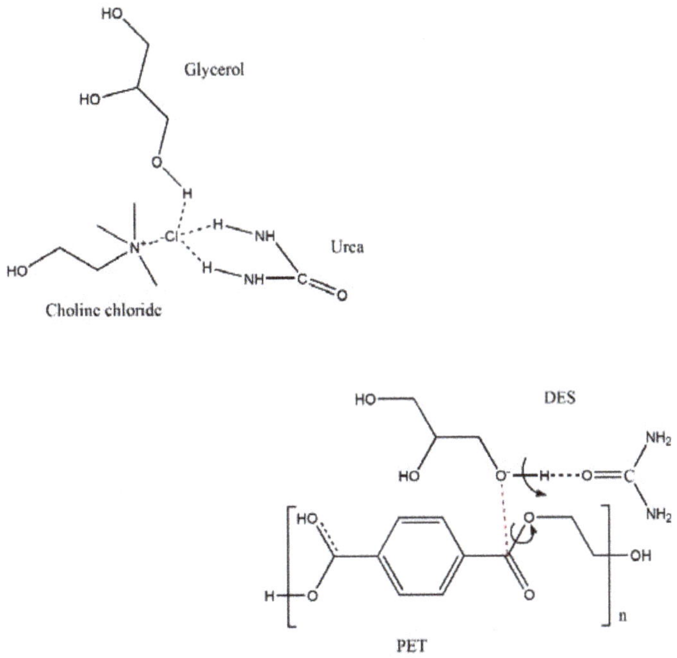

Figure 1. Schematic diagram for the proposed interactions of DES (composed of choline chloride/urea/glycerol in the ratio of 1:1:1) with PET via hydrogen bonding.

Based on previous reports, DESs with high pH values and low viscosity can contribute to the catalytic activity and influence the reaction rate [39,40]. The pH and density values of the proposed ternary DES were found to be 9.95 and 1.16 g/mL, respectively, giving rise to a suitable pH for depolymerization reaction to occur. Moreover, the employment of MW irradiation in the current depolymerization technique led to the production of expeditious heating, in particular with DES due to its high electric conductivity [41]. Therefore, initial depolymerization of PET can be achieved very efficiently due to the synergistic effects of MW irradiation, glycolysis due to glycerol, and the catalytic activities of ChCl, urea alone, and in DES.

Prior to the depolymerization process, the prepared DES was also characterized using FTIR, as shown in Figure 2. In the typical FTIR spectrum of pure urea, the characteristic C=O, N–H and C–N stretching peaks appeared at 1677, 3427.12, and 1459.79 cm^{-1}, respectively, while the N–H deformation peak appeared at 1590.21 cm^{-1}. Pure glycerol's spectrum showed a C–O stretching peak at 1043.55 and 1111.08 cm^{-1} and O–H stretching peak at 3339.48 cm^{-1}. The FTIR spectrum of pure ChCl had O–H and C–H stretching peaks at 3220.59 and 3006.97 cm^{-1}, respectively, and the asymmetric and symmetric stretching peaks of C–N linkage were observed at 954.54 and 892.96 cm^{-1}, respectively. The FTIR spectrum of the prepared DES showed the formation of new bonds at 1668.38 and 1624.66 cm^{-1} compared to the FTIR spectra of pure urea, glycerol, and ChCl. The C=O linkage frequency peak at 1677 cm^{-1} of urea was shifted to lower side at 1668 cm^{-1}, indicating the formation of more hydrogen bonds, and the N–H stretching frequency of urea at 3427.12 cm^{-1} was masked by the O–H stretching peak [42]. The C–O stretching peak of glycerol at 1043.55 cm^{-1} was also shifted to higher wavenumber in the DES, indicating the successful formation of DES.

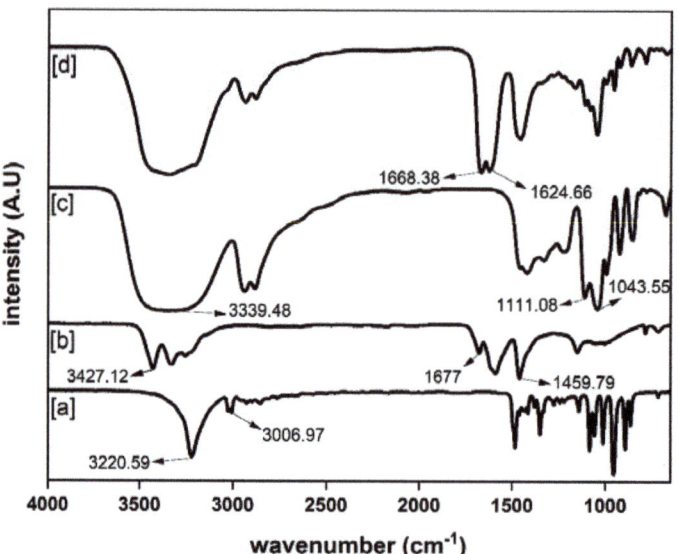

Figure 2. FTIR spectra of: (a) choline chloride, (b) urea, (c) glycerol, and (d) DES (choline chloride/urea/glycerol in the ratio of 1:1:1).

3.2. Experimental Design Results

The model of PET MW treatment was studied using response surface methodology. In the current study, the experimental runs were carried out based on the design plan proposed for the studied parameters (MW irradiation time, MW power, and volume of DES). After each run, the crystallinity index, carbonyl index, and weight loss (%) at T_0 of degradation of treated PET were calculated. The results are presented as responses for each run in Table 2.

Table 2. Experimental matrix and observed responses for PET pretreatment in BBD.

Run	Independent Variable			Dependent Variable		
	X_1 (min)	X_2 (W)	X_3 (mL)	Y_1	Y_2	Y_3 (%)
1	3	400	35	54.10	3.47	8.90
2	2	100	50	18.57	4.41	2.17
3	2	250	35	48.30	3.61	4.80
4	1	250	20	19.00	3.91	1.20
5	3	250	50	41.67	3.79	4.91
6	2	400	20	45.07	4.16	7.39
7	2	250	35	47.80	3.62	5.00
8	2	400	50	50.29	4.30	5.63
9	3	100	35	12.83	3.95	0.90
10	3	250	20	32.79	4.17	6.20
11	2	100	20	20.21	4.28	0.94
12	1	400	35	19.86	4.15	1.89
13	1	250	50	12.86	4.55	2.20
14	2	250	35	49.32	3.59	5.20
15	1	100	35	9.52	3.87	0.75

X_1: MW irradiation time, X_2: MW power, X_3: volume of DES, Y_1: crystallinity index, Y_2: carbonyl index, and Y_3: weight loss (%) at T_0 of degradation.

The studied responses were then tested against different regression models to determine the best-fitting mathematical model and the significance of varying the process parameters. The quadratic model was chosen as the best fitting model for the studied responses in comparison to the other models. The relationship between the crystallinity index (Y_1), carbonyl index (Y_2), and weight loss of PET at T_0 of degradation (Y_3) and the studied parameters of MW irradiation time (X_1), MW power (X_2), and volume of DES (X_3) is demonstrated in Table 3.

For the crystallinity index (Y_1), the coefficients of the quadratic model equation indicated that the increase in all the studied factors led to a significant increase in the crystallinity index of residual PET except for the volume of DES, where the p-value was more than 0.05. The interaction between MW irradiation time and volume of DES showed a positive effect on the crystallinity index as well revealed the significant effect both factors had on the PET residues. Moreover, the MW power interactions with both the MW irradiation time and volume of DES also showed a significant positive efficacy on the crystallinity index. Such results indicate that all the studied factors and their interactions had positive effects on the crystallinity index of treated PET, with the increase in the studied factors leading to an increase in the degradation of PET and causing an increase in the crystallinity index of the treated samples [43].

For the carbonyl index (Y_2), as demonstrated in Table 2, all treated PET residues had higher carbonyl index than that of the untreated PET (2.80). Both MW irradiation time and power showed a significant negative effect on the carbonyl index values, while the volume of DES showed a positive effect. The MW irradiation time interactions with both MW power and volume of DES also showed significant negative effects on the carbonyl index of PET residue. Thus, based on the obtained results and the carbonyl index of untreated PET, the increase in both MW irradiation time and power led to a degree of PET depolymerization, which was observed through the low values of the carbonyl index of the residual PET. On the other hand, high levels of DES volume with low levels of MW irradiation time and power caused a significant increase in the carbonyl groups on the surface of the PET as a result of surface oxidation rather than complete depolymerization of treated PET.

Table 3. Statistical analysis of measured responses for PET pretreatment.

Fitting Model	Factors	Coefficient	p-Value	ANOVA
PET crystallinity index (Y_1)	Intercept	48.47		
	X_1	10.02	<0.0001	
	X_2	13.52	<0.0001	
	X_3	0.79	0.1290	$F = 276.92$,
	X_1X_2	7.73	<0.0001	$R^2 = 0.9944$, Model p-value < 0.0001,
	X_1X_3	3.76	0.0017	p-value of lack of fit = 0.228
	X_2X_3	1.71	0.0385	
	X_1^2	−15.68	<0.0001	
	X_2^2	−8.72	<0.0001	
	X_3^2	−6.22	0.0002	
PET carbonyl index (Y_2)	Intercept	3.61		
	X_1	−0.14	<0.0001	
	X_2	−0.054	<0.0001	
	X_3	0.066	0.0005	$F = 461.34$,
	X_1X_2	−0.19	0.0002	$R^2 = 0.9966$, Model p-value < 0.0001,
	X_1X_3	−0.25	<0.0001	p-value of lack of fit = 0.355
	X_2X_3	2.5×10^{-3}	<0.0001	
	X_1^2	0.035	0.8048	
	X_2^2	0.22	0.0165	
	X_3^2	0.46	<0.0001	
Weight loss of PET at T_0 of degradation (Y_3)	Intercept	5.00		
	X_1	1.86	<0.0001	
	X_2	2.38	<0.0001	
	X_3	−0.10	0.2060	$F = 264.71$,
	X_1X_2	1.72	<0.0001	$R^2 = 0.9941$, Model p-value < 0.0001,
	X_1X_3	−0.57	0.0023	p-value of lack of fit = 0.537
	X_2X_3	−0.75	0.0007	
	X_1^2	−1.15	0.0001	
	X_2^2	−0.74	0.0008	
	X_3^2	−0.23	0.0825	

X_1: MW irradiation time, X_2: MW power, X_3: volume of DES, Y_1: crystallinity index, Y_2: carbonyl index, and Y_3: weight loss (%) at T_0 of degradation.

For PET weight loss at T_0 of degradation (Y_3), as elaborated in Table 3, the coefficients of the model equation showed that MW irradiation time and power and their interaction with each other had positive effect on PET weight loss. On the other hand, the volume of DES and its interactions with both MW irradiation time and power showed negative efficacy on PET weight loss at T_0 of degradation. These results indicate that increased levels of MW irradiation time and power and low levels of DES induce a decrease in thermal stability of treated PET, leading to a greater degree of PET depolymerization at the T_0 of degradation.

The adequacy of the proposed model to describe the crystallinity index, carbonyl index, and weight loss of treated PET at T_0 of degradation was evaluated, and the results are demonstrated in Table 3. Based on the statistics test, high coefficients of determination were observed for all responses. The adjusted R^2 values were calculated to be 0.9944 for the crystallinity index, 0.9966 for the carbonyl index, and 0.9941 for the percentage weight loss of PET at T_0 of degradation.

Analysis of variance (ANOVA) was also applied to determine the significance of the model at a 95% confidence interval. A model is said to be significant if the probability value (p-value) is <0.05. The p-values demonstrated in Table 3 indicate that all the studied responses fitted the model well. From the lack-of-fit test. the response showed a highly desirable nonsignificant lack-of-fit ($p > 0.1$) with p-values of 0.228 for the crystallinity index, 0.355 for the carbonyl index, and 0.537 for the percentage weight loss of PET at T_0 of degradation.

3.3. Response Surface Analysis

Response surface graphical plots were generated between the responses obtained for PET MW treatment and the studied independent variables to estimate the effect of combinations of these variables on the studied responses. The 3D and contour plots for the crystallinity index, carbonyl index, and weight loss of PET at T_0 of degradation are demonstrated in Figures 3–5, respectively.

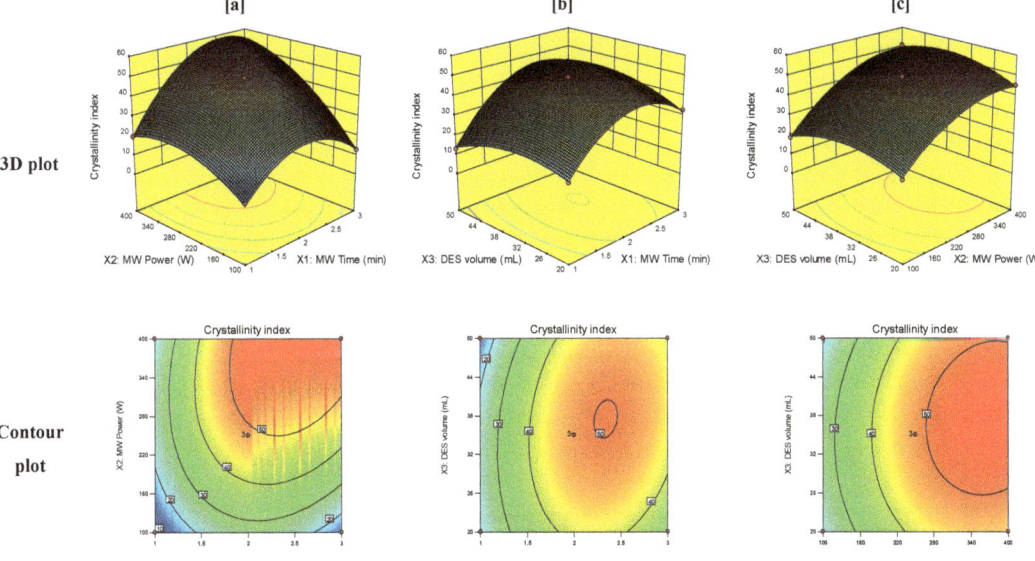

Figure 3. 3D and contour plots of the effect of the interaction of (**a**) MW time (X_1) and MW power (X_2), (**b**) MW time (X_1) and volume of DES (X_3), and (**c**) MW power (X_2) and volume of DES (X_3) on the crystallinity index.

As shown in Figure 3, high levels of both MW irradiation time and power caused an increase in the crystallinity index of PET residue. Such result can be attributed to the initial degradation of PET upon treatment with the MW-assisted DES technique, which usually occurs in the amorphous phase of PET, causing an increase in the overall crystallinity of the polymer [43]. A significant increase in the crystallinity index was also observed with the increase in DES volume until 35 mL. Further increase in DES volume did not show a profound effect on the crystallinity index, indicating that low volumes of DES are more effective in the initial depolymerization of PET using the proposed MW treatment technique.

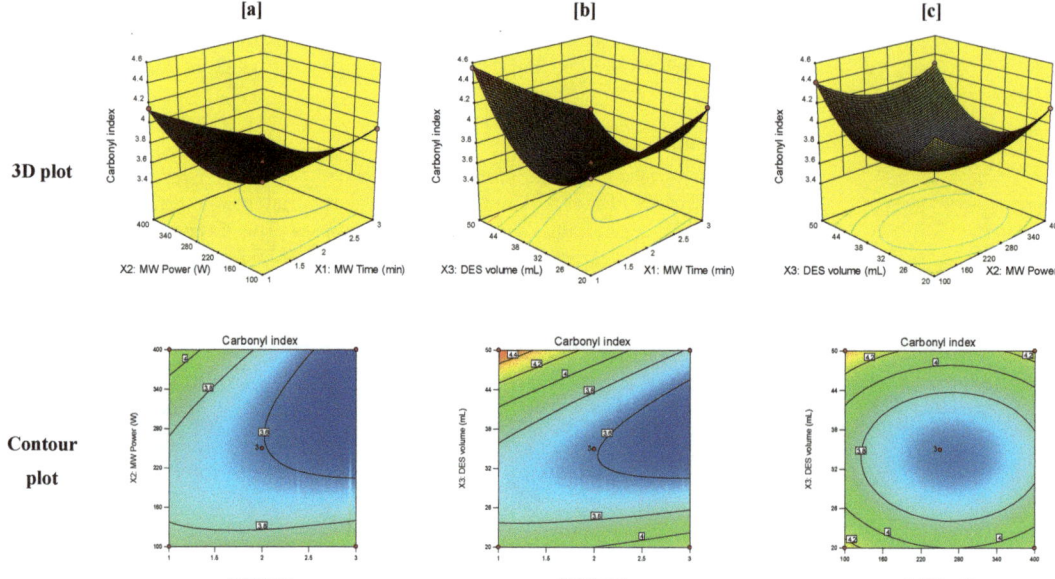

Figure 4. 3D and contour plots of the effect of the interaction of (**a**) MW time (X_1) and MW power (X_2), (**b**) MW time (X_1) and volume of DES (X_3), and (**c**) MW power (X_2) and volume of DES (X_3) on the carbonyl index.

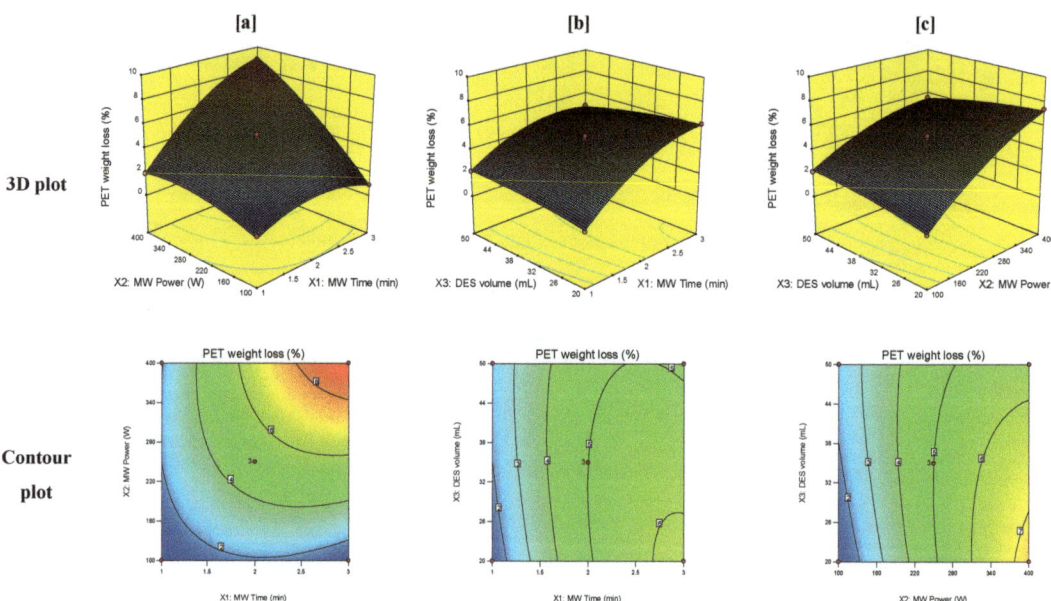

Figure 5. 3D and contour plots of the effect of the interaction of (**a**) MW time (X_1) and MW power (X_2), (**b**) MW time (X_1) and volume of DES (X_3), and (**c**) MW power (X_2) and volume of DES (X_3) on PET weight loss (%).

The carbonyl index is considered one of the critical responses used in the evaluation of the hydrophilic nature of treated polymers. Hydrophilic polymers with high values of carbonyl index are more amenable for microbial depolymerization than hydrophobic ones. Figure 4 illustrates the dependence of the carbonyl index on the studied factors of MW irradiation time, MW power, and volume of DES. It can be observed that the increase in both MW irradiation time and power did not cause a significant increase in the carbonyl index of treated PET samples. Thus, it can be assumed that high levels of MW irradiation time and power result in PET depolymerization rather than surface oxidation, which is in accordance with the crystallinity index results. On the other hand, the interactions of high levels of DES volume with low levels of both MW irradiation time and power showed a significant increase in the carbonyl index, reaching a value of 4.55 and confirming the process of PET surface oxidation at these levels of the studied factors.

PET weight loss at T_0 of degradation is also an important response for assessing the initial depolymerization of PET. The higher the value of weight loss, the greater was the decrease in the thermal stability of the treated polymer, which confirmed the initial depolymerization of the treated PET samples. In Table 3, it can be observed that all the independent variables and their interactions influenced PET weight loss significantly (term p-value < 0.05) except for the volume of DES. As indicated in Figure 5, the interaction between MW irradiation time and MW power resulted in a significant increase in PET weight loss at T_0 of degradation to reach a value of 9%, whereas the value of weight loss of untreated PET was measured to be only 0.44%. In addition, it should be noted that low levels of DES volume showed higher PET weight loss percentage than high levels of DES volume, which means that initial depolymerization of PET occurs better at low volumes of DES.

3.4. Optimization of the MW-Assisted DES Technique

All three responses were optimized simultaneously using BBD optimization. Optimum MW treatment conditions were chosen with the aim of obtaining maximum initial depolymerization of PET and enhancing the biodegradation of residual PET after MW treatment. As previously described, maximum initial depolymerization of PET was observed with PET residues of increased percentage of weight loss at T_0 of degradation. Additionally, based on literature review, enhanced PET biodegradation can be achieved through low crystallinity and high carbonyl index of residual PET [44]. Thus, the MW treatment conditions were adjusted to attain minimum crystallinity index and maximum carbonyl index and percentage weight loss of PET at T_0, as shown in Table 4. Based on BBD, a total of 63 optimized solutions were obtained. The selected solution was determined according to its success in attaining an acceptable desirability of >0.5 for the studied responses and in fulfilling the low carbon footprint goal with the lowest energy consumption concerning MW irradiation time and power. A batch experiment was carried out for PET MW treatment using the optimized conditions while keeping PET concentration at 1.0 g, and the three responses were evaluated to validate the predicted model factors and responses. The response values (predicted and observed) for the optimized conditions are recorded in Table 4. The model was proven to be validated as a fine agreement existed between the predicted and observed results. This indicates the success of BBD for the evaluation and optimization of the proposed PET treatment process.

Table 4. The optimized PET pretreatment process with observed and predicted response values.

Independent Variable		Optimized Level	
X_1: MW time (min)		3.0	
X_2: MW power (W)		260	
X_3: Volume of DES (mL)		20.0	
Over all desirability		0.59	
Dependent variables	Desirability	Expected	Observed
Y_1: PET crystallinity index	Minimize	33.39	32.98
Y_2: PET carbonyl index	Maximize	4.14	4.22
Y_3: PET weight loss (%)	Maximize	6.47	6.25

3.5. Enzymatic Depolymerization of PET Materials

Combination of green treatments with enzymatic hydrolysis as means for enhanced plastic recycling has been very limited within the literature. In this work, a multistep depolymerization process for PET comprising a green treatment process followed by enzymatic hydrolysis using a highly PET-active enzyme was performed. The total depolymerization of PET obtained after the optimized MW treatment and enzymatic hydrolysis was compared against that of untreated virgin PET biodegradation. The polymer's biodegradability was assessed based on the percentage weight loss of the material and the amount of the produced water-soluble monomers, namely TPA, mono-(2-hydroxyethyl) terephthalate (MHET), and bis-(2-hydroxyethyl) terephthalate (BHET), analyzed via HPLC (Figure S1). Following the MW treatment process, the optimized PET sample showed 20.2 ± 1.4% weight loss, and 15.76% (w/w) of the treated PET sample corresponded to a mixture of the released soluble monomers TPA, MHET, and BHET. Alternatively, the average percentage weight loss (%) for untreated and MW-treated PET samples was estimated to be 1.5 ± 0.3 and 1.8 ± 0.3%, respectively, after enzymatic hydrolysis. Such low percentage of weight loss can be attributed to the crystallinity of the materials where there are a few accessible amorphous regions for the enzyme to act on. Slightly higher weight loss percentage was observed for the MW-treated PET sample, which indicated the efficiency of MW treatment in enhancing the biodegradability of PET.

As detailed in Table 5, the total amount of monomers released after enzymatic hydrolysis was slightly higher for the virgin PET than for MW-treated PET residue (0.82 versus 0.55 mM). In both cases, 60–70% of the total products released was in the form of TPA and 30–35% as MHET, while only 1% remained as BHET. The higher product release for virgin PET could be explained by its slightly lower crystallinity index (31.40) compared to MW-treated PET residue (32.98), a factor that enhances enzyme action.

Table 5. Monomer concentration after a four-day incubation of LCCv enzyme with the untreated and treated PET materials.

Material	TPA (µM)	MHET (µM)	BHET (µM)
Untreated: Control	0.55 ± 0.04	0.00 ± 0.00	0.00 ± 0.00
Untreated: LCCv	521.13 ± 23.22	287.04 ± 7.63	7.07 ± 0.36
Treated: Control	0.83 ± 0.04	0.20 ± 0.00	0.00 ± 0.00
Treated: LCCv	384.79 ± 4.91	158.83 ± 4.52	4.21 ± 0.06

Moreover, the proposed multistep depolymerization approach resulted in the production of monomers during both the MW treatment and enzymatic hydrolysis processes, giving rise to an average PET weight loss of 22 ± 1.7% and a total monomer conversion of ≈16% (w/w). Thus, the obtained results for the suggested depolymerization protocol were considerably higher than the virgin PET undergoing enzymatic hydrolysis only.

4. Conclusions

A stepwise depolymerization process for PET comprising an all-green, fast, low-energy, MW-assisted DES technique without the use of additional depolymerizing agents followed by enzymatic hydrolysis using LCCv enzyme was demonstrated. Compared to virgin PET biodepolymerization, the demonstrated combined approach was able to achieve increased PET depolymerization with a total of ≈16% (w/w) monomer conversion. The developed MW treatment process was optimized using BBD, where the volume of DES, MW power, and MW irradiation time were studied as independent variables. FTIR, TGA, and DSC spectra of the residual PET obtained after treatment with the MW-assisted DES technique showed a significant increase in residual PET carbonyl index and percentage weight loss at T_0 of degradation and maintenance of PET crystallinity percentage. Furthermore, optimum MW treatment was obtained at low DES volume (20 mL), 260 W MW power, and 3 min MW irradiation time. The enzymatic hydrolysis of treated PET demonstrated 1.8% weight loss and 0.55 mM monomers released after enzymatic hydrolysis, while 1.5% weight loss and 0.82 mM monomers were recorded for virgin PET. Analysis of the recycled monomers using HPLC confirmed the presence of TPA, MHET, and BHET as the monomers produced in the treated samples. The isolation of these monomers will be done in future work. The combination all-green treatments, which operated under mild, low-energy conditions without the use of additional depolymerization agents, produced an average PET weight loss of 22 ± 1.7% and a total monomer conversion of ≈16% (w/w). This MW-assisted DES followed by enzymatic hydrolysis methodology shows strong potential to achieve high conversion rates and is amenable to the incorporation of additional and sequential green approaches.

Moreover, large-scale applications of MW-assisted depolymerization in a continuous manner has recently shown great potential for recycling as it facilitates depolymerization of a large amount of materials in relatively mild conditions (lower temperature and frequencies) [45]. Nevertheless, there are certain challenges with respect to high-cost reactor designs, emission of volatile degradation products, unequal irradiations due to hot spots, and nonuniform heating that still need to be resolved [46].

In conclusion, the promise of ultramild routes with no requirement for additional depolymerization agent is demonstrated herein for their capacity to play an instrumental role in highly sustainable degradation and depolymerization processes for PET and other polyesters, thus serving as a key step in delivering ultrasustainable all-green routes for circular plastic value chains.

Supplementary Materials: The following supporting information can be downloaded at: https://www.mdpi.com/article/10.3390/polym14010109/s1, Figure S1: HPLC chromatograms of post-enzymatic hydrolysis products; Table S1: Different compositions of DESs used in PET recycling.

Author Contributions: Conceptualization, O.A.A., M.A. and M.B.F.; methodology, M.A., O.A.A. and E.N.; software, O.A.A. and M.A.; validation, M.A., O.A.A. and E.N.; formal analysis, M.A., O.A.A., E.N. and E.T.; investigation, M.A., E.N. and E.T.; resources, M.B.F.; data curation O.A.A., M.A. and E.N.; writing—original draft preparation, O.A.A., M.A. and E.N.; writing—review and editing, O.A.A., E.T. and M.B.F.; visualization, M.A. and O.A.A.; supervision, M.B.F., O.A.A. and E.T.; project administration, M.B.F.; funding acquisition, M.A. and M.B.F. All authors have read and agreed to the published version of the manuscript.

Funding: This project received funding from the European Union's Horizon 2020 research, the Irish Research Council (GOIPG/2021/1739), and innovation program under grant agreement No. 870292 (BIOICEP) and was supported by the National Natural Science Foundation of China (grant numbers: Institute of Microbiology, Chinese Academy of Sciences: 31961133016; Beijing Institute of Technology: 31961133015; Shandong University: 31961133014).

Institutional Review Board Statement: Not applicable.

Data Availability Statement: All data generated or analyzed during this study are included in the article (and its Supplementary Information files).

Conflicts of Interest: The authors declare no conflict of interest.

References

1. Gómez, A.V.; Biswas, A.; Tadini, C.C.; Furtado, R.F.; Alves, C.R.; Cheng, H.N. Use of Natural Deep Eutectic Solvents for Polymerization and Polymer Reactions. *J. Braz. Chem. Soc.* **2019**, *30*, 717–726. [CrossRef]
2. George, N.; Kurian, T. Recent Developments in the Chemical Recycling of Postconsumer Poly(Ethylene Terephthalate) Waste. *Ind. Eng. Chem. Res.* **2014**, *53*, 14185–14198. [CrossRef]
3. Rorrer, N.A.; Nicholson, S.; Carpenter, A.; Biddy, M.J.; Grundl, N.J.; Beckham, G.T. Combining Reclaimed PET with Bio-Based Monomers Enables Plastics Upcycling. *Joule* **2019**, *3*, 1006–1027. [CrossRef]
4. Singh, N.; Hui, D.; Singh, R.; Ahuja, I.P.S.; Feo, L.; Fraternali, F. Recycling of Plastic Solid Waste: A State of Art Review and Future Applications. *Compos. Part B Eng.* **2017**, *115*, 409–422. [CrossRef]
5. Filella, M. Antimony and PET Bottles: Checking Facts. *Chemosphere* **2020**, *261*, 127732. [CrossRef]
6. Khoonkari, M.; Haghighi, A.H.; Sefidbakht, Y.; Shekoohi, K.; Ghaderian, A. Chemical Recycling of PET Wastes with Different Catalysts. *Int. J. Polym. Sci.* **2015**, *2015*, 124524. [CrossRef]

7. Yue, Q.F.; Xiao, L.F.; Zhang, M.L.; Bai, X.F. The Glycolysis of Poly(Ethylene Terephthalate) Waste: Lewis Acidic Ionic Liquids as High Efficient Catalysts. *Polymers* **2013**, *5*, 1258–1271. [CrossRef]
8. Chen, W.; Yang, Y.; Lan, X.; Zhang, B.; Zhang, X.; Mu, T. Biomass-Derived γ-Valerolactone: Efficient Dissolution and Accelerated Alkaline Hydrolysis of Polyethylene Terephthalate. *Green Chem.* **2021**, *23*, 4065–4073. [CrossRef]
9. Wang, Q.; Yao, X.; Geng, Y.; Zhou, Q.; Lu, X.; Zhang, S. Deep Eutectic Solvents as Highly Active Catalysts for the Fast and Mild Glycolysis of Poly(Ethylene Terephthalate)(PET). *Green Chem.* **2015**, *17*, 2473–2479. [CrossRef]
10. Rubio Arias, J.J.; Thielemans, W. Instantaneous Hydrolysis of PET Bottles: An Efficient Pathway for the Chemical Recycling of Condensation Polymers. *Green Chem.* **2021**. [CrossRef]
11. Nikolaivits, E.; Pantelic, B.; Azeem, M.; Taxeidis, G.; Babu, R.; Topakas, E.; Brennan Fournet, M.; Nikodinovic-Runic, J. Progressing Plastics Circularity: A Review of Mechano-Biocatalytic Approaches for Waste Plastic (Re)Valorization. *Front. Bioeng. Biotechnol.* **2021**, *9*, 535. [CrossRef] [PubMed]
12. Musale, R.M.; Shukla, S.R. Deep Eutectic Solvent as Effective Catalyst for Aminolysis of Polyethylene Terephthalate (PET) Waste. *Int. J. Plast. Technol.* **2016**, *20*, 106–120. [CrossRef]
13. Barnard, E.; Rubio Arias, J.J.; Thielemans, W. Chemolytic Depolymerisation of PET: A Review. *Green Chem.* **2021**, *23*, 3765–3789. [CrossRef]
14. Imran, M.; Kim, B.K.; Han, M.; Cho, B.G.; Kim, D.H. Sub-and Supercritical Glycolysis of Polyethylene Terephthalate (PET) into the Monomer Bis (2-Hydroxyethyl) Terephthalate (BHET). *Polym. Degrad. Stab.* **2010**, *95*, 1686–1693. [CrossRef]
15. Choi, H.M.; Choi, H. Eco-Friendly, Expeditious Depolymerization of PET in the Blend Fabrics by Using a Bio-Based Deep Eutectic Solvent under Microwave Irradiation for Composition Identification. *Fibers Polym.* **2019**, *20*, 752–759. [CrossRef]
16. Attallah, O.A.; Janssens, A.; Azeem, M.; Fournet, M.B. Fast, High Monomer Yield from Post-Consumer Polyethylene Terephthalate via Combined Microwave and Deep Eutectic Solvent Hydrolytic Depolymerization. *ACS Sustain. Chem. Eng.* **2021**. [CrossRef]
17. Kawai, F.; Kawabata, T.; Oda, M. Current Knowledge on Enzymatic PET Degradation and Its Possible Application to Waste Stream Management and Other Fields. *Appl. Microbiol. Biotechnol.* **2019**, *103*, 4253–4268. [CrossRef]
18. Park, S.H.; Alammar, A.; Fulop, Z.; Pulido, B.A.; Nunes, S.P.; Szekely, G. Hydrophobic Thin Film Composite Nanofiltration Membranes Derived Solely from Sustainable Sources. *Green Chem.* **2021**, *23*, 1175–1184. [CrossRef]
19. Attallah, O.A.; Mojicevic, M.; Garcia, E.L.; Azeem, M.; Chen, Y.; Asmawi, S.; Brenan Fournet, M. Macro and Micro Routes to High Performance Bioplastics: Bioplastic Biodegradability and Mechanical and Barrier Properties. *Polymers* **2021**, *13*, 2155. [CrossRef] [PubMed]
20. Wei, R.; Zimmermann, W. Biocatalysis as a Green Route for Recycling the Recalcitrant Plastic Polyethylene Terephthalate. *Microb. Biotechnol.* **2017**, *10*, 1302–1307. [CrossRef] [PubMed]
21. Pestana, S.C.; Machado, J.N.; Pinto, R.D.; Ribeiro, B.D.; Marrucho, I.M. Natural Eutectic Solvents for Sustainable Recycling of Poly (Ethyleneterephthalate): Closing the Circle. *Green Chem.* **2021**, *23*, 9460–9464. [CrossRef]
22. Kawai, F. The Current State of Research on PET Hydrolyzing Enzymes Available for Biorecycling. *Catalysts* **2021**, *11*, 206. [CrossRef]
23. Ali, S.S.; Elsamahy, T.; Koutra, E.; Kornaros, M.; El-Sheekh, M.; Abdelkarim, E.A.; Zhu, D.; Sun, J. Degradation of Conventional Plastic Wastes in the Environment: A Review on Current Status of Knowledge and Future Perspectives of Disposal. *Sci. Total Environ.* **2021**, *771*, 144719. [CrossRef] [PubMed]
24. Donelli, I.; Taddei, P.; Smet, P.F.; Poelman, D.; Nierstrasz, V.A.; Freddi, G. Enzymatic Surface Modification and Functionalization of PET: A Water Contact Angle, FTIR, and Fluorescence Spectroscopy Study. *Biotechnol. Bioeng.* **2009**, *103*, 845–856. [CrossRef] [PubMed]
25. Falkenstein, P.; Gräsing, D.; Bielytskyi, P.; Zimmermann, W.; Matysik, J.; Wei, R.; Song, C. UV Pretreatment Impairs the Enzymatic Degradation of Polyethylene Terephthalate. *Front. Microbiol.* **2020**, *11*, 689. [CrossRef]
26. Falah, W.; Chen, F.J.; Zeb, B.S.; Hayat, M.T.; Mahmood, Q.; Ebadi, A.; Toughani, M.; Li, E.Z. Polyethylene Terephthalate Degradation by Microalga Chlorella Vulgaris along with Pretreatment. *Mater. Plast.* **2020**, *57*, 260–270. [CrossRef]
27. Quartinello, F.; Vajnhandl, S.; Volmajer Valh, J.; Farmer, T.J.; Vončina, B.; Lobnik, A.; Herrero Acero, E.; Pellis, A.; Guebitz, G.M. Synergistic Chemo-Enzymatic Hydrolysis of Poly(Ethylene Terephthalate) from Textile Waste. *Microb. Biotechnol.* **2017**, *10*, 1376–1383. [CrossRef]
28. Gong, J.; Kong, T.; Li, Y.; Li, Q.; Li, Z.; Zhang, J. Biodegradation of Microplastic Derived from Poly(Ethylene Terephthalate) with Bacterial Whole-Cell Biocatalysts. *Polymers* **2018**, *10*, 1326. [CrossRef]
29. Liu, B.; Fu, W.; Lu, X.; Zhou, Q.; Zhang, S. Lewis Acid-Base Synergistic Catalysis for Polyethylene Terephthalate Degradation by 1,3-Dimethylurea/Zn(OAc)2 Deep Eutectic Solvent. *ACS Sustain. Chem. Eng.* **2019**, *7*, 3292–3300. [CrossRef]
30. Didaskalou, C.; Kupai, J.; Cseri, L.; Barabas, J.; Vass, E.; Holtzl, T.; Szekely, G. Membrane-Grafted Asymmetric Organocatalyst for an Integrated Synthesis-Separation Platform. *ACS Catal.* **2018**, *8*, 7430–7438. [CrossRef]
31. Marichamy, S.; Stalin, B.; Ravichandran, M.; Sudha, G.T. Optimization of Machining Parameters of EDM for α-β Brass Using Response Surface Methodology. *Mater. Today Proc.* **2020**, *24*, 1400–1409. [CrossRef]
32. Kadhom, M.A.; Abdullah, G.H.; Al-Bayati, N. Studying Two Series of Ternary Deep Eutectic Solvents (Choline Chloride–Urea–Glycerol) and (Choline Chloride–Malic Acid–Glycerol), Synthesis and Characterizations. *Arab. J. Sci. Eng.* **2017**, *42*, 1579–1589. [CrossRef]

33. Tournier, V.; Topham, C.M.; Gilles, A.; David, B.; Folgoas, C.; Moya-Leclair, E.; Kamionka, E.; Desrousseaux, M.-L.; Texier, H.; Gavalda, S.; et al. An Engineered PET Depolymerase to Break down and Recycle Plastic Bottles. *Nature* **2020**, *580*, 216–219. [CrossRef] [PubMed]
34. Djapovic, M.; Milivojevic, D.; Ilic-Tomic, T.; Lješević, M.; Nikolaivits, E.; Topakas, E.; Maslak, V.; Nikodinovic-Runic, J. Synthesis and Characterization of Polyethylene Terephthalate (PET) Precursors and Potential Degradation Products: Toxicity Study and Application in Discovery of Novel PETases. *Chemosphere* **2021**, *275*, 130005. [CrossRef]
35. Chelliah, A.; Subramaniam, M.; Gupta, R.; Gupta, A. Evaluation on the Thermo-Oxidative Degradation of PET Using Prodegradant Additives. *Indian J. Sci. Technol.* **2017**, *10*, 2–5. [CrossRef]
36. Li, W.; Zhang, C.; Chi, H.; Li, L.; Lan, T.; Han, P.; Chen, H.; Qin, Y. Development of Antimicrobial Packaging Film Made from Poly(Lactic Acid) Incorporating Titanium Dioxide and Silver Nanoparticles. *Molecules* **2017**, *22*, 1170. [CrossRef] [PubMed]
37. Fosse, C.; Bourdet, A.; Ernault, E.; Esposito, A.; Delpouve, N.; Delbreilh, L.; Thiyagarajan, S.; Knoop, R.J.I.; Dargent, E. Determination of the Equilibrium Enthalpy of Melting of Two-Phase Semi-Crystalline Polymers by Fast Scanning Calorimetry. *Thermochim. Acta* **2019**, *677*, 67–78. [CrossRef]
38. Alomar, M.K.; Hayyan, M.; Alsaadi, M.A.; Akib, S.; Hayyan, A.; Hashim, M.A. Glycerol-Based Deep Eutectic Solvents: Physical Properties. *J. Mol. Liq.* **2016**, *215*, 98–103. [CrossRef]
39. Wang, Q.; Yao, X.; Tang, S.; Lu, X.; Zhang, X.; Zhang, S. Urea as an Efficient and Reusable Catalyst for the Glycolysis of Poly (Ethylene Terephthalate) Wastes and the Role of Hydrogen Bond in This Process. *Green Chem.* **2012**, *14*, 2559–2566. [CrossRef]
40. Sert, E.; Yılmaz, E.; Atalay, F. Chemical Recycling of Polyethlylene Terephthalate by Glycolysis Using Deep Eutectic Solvents. *J. Polym. Environ.* **2019**, *27*, 2956–2962. [CrossRef]
41. Choi, H.M.; Cho, J.Y. Microwave-Mediated Rapid Tailoring of PET Fabric Surface by Using Environmentally-Benign, Biodegradable Urea-Choline Chloride Deep Eutectic Solvent. *Fibers Polym.* **2016**, *17*, 847–856. [CrossRef]
42. Çabuk, H.; Yılmaz, Y.; Yıldız, E. A Vortex-Assisted Deep Eutectic Solvent-Based Liquid-Liquid Microextraction for the Analysis of Alkyl Gallates in Vegetable Oils. *Acta Chim. Slov.* **2019**, *66*, 385–394. [CrossRef]
43. Beltrán-Sanahuja, A.; Casado-Coy, N.; Simó-Cabrera, L.; Sanz-Lázaro, C. Monitoring Polymer Degradation under Different Conditions in the Marine Environment. *Environ. Pollut.* **2020**, *259*, 113836. [CrossRef] [PubMed]
44. Cho, J.Y.; Choi, H.M.; Oh, K.W. Rapid Hydrophilic Modification of Poly(Ethylene Terephthalate) Surface by Using Deep Eutectic Solvent and Microwave Irradiation. *Text. Res. J.* **2016**, *86*, 1318–1327. [CrossRef]
45. Winter, R. Method and Apparatus for Recycling of Organic Waste Preferably Used Tires Using Microwave Technique. *Eur. Pat. Appl.* **2013**, *1*, 1–23.
46. Formela, K.; Hejna, A.; Zedler; Colom, X.; Cañavate, J. Microwave Treatment in Waste Rubber Recycling – Recent Advances and Limitations. *Express Polym. Lett.* **2019**, *13*, 565–588. [CrossRef]

Article

Thermo-Oxidative Destruction and Biodegradation of Nanomaterials from Composites of Poly(3-hydroxybutyrate) and Chitosan

Anatoly A. Olkhov [1,2,3], Elena E. Mastalygina [1,2], Vasily A. Ovchinnikov [1,3], Tatiana V. Monakhova [2], Alexandre A. Vetcher [4,5,*] and Alexey L. Iordanskii [3]

[1] Scientific Laboratory "Advanced Composite Materials and Technologies", Plekhanov Russian University of Economics, 36 Stremyanny Ln, 117997 Moscow, Russia; aolkhov72@yandex.ru (A.A.O.); elena.mastalygina@gmail.com (E.E.M.); fizhim@rambler.ru (V.A.O.)
[2] N.M. Emanuel Institute of Biochemical Physics, Russian Academy of Sciences, 4 Kosygin St., 119991 Moscow, Russia; tvmonakhova@ya.ru
[3] N.N. Semenov Federal Research Center for Chemical Physics, Russian Academy of Sciences, 4 Kosygin St. 4, 119334 Moscow, Russia; aljordan08@gmail.com
[4] Institute of Biochemical Technology and Nanotechnology (IBTN), Peoples' Friendship University of Russia (RUDN), 6 Miklukho-Maklaya St., 117198 Moscow, Russia
[5] Complementary and Integrative Health Clinic of Dr. Shishonin, 5 Yasnogorskaya St., 117588 Moscow, Russia
* Correspondence: avetcher@gmail.com

Citation: Olkhov, A.A.; Mastalygina, E.E.; Ovchinnikov, V.A.; Monakhova, T.V.; Vetcher, A.A.; Iordanskii, A.L. Thermo-Oxidative Destruction and Biodegradation of Nanomaterials from Composites of Poly(3-hydroxybutyrate) and Chitosan. *Polymers* **2021**, *13*, 3528. https://doi.org/10.3390/polym13203528

Academic Editor: Andrea Sorrentino

Received: 9 September 2021
Accepted: 7 October 2021
Published: 14 October 2021

Publisher's Note: MDPI stays neutral with regard to jurisdictional claims in published maps and institutional affiliations.

Copyright: © 2021 by the authors. Licensee MDPI, Basel, Switzerland. This article is an open access article distributed under the terms and conditions of the Creative Commons Attribution (CC BY) license (https://creativecommons.org/licenses/by/4.0/).

Abstract: A complex of structure-sensitive methods of morphology analysis was applied to study film materials obtained from blends of poly(3-hydroxybutyrate) (PHB) and chitosan (CHT) by pouring from a solution, and nonwoven fibrous materials obtained by the method of electrospinning (ES). It was found that with the addition of CHT to PHB, a heterophase system with a nonequilibrium stressed structure at the interface was formed. This system, it undergone accelerated oxidation and hydrolysis, contributed to the intensification of the growth of microorganisms. On the other hand, the antimicrobial properties of CHT led to inhibition of the biodegradation process. Nonwoven nanofiber materials, since having a large specific surface area of contact with an aggressive agent, demonstrated an increased ability to be thermo-oxidative and for biological degradation in comparison with film materials.

Keywords: poly(3-hydroxybutyrate); chitosan; electrospinning; thermal oxidation; biodegradation; Sturm's method

1. Introduction

Nowadays, much attention in world science is devoted to the creation of a new class of functional biodegradable highly porous materials based on ultrathin and nanoscale fibrous fibrillar structures with a wide variety of specific characteristics: physical, mechanical, sorption, and diffusion properties [1–3]. Such materials, based on synthetic and biopolymers, are widely used in biology, medicine, cell engineering, separation and filtration processes, reinforced composites, electronics, analytics, sensor diagnostics, as eco-sorbents for cleaning the environment from emergency spills of oil products and heavy metal compounds, and in many other innovative applications [4–6].

With the employment of traditional technologies, such as extrusion of polymer melts and solutions, spinneret drawing, etc., technological difficulties arise in obtaining fibers with a diameter of less than 1 μm [7]. Namely for this reason, currently, an electrospinning (ES) method, as a versatile and relatively simple method for forming ultra-thin and nanofibers with a diameter in the range from 10 nm to 10 μm from polymer solutions and melts, is employed [8]. The process is based on pulling a drop of solution or melt, formed at the end of a capillary, by the action of electrostatic forces applied to the solution or polymer melt, resulting in the formation of an ultra-thin nonwoven fiber [9,10].

Earlier, on the example of PHB and a number of its compositions with other polymers [11–13], the physicochemical, dynamic, and transport characteristics of film biodegradable matrices, microparticles, and microcapsules of PHB were studied for prolonged and targeted drug delivery [14,15]. High biocompatibility, controlled biodegradation into environmentally friendly products (carbon dioxide and water), and satisfactory mechanical characteristics make it possible to consider this polymer for the creation of innovative disposable products in food, packaging, medical, and environmental areas [11,16].

For the effective regulation of operational properties, combined (composite or mixed) materials based on PHB and other biopolymers, various organic and inorganic dispersed and fibrous fillers are currently available. Such composites, due to their heterophase and high heterogeneity, create, for example, a multiform dynamic of transport of low-molecular substances, due to the cooperation of the processes of structure formation (crystallization, plasticization, and swelling) and the kinetics of biodegradation of one of the components [17,18].

The most promising system from this view is a blend of PHB and CHT [19]. As was shown in the example of film matrices [20], the combination of PHB and CHT allowed to create of a new generation of amphiphilic fully biodegradable composite systems with an increased sorption capacity, a controlled rate of biodegradation, and the ability to implement various kinetic profiles of prolonged release of a low-molecular-weight substance in a wide time range (from weeks to months).

The structural organization and resistance to aggressive media of ultra-fibrous materials based on PHB and ultra-low concentrations of CHT (0.05–0.3 wt%) have been already analyzed [21]. The effect of CHT on both the crystalline and amorphous phases was shown: the addition of 0.1–0.2 wt% of CHT to the PHB structure increases the enthalpy of melting of PHB crystallites by almost 20%, which, in our opinion, is associated with the nucleating effect of chitosan, while the proportion of dense amorphous regions of PHB increases. The oxidation of PHB-CHT fibrous matrices with an ozone-oxygen mixture for up to four hours contributes to the compaction of amorphous regions, restricting molecular mobility, due to the predominant process of crosslinking of molecules over their oxidative destruction.

Hence, we would like to study in detail the structural organization and processes of thermo-oxidative and biodegradation of composite film and nonwoven fiber materials based on PHB with a high (40–70 wt%) content of CHT.

2. Materials and Methods

PHB (16F series) was obtained by microbiological synthesis by Biomer (Schwalbach am Taunus, Germany) with a viscosity-average molecular mass (M_v) of 4.6×10^5 g·mol^{-1}, with a degree of crystallinity around 59%, and CHT in the form of a fine powder with an average particle size of 1 ÷ 5 μm (Bioprogress CJSC, Shchyolkovo, Russian Federation) with the M_v of 4.4×10^5 g·mol^{-1} and the degree of deacetylation of 82.3% were employed.

According to the available publications, PHB is highly soluble in chloroform, dichloroethane, dimethylformamide, pyridine, dioxane, 1 M NaOH solution, higher alcohols, camphor. The more accessible and less toxic of these solvents are chloroform and dimethylformamide (DMF). The calculation of the compatibility of these solvents with PHB was carried out on the basis of a comparison of the solubility parameters by three different methods (according to Small, according to Hoya, and according to Van Krevelen). The average solubility parameter for PHB was 9.23, for chloroform 10.2, for DMF 12.5. Based on the data obtained, the solvent chloroform was selected as the most suitable in terms of solubility and toxicity.

To obtain the fibers, forming solutions of PHB in chloroform were prepared. The concentration of PHB in the solution was 7 wt%. The content of CHT in the forming solution was 40, 50, and 60 wt%, relative to the mass of PHB. Forming solutions of PHB with CHT were prepared at a temperature of 60 °C using an automatic high-speed agitator and an ultrasonic bath. The fibers were obtained by ES using a single-capillary laboratory setup with a capillary diameter of 0.1 mm, voltage of 12 kV, distance between the electrodes of 18 cm, and electrical conductivity of 10 μScm^{-1} [5,22].

In a comparison, samples of PHB/CHT were prepared in the form of films in several steps. First, PHB film was compressed from granules on the glass surface at a temperature of (180 ± 5) °C and a pressure of 10^7 Pa. The resulting film was crushed and dissolved in chloroform. CHT powder was added to the solution with vigorous stirring using a high-speed mechanical stirrer at 1000 rpm, Disperser IKA T18 digital ULTRA TURRAX (IKA-Werke GmbH & Co. KG, Staufen, Germany), until a homogeneous suspension was obtained. Then, the suspension was subjected to exposure by simultaneous action of stirring at 500 rpm, 50 °C, 40 kHz for 30 min on an overhead stirrer LABTEH OS-20LT (Labteh Ltd., Moscow, Russian Federation) in ultrasonic bath Vilitek VBS-10DS (Vilitek Ltd., Moscow, Russian Federation). The resulting suspension was poured into Petri dishes, covered with a lid, and conditioned at room temperature until the films were completely dry.

The morphology of fibrous materials was studied by scanning electron microscopy (SEM) on Hitachi TM-1000 scanning electron microscope (Hitachi, Ltd., Tokyo, Japan). Changes in the microstructure of the samples before, during, and after the biological degradation was recorded with an Axio Imager Z2m optical microscope (Carl Zeiss, Jena, Germany) in transmitted (TL) and reflected (RL) light at magnifications of 50, 200, and 500×.

The study of thermophysical characteristics (temperature and enthalpy of melting) was carried out with a differential scanning calorimeter (DSC) DSC 214 Polyma (NETZSCH-Geratebau GmBH, Selb, Germany) according to ISO 11357-3:2018. The heating of the samples was carried out in the temperature range of 25–180 °C at a scanning speed of 10 °C/min. The weight of the sample was 10 ± 0.1 mg. The temperature scale and enthalpy of melting were calibrated against an indium standard sample.

The analysis of the chemical composition was carried out according to the manufacturer's recommendation on the Bruker Lumos IR Fourier microscope (Bruker Corp., Bremen, Germany) at the temperature of (24 ± 2) °C in the range of wave numbers 400–4000 cm^{-1}. The analysis was carried out by attenuated total reflection (ATR) using a diamond crystal.

The study of biological degradation was carried out under the action of soil microorganisms for the release of CO_2 (Sturm method) according to ISO 14855-1:2012. To obtain an inoculum, compost was prepared according to Russian Standard GOST 9.060-75: equal proportions of sand, garden soil, and horse manure were mixed and conditioned for 60 days. Then, 860 g of prepared compost was mixed with 2 L of double-distilled water and filtered through a sieve with a mesh size of 0.5 mm. Subsequently, to purify the inoculum, the liquid part was subjected to separation in a laboratory centrifuge Hettich EBA 200 (Andreas Hettich GmbH & Co. KG, Tuttlingen, Germany) for 5 min at 1500 rpm, after which the liquid fraction was taken. Thereafter, the resulting suspension was passed through filter paper three times. The resulting soil concentrate was diluted with double distilled water to a volume of 9 L. The volume of round-bottomed test flasks was 500 mL. Six test flasks with inoculum without samples were bubbled with CO_2-free air for 72 h before the start of the biodegradation experiment. The viability of bacteria in the soil concentrate (inoculum) was analyzed using a Polar 3 ToupCam 5.1 MP optical microscope (Micromed, Shenzhen, China) at 400 × magnification. In the process of biodegradation of materials by the microbiota of the inoculum, carbon dioxide was released. A 0.0125 M Ba(OH)$_2$ solution was used as an absorbent. The amount of evolved carbon dioxide was estimated by titration with 0.05 M HCl solution and was calculated by the formula: CO_2 (mg) = 1.1 × V(HCl), where V(HCl) is the amount of hydrochloric acid used for titration (ml). For alkalimetric testing, a set for automatic titration Titrion (Econix-Expert, Ltd., Moscow, Russia) was used. With a known initial content of total organic carbon in the samples, the biodegradation index was calculated. The analysis of the accumulation of carbon dioxide was carried out with an interval of 2–4 days. The biodegradation test was carried out for 60 days, the number of test flasks was 6: flasks 1, 2, 3, and 4 containing samples of the test materials and 500 mL of inoculum, as well as flasks 4 and 5 containing only inoculum (zero control).

The kinetics of oxidation of the samples was investigated by determining the amount of absorbed oxygen by the manometric method using a special manometric device [23]. The analysis was carried out at an elevated temperature (140 ± 2) °C and an oxygen pressure of 500 torr on a manometric device with the absorption of volatile oxidation products with solid KOH.

3. Results

3.1. Morphology Analysis

At the first stage of our work, we studied the features of the morphology and kinetics of biodegradation of fibrous and film materials based on PHB/CHT blends. All film and fibrous samples in each group, regardless of the PHB/CHT ratio, showed approximately the same patterns of morphology formation. Therefore, to consider the comparative kinetics of biodegradation, we selected two samples of film and fibrous materials without CHT and with a CHT weight % of 40–50 (all over the text (X0/Y0) means PHB/CHT (w/w)%).

To describe the morphology of film composites and ultrathin fibers based on a mixed composition of PHB and CHT, micrographs of surfaces and edge cleavages were obtained, recorded by the SEM method (Figure 1).

Figure 1. Images of the morphology of samples of nonwoven fibrous and film materials obtained by SEM: (**a**) PHB—fibers, (**b**) PHB/CHT—fibers (60/40), (**c**) PHB/CHT—film (50/50), and (**d**) PHB—film. The size of the scale bar—100 μm.

The analysis of micrographs of surfaces and chips for mixed films demonstrates that as the content of PHB increases, the globularity of the structure becomes more and more pronounced. The globules look "inserted" into the CHT matrix, and, in general, both components form a heterophase structure. Note that, despite the heterophase nature, almost all ratios of polymer components retain the integrity of the PHB matrix with an increase in the concentration of CHT globules in it. It should be noted that the morphology of films with an equal polymer ratio is characterized by the separation of components with the formation of a two-layer structure. At a high content of CHT in the layer, an insignificant amount of PHB globules can be seen; however, in general, phase separation occurs quite noticeably. The formation of a bilayer structure should affect the diffusion, mechanical, and other physico-mechanical characteristics of the polymer composition.

The fibrous samples analysis reveals the morphology difference of the nonwoven materials from PHB and PHB/CHT. PHB fibers are randomly located, individual filaments of circular cross-sections with a diameter of 3.5 ± 2.5 µm. This diameter distribution is a consequence of the splitting of the primary jet during the ES of the polymer solution. The fibers contain single defects in the form of extended thickenings of arbitrary geometry. The origin of these defects is associated with the suboptimal electrical conductivity of the polymer solution [24]. PHB/CHT fibers, on the other hand, consist of cylindrical sections with an average diameter of 3 ± 1 µm and spherical thickenings slightly elongated along the fiber axis with an average size of 15 ± 5 µm. The presence of thickenings in the fiber structure is associated with particles of the dispersed phase of CHT. Dispersion of CHT in the form of relatively large particles is associated with the lack of solubility in $CHCl_3$ during the preparation of ES solutions based on PHB. The solutions were instead suspensions.

3.2. Study of Biological Degradation

The kinetics of biodegradation is demonstrated in Figure 2. In comparison with films, nonwoven materials were characterized by a higher rate of biodegradation, which is associated with a higher specific surface area and, therefore, increased accessibility for attack by microorganisms. Similar patterns have been described for nonwoven materials based on polylactic acid [25].

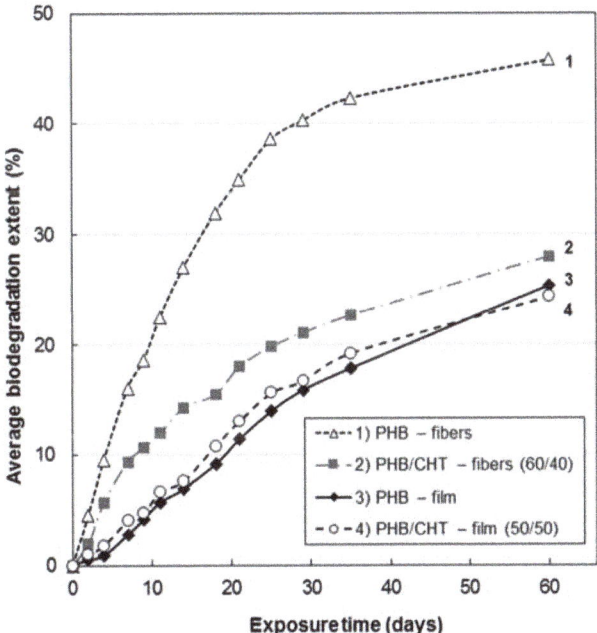

Figure 2. Kinetics of biodegradation (according to the Sturm method) of samples of nonwoven fibrous and film materials: (**1**) PHB—fibers; (**3**) PHB—film; (**2**) PHB/CHT—fibers (60/40); and (**4**) PHB/CHT—film (50/50).

Analysis of the kinetic curves in Figure 2 suggests that the introduction of CHT into PHB leads to inhibition of the biodegradation process. This can be explained by the antibacterial properties of CHT, which are a consequence of the interaction of its positively charged amino groups with negatively charged phosphoryl groups of phospholipids of the bacterial cell wall, a violation of its integrity with changes in metabolism, which eventually leads to cell death [26].

In addition, CHT is capable of interacting with nucleic acids, thereby disrupting the biosynthesis of proteins and damaging the structural and functional complexes of the bacterial cell [27]. The fungicidal properties of CHT are explained by similar mechanisms.

Analysis of the nature of the kinetic dependences in Figure 2 allows distinguishing two characteristic areas in them. The first section up to 30 days is characterized by an intensive course of the process of hydrolysis of film and fibrous materials based on PHB. This feature is associated with the processes of degradation of the loose amorphous phase of PHB, intercrystalline regions, stressed regions at the PHB/CHT interface, and the CHT phase itself. The second section is characterized by a low rate of processes of hydrolytic degradation of materials, which we associate with the destruction of directly crystalline formations of polymers. It is known that crystalline grains have a high packing density of macromolecules and are resistant to aggressive environmental factors [28]. Their degradation is associated with low permeability for all types of low molecular weight liquids and gases [29].

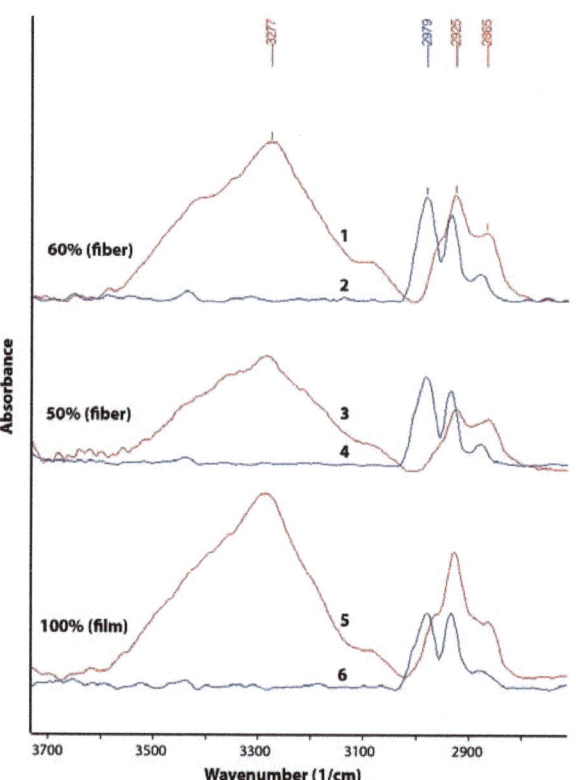

Figure 3. IR spectra (ATR method) in the wavelength range of 2700–3700 cm^{-1} of the initial samples (2,4,6) and samples after 60 days of testing by the Sturm method (1,3,5): PHB/CHT (fibers)—(60/40) (1 and 2), PHB/CHT (film)—(50/50) (3 and 4), PHB (film) (5 and 6).

The change in chemical composition during biodegradation was studied by FTIR (Figures 3 and 4). An analysis was made of changes in the chemical composition of samples of materials PHB (film), PHB/CHT (film) (50/50), PHB/CHT (fibers) (60/40) after 60 days of tests for biodegradation under the influence of soil microflora. The PHB-based samples obtained by ES on a substrate (PHB—fibers) were severely damaged after testing, which made them difficult to analyze.

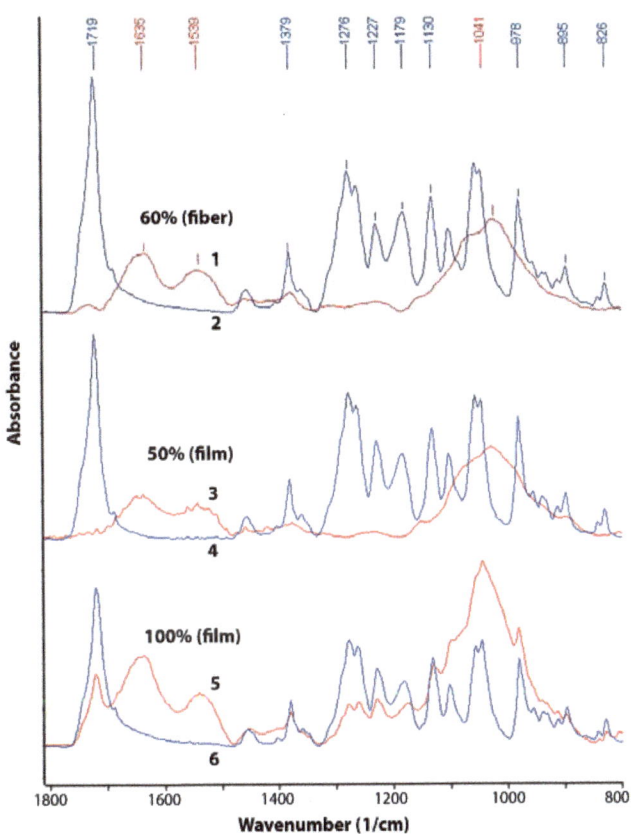

Figure 4. IR spectra (ATR method) in the wavelength range 800–1800 cm^{-1} of the initial samples (2, 4, 6) and samples after 60 days of testing by the Sturm method (1, 3, 5): PHB/CHT (fiber)—(60/40) (1 and 2), PHB/CHT (film)—(50/50) (3 and 4), PHB (film) (5 and 6).

Since the analysis was carried out by the ATR method, changes in the outer layers of the material were monitored. It should be noted that the IR spectra of the initial samples of PHB (film) and PHB/CHT (film) (50/50) have similar absorption peaks in the entire wavelength range corresponding to pure PHB. For example, in the near-surface layers of the composite material with the equal content of components, PHB is predominantly located.

After 60 days of biodegradation in the soil inoculum, changes in the chemical composition were observed both in materials based on pure PHB and in composites based on PHB and CHT. Moreover, composite nonwoven materials were characterized by greater changes than film samples. When comparing the IR spectra of PHB (film) of the original and subjected to biodegradation, a significant increase in absorption bands was found in the regions of 3100–3600, 1580–1700, 1490–1580, and 950–1150 cm^{-1}. In this case, the intensity of the absorption peak with a maximum at 1720 cm^{-1}, characteristic of PHB and corresponding to stretching vibrations of carbonyl groups (C=O), significantly decreased [30].

The appearance of a diffuse absorption peak in the region of 3100–3600 cm^{-1} indicates the accumulation of hydroxyl groups (–OH) and amide groups (–NH$_2$) in the near-surface layers of the material [31]. The emerging absorption peaks at 1530 and 1635 cm^{-1} correspond to bending vibrations of amide groups (–NH$_2$) in primary amides [32]. There was also a significant increase in the absorption band at 1041 cm^{-1}. This band corresponds to vibrations of C-O-C bonds and is characteristic of carbohydrates [33].

Changes in the microstructure of the samples were examined by optical microscopy. Figure 5 demonstrates that the PHB sample (film) (Figure 5a) has a fairly uniform structure, with some roughness on the surface. There are no violations of the film continuity. After tests by the Sturm method, violations of the integrity of the sample and numerous through-holes are visible (Figure 5b). The damage to the material is visible under the microscope—darkening of the material in the appearance of the micromycetes' mycelium.

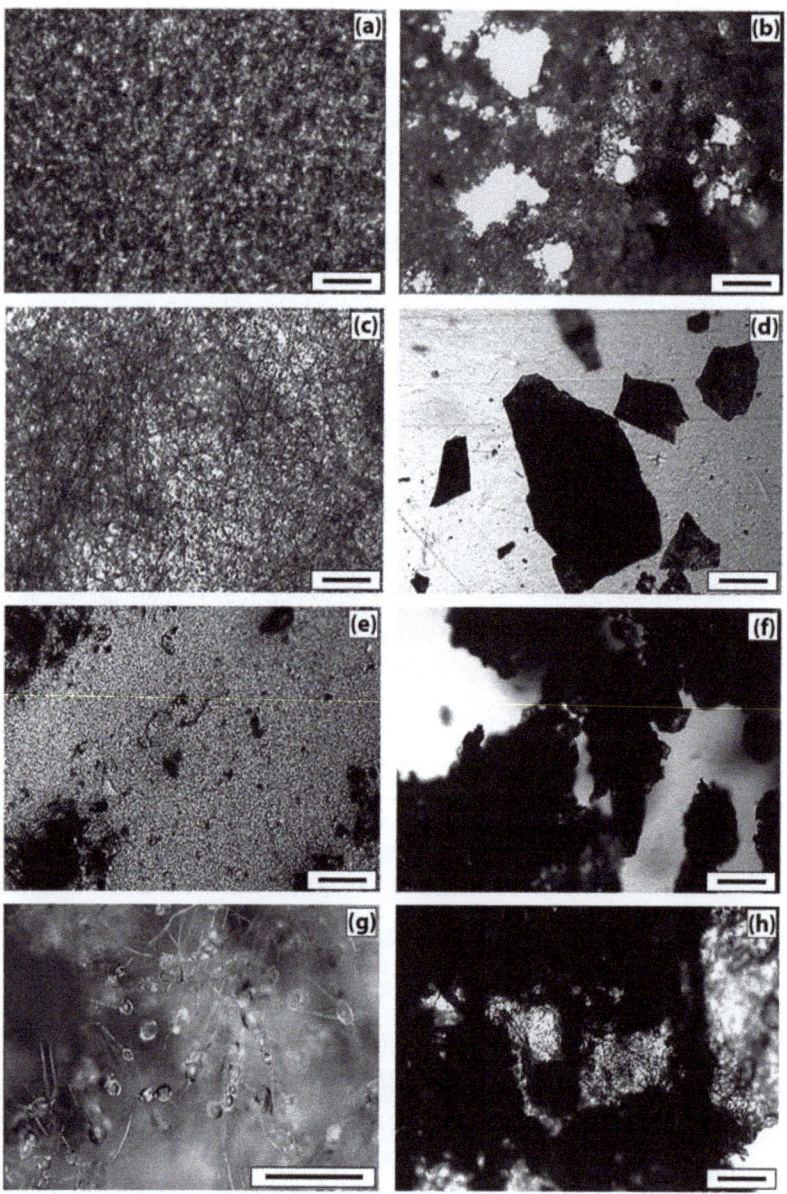

Figure 5. Optical micrographs of PHB (film) samples—(**a**) initial and (**b**) after 60 days of tests by the Sturm method; PHB (fiber)—(**c**) initial and (**d**) after 60 days of tests by the Sturm method; PHB/CHT (film) —(50/50)—(**e**) initial and (**f**) after 60 days of tests by the Sturm method; and PHB/CHT (fiber)—(60/40)—(**g**) initial and (**h**) after 60 days of tests by the Sturm method). The size of the scale bar—100 μm.

The original PHB sample (nonwoven) (Figure 5c) is characterized by a uniform fibrous structure. The average fiber diameter is 4.5 ± 1.5 μm. As a result of tests by the Sturm method, significant damage to the material is observed. The fibrous structure of the sample becomes almost indistinguishable, numerous fiber breaks and signs of microbiological damage are visible.

The original film sample of PHB/CHT (50/50) (Figure 5e) differs from the PHB film by a more heterogeneous structure. Granular areas of the structure and through defects are visible, which can adversely affect the physical and mechanical properties of the material. On the other hand, the loose structure provides access to the material for degradation agents. After biodegradation by the Sturm method, the number of defects increases, and the darkening of the material is observed.

The original nonwoven sample PHB/CHT (60/40) (Figure 5g) is characterized by an uneven fibrous structure, thickness differences, and spindle-shaped thickenings of 15–20 μm are visible. As a result of tests by the Sturm method, significant damage to the material is observed; however, the fibrous structure remains. Blind cavities and traces of microbiological damage appear.

3.3. Thermo-Oxidative Destruction

Under real conditions of biodegradation of products, the processes of hydrolysis and oxidation proceed simultaneously. In our experiment with the Sturm method, the samples were incubated in an aqueous inoculum, which led to the predominant contact of materials with water and a relatively small amount of oxygen dissolved in water. In this case, the processes of hydrolytic degradation of polymers predominantly took place. Therefore, in the second part of our study, we will separately consider the processes of oxidative degradation of PHB/CHT samples. For acceleration, the oxidation process was carried out at 140 °C in an oxygen atmosphere. This is the maximum possible temperature at which no noticeable changes in the chemical composition and supramolecular structure occurred in the samples.

Figure 6 demonstrates the kinetic dependences of the oxidation of PHB/CHT fibrous materials.

Figure 6 demonstrates the kinetic curve for PHB fibers has an induction period of about 90 min. This indicates the oxidation of the material in the diffusion mode. The sample oxidizes extremely slowly. The same effect was previously noted in the study of thermal oxidation of extrusion films based on blends of polyethylene/PHB [34], polyvinyl alcohol/PHB [35], and ethylene-propylene copolymer/PHB [36]. The increased resistance of PHB to oxidation is due to the high degree of crystallinity and the dense structure of the amorphous regions, which impede the diffusion of oxygen [37].

The kinetic curves of oxidation of mixed fibers containing 40–50 wt% of CHT are intermediate between pure PHB and CHT, but are relatively close to the kinetic curve of the latter. It should be noted that CHT is oxidized better than PHB. This is due to the better oxygen permeability. Such closeness of the oxidation curves of blends to CHT is explained by the fact that, in the range of CHT content of 40–50 wt%, phase inversion occurs in the mixed fibers and both components form a continuous network. It should also be noted that the analysis of the IR spectra (ATR) of blended fibers with a CHT content of 40–50 wt%, presented above, exhibited the presence of only the PHB phase in the surface layers. It is known that crystallization does not occur in the surface layers of films and fibers [38,39]. In this case, oxygen diffuses faster through the amorphous surface layer of PHB to the more reactive phase of CHT, and the oxidation of blends proceeds in a kinetic mode without an induction period typical for PHB fibers.

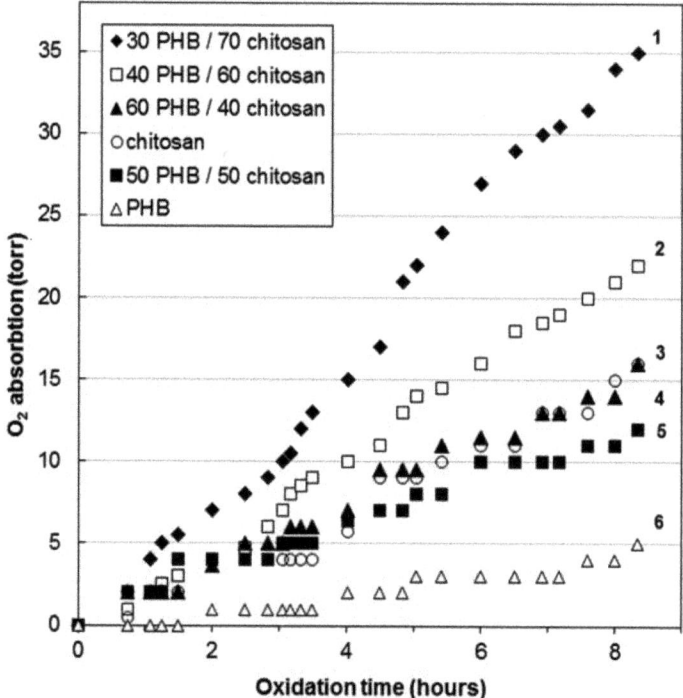

Figure 6. Kinetic curves of oxygen absorption by nonwoven fibrous materials PHB/CHT: (30/70)—1, (40/60)—2, (60/40)—3, (0/100)—4, (50/50)—5, and (100/0)—6.

With an increase in the content of CHT in blends to 60–70 wt%, the structure of the fibers changes. The PHB phase becomes dispersed and acts as a binder for CHT particles during the formation of a nonwoven fibrous material. The kinetic curves of oxidation of these materials are significantly higher than of PHB and CHT themselves.

When heated, pure CHT is characterized by an endothermic peak at 50–100 °C, which is related to the process of moisture evaporation. In this case, CHT is thermostable up to 250 °C [40]. Therefore, CHT should not have a significant effect on the endothermic peak of PHB melting in the range of 160–170 °C. However, upon heating, a complex bimodal peak is observed on the thermograms of PHB/CHT blends. Therefore, when calculating the enthalpies of the effect for the initial materials, an apparent increase in the enthalpy of melting of PHB is observed. This is associated with the interaction of PHB and CHT molecules with the formation of hydrogen bonds, which are destroyed when the blends are heated, which contributes to the endothermic effect [41].

From the data presented in Table 1, at 500 min of thermal oxidation, the enthalpy drops by 30–40% in blends with a content of 60–70 wt% CHT and by 10% in blends with 40 wt% CHT. At the same time, the enthalpy of melting in PHB fibers increased by 30%. The increase in the enthalpy of melting is most likely associated with the processes of oxidative degradation of overstressed covalent bonds of the PHB main chain localized in the amorphous regions, which leads to additional crystallization. The decrease in the enthalpy of mixed samples can be explained by the processes of CHT oxidation, as a result of which the number of intermolecular hydrogen bonds with PHB macromolecules decreases.

Table 1. Changes in thermophysical parameters melting temperature (T_m) and enthalpy (ΔH_m) of PHB/CHT fibrous materials during oxidation for 500 min at 140 °C.

(*w/w*) Ratio of PHB/CHT Samples	Initial		Oxidazed	
	T_m, °C ($\delta \pm 1°C$)	ΔH_m, J·g^{-1} ($\delta \pm 5$ J·g^{-1})	T_m, °C ($\delta \pm 1°C$)	ΔH_m, J·g^{-1} ($\delta \pm 5$ J·g^{-1})
30/70	169	106	160	79
40/60	163	125	166	72
60/40	162	69	159	63
100/0	166	65	162	87

At the same time, it should be noted that the change in the melting temperatures of the PHB phase in the mixed fibers changes insignificantly (3–4 °C), which indicates the invariability of the average crystallite size. According to Table 1, as the content of CHT increases, the fraction of the crystalline phase in PHB also increases, as evidenced by the value of the enthalpy of melting. Upon thermal oxidation for 500 min, the melting point of PHB crystallites practically does not change (3–4 °C), while the amount of the crystalline phase decreases significantly, especially in fibers containing 60–70 wt% of CHT.

4. Discussion

The interaction of PHB and CHT is contingent on forming hydrogen bonds. According to the available published data, the interaction of PHB and CT was proved through the formation of hydrogen bonds between the ester groups of PHB and the mobile protons of the amino and hydroxyl groups of CHT [42]. These interactions between polymers, in turn, affect the crystallization of PHB. The combination of the results of FTIR spectroscopy and the results of DSC, as well as SEM micrographs of the edge cleavages, also confirm that PHB and CHT interact with each other through the formation of hydrogen bonds between the ester groups of PHB and the amine groups of CHT.

The biodegradability of PHB-based materials is an important parameter. One of the objectives of this work was to study the kinetics of biodegradation of PHB/CHT materials under the influence of soil microorganisms and to reveal the effect of CHT on the biodegradation of PHB. It is well known that PHB is capable of hydrolysis [43,44], while enzymes secreted by microorganisms of the medium are capable of accelerating this process. In addition, an important factor for the biodegradation process is the availability of the material for attack by microorganisms, including its contact area with the environment.

There are several publications devoted to the degradation of PHB composites with other polymers in model media. As a rule, water or aqueous solutions of acids and alkalines are used as media. For example, the kinetics of degradation of fibrous scaffolds based on PHB with additions of CHT and bioglass were studied in the work [45]. The fibrous PHB/CHT (50/50) material was characterized by weight loss of 25% after conditioning in water at 37 °C for 60 days. The other researchers [46] demonstrated that PHB/CHT (85/15) blend experienced a weight loss of 10% after conditioning in water at 37 °C for 90 days. No studies are available on the biodegradability of PHB/CHT composites under the action of a combined microbiota that simulate the real conditions of materials degradation.

In this work, the study of biological degradation was carried out under the action of soil microorganisms for the release of carbon dioxide. The amount of emitted carbon dioxide, measured using the Sturm method, is the most reliable criterion for the biodegradation of organic molecules. The results of this study showed that PHB-based fibrous material was characterized by the highest biodegradability (biodegradation rate was 45.8% after 60 days). Pure chitosan has slow biodegradation compared to PHB, about 12% for 30 days [47]. The addition of chitosan inhibited the biological degradation of PHB due to the antimicrobial properties of the former. Thus, for the fibrous material based on the blend PHB/CHT (60/40), the degree of biodegradation was 28.0% after 60 days of the experiment.

After the biodegradation process under the action of soil microbiota, significant changes in chemical composition and microstructure were observed. The emerging absorption peaks at 1530 and 1635 cm^{-1} attributed to amide groups and a significant increase in the absorption band at 1041 cm^{-1} (C-O-C bonds in carbohydrates) is likely related to the increased number of chitin molecules contained in fungal cell walls [48]. This fact once again confirms the biofouling of the samples by microorganisms with their subsequent destruction.

The complete disappearance of the peak at 1720 cm^{-1} for the blends PHB/CHT and a decrease in the intensity for the pure PHB after biodegradation test indicate the destruction of PHB followed by forming microdefects and integrity damage. According to optical microscopy analysis, the materials based on pure PHB were characterized by biodeterioration throughout the samples. The film and fibrous materials based on the blends PHB/CHT had numerous sites of biodeterioration.

The reports also revealed the effect of CHT on the ability of PHB to oxidative destruction. Some publications confirmed that during the mild oxidation of CHT, hydroxyl groups (C3 and C6 sites) are oxidized to carbonyl ones without the noticeable effect of the polymer backbone structure. The number of amino groups may also decrease [49]. However, CHT is not an oxidation catalyst. Usually, catalysts and photosensitizers are used to oxidize chitosan itself.

In this case, the supramolecular structure of PHB/CHT blends plays an important role in the process of oxygen sorption. As shown earlier [50], the amorphous phase of PHB fibers consists of dense and relatively "loose" regions. The presence of dense areas makes it difficult to diffuse low molecular weight substances and gases. However, when low-molecular substances are added to the forming solution, it leads to a change in the processes of structure formation of PHB, as a result of which an amorphous phase with an increased free volume is formed. The formation of a looser structure of the amorphous phase is a direct consequence of the inhibition of crystallization processes as a result of adsorption or electrostatic interaction with polar groups of low molecular weight substances. Because film and fibrous samples were formed from the solution, the increase of PHB crystallinity after the oxidative degradation could be explained by isothermal crystallization during the experiment. Thermal oxidation of PHB leads to the destruction of amorphous regions and an increase in the mobility of macromolecules, which facilitates the recrystallization of the most ordered PHB regions. An increase in free volume in composite fibers converts the diffusion mode of oxidation into a kinetic one. Accelerated oxidation of blended fibers with a CHT content of more than 60 wt% can also be a consequence of the presence of intermolecular interaction between PHB and CHT, which have polar groups in the main chain (oxygen-containing—PHB, amine—CHT). In this case, the resulting physical bonds with CHT particles prevent the convergence of macromolecule regions in the amorphous phase of PHB, creating an additional free volume. Due to intermolecular interaction at the phase boundary of PHB/CHT, a layer more reactive to oxidation is formed, characterized by a stressed state of a portion of macromolecules [51]. This occurs during the ES. When a fiber is pulled in the field of action of an electrostatic force in the PHB phase, stresses arise at the interface with CHT due to the inability of the latter to perform deformation stretching.

5. Conclusions

This work analyzes the patterns of structure formation of film and fibrous biocomposites based on blends of natural polymers—poly(3-hydroxybutyrate) and chitosan. Chitosan having antimicrobial activity due to its cationic residue is an efficient additive for biomedical materials based on fully biodegradable PHB. Since chitosan possesses hydroxy and amino groups, PHB can be chemically bonded with it by blending. The use of a complex of such structure-sensitive methods like FTIR spectroscopy, DSC, and SEM made it possible to

31. López, J.A.; Naranjo, J.M.; Higuita, J.C.; Cubitto, M.A.; Cardona, C.A.; Villar, M.A. Biosynthesis of PHB from a new isolated Bacillus megaterium strain: Outlook on future developments with endospore forming bacteria. *Biotechnol. Bioproc. Eng.* **2012**, *17*, 250–258. [CrossRef]
32. Negrea, P.; Caunii, A.; Sarac, I.; Butnariu, M. The study of infrared spectrum of chitin and chitosan extract as potential sources of biomass. *Dig. J. Nanomater. Biostruct.* **2015**, *10*, 1129–1138.
33. Fan, M.; Dai, D.; Huang, B. Fourier Transform Infrared Spectroscopy for Natural Fibres. In *Fourier Transform—Materials Analysis*; Salih, S., Ed.; InTech: Rijeka, Croatia, 2012; pp. 45–68.
34. Olkhov, A.A.; Vlasov, S.V.; Iordanskii, A.L.; Zaikov, G.E.; Lobo, V.M.M. Water transport, structure features and mechanical behavior of biodegradable PHB/PVA blends. *J. Appl. Polym. Sci.* **2003**, *90*, 1471–1476. [CrossRef]
35. Ol'Khov, A.A.; Iordanskii, A.L.; Danko, T. Morphology of poly(3-hydroxybutyrate)–polyvinyl alcohol extrusion films. *J. Polym. Eng.* **2015**, *35*, 765–771. [CrossRef]
36. Ol'khov, A.A.; Shibryaeva, L.S.; Tertyshnaya, Y.V.; Kovaleva, A.N.; Kucherenko, E.L.; Zhul'kina, A.L.; Iordanskii, A.L. Resistance to thermal oxidation of ethylene propylene rubber and polyhydroxybutyrate blends. *Int. Polym. Sci. Technol.* **2017**, *44*, T/11–T/14.
37. Karpova, S.G.; Olkhov, A.A.; Bakirov, A.V.; Chvalun, S.N.; Shilkina, N.G.; Popov, A.A. Poly(3-Hydroxybutyrate) Matrices Modified with Iron(III) Complexes with Tetraphenylporphyrin. Analysis of the Structural Dynamic Parameters. *Russ. J. Phys. Chem. B* **2018**, *12*, 142–154. [CrossRef]
38. Doye, J.P.K.; Frenkel, D. Crystallization of a polymer on a surface. *J. Chem. Phys.* **1998**, *109*, 10033–10041. [CrossRef]
39. Zhang, M.C.; Guo, B.-H.; Xu, J. A Review on Polymer Crystallization Theories. *Crystals* **2016**, *7*, 4. [CrossRef]
40. Wan, Y.; Lu, X.; Dalai, S.; Zhang, J. Thermophysical properties of polycaprolactone/chitosan blend membranes. *Thermochim. Acta* **2009**, *487*, 33–38. [CrossRef]
41. Medvecky, L. Microstructure and Properties of Polyhydroxybutyrate-Chitosan-Nanohydroxyapatite Composite Scaffolds. *Sci. World J.* **2012**, *2012*, 1–8. [CrossRef]
42. Bonartsev, A.; Boskhomdzhiev, A.; Voinova, V.; Makhina, T.; Myshkina, V.; Yakovlev, S.; Zharkova, I.; Filatova, E.; Zernov, A.; Bagrov, D.; et al. Degradation of poly(3-hydroxybutyrate) and its derivatives: Characterization and kinetic behavior. *Chem. Chem. Technol.* **2012**, *6*, 385–392. [CrossRef]
43. Bonartsev, A.; Boskhomodgiev, A.P.; Iordanskii, A.; Bonartseva, G.A.; Rebrov, A.V.; Makhina, T.; Myshkina, V.; Yakovlev, S.; Filatova, E.A.; Ivanov, E.A.; et al. Hydrolytic degradation of poly(3-hydroxybutyrate), polylactide and their derivatives: Kinetics, crystallinity, and surface morphology. *Mol. Cryst. Liq. Cryst.* **2012**, *556*, 288–300. [CrossRef]
44. Yu, J.; Plackett, D.; Chen, L.X. Kinetics and mechanism of the monomeric products from abiotic hydrolysis of poly[(R)-3-hydroxybutyrate] under acidic and alkaline conditions. *Polym. Degrad. Stab.* **2005**, *89*, 289–299. [CrossRef]
45. Foroughi, M.R.; Karbasi, S.; Khoroushi, M.; Khademi, A.A. Polyhydroxybutyrate/chitosan/bioglass nanocomposite as a novel elec-trospun scaffold: Fabrication and characterization. *J. Porous Mat.* **2017**, *24*, 1447–1460. [CrossRef]
46. Keikhaei, S.; Mohammadalizadeh, Z.; Karbasi, S.; Salimi, A. Evaluation of the effects of β-tricalcium phosphate on physical, mechanical and biological properties of poly(3-hydroxybutyrate)/chitosan electrospun scaffold for cartilage tissue engineering applications. *Mater. Technol.* **2019**, *34*, 615–625. [CrossRef]
47. Ikejima, T.; Inoue, Y. Crystallization behavior and environmental biodegradability of the blend films of poly(3-hydroxybutyric acid) with chitin and chitosan. *Carbohydr. Polym.* **2000**, *41*, 351–356. [CrossRef]
48. Garcia-Rubio, R.; De Oliveira, H.C.; Rivera, J.; Trevijano-Contador, N. The Fungal Cell Wall: Candida, Cryptococcus, and Aspergillus Species. *Front. Microbiol.* **2020**, *10*, 2993. [CrossRef] [PubMed]
49. Jawad, A.H.; Nawi, M.A.; Mohamed, M.H.; Wilson, L.D. Oxidation of chitosan in solution by photocatalysis and product characterization. *J. Polym. Environ.* **2017**, *25*, 828–835. [CrossRef]
50. Karpova, S.G.; Ol'Khov, A.A.; Krivandin, A.V.; Shatalova, O.V.; Lobanov, A.V.; Popov, A.A.; Iordanskii, A.L. Effect of Zinc–Porphyrin Complex on the Structure and Properties of Poly(3-hydroxybutyrate) Ultrathin Fibers. *Polym. Sci. Ser. A* **2019**, *61*, 70–84. [CrossRef]
51. Tertyshnaya, Y.V.; Shibryaeva, L.S.; Ol'khov, A.A. The structure and properties of blends of poly(3-hydroxybutyrate) and an ethylene-propylene copolymer. *Polym. Sci. Ser. B+* **2002**, *44*, 287–290.

Article

Characteristics of Biodegradable Gelatin Methacrylate Hydrogel Designed to Improve Osteoinduction and Effect of Additional Binding of Tannic Acid on Hydrogel

Ji-Bong Choi [1,†], Yu-Kyoung Kim [1,†], Seon-Mi Byeon [2], Jung-Eun Park [1], Tae-Sung Bae [1], Yong-Seok Jang [1,*] and Min-Ho Lee [1,*]

1. Department of Dental Biomaterials, Institute of Biodegradable Materials, School of Dentistry, Jeonbuk National University, Jeonju-si 54896, Jeollabuk-do, Korea; submissi@naver.com (J.-B.C.); yk0830@naver.com (Y.-K.K.); pje312@naver.com (J.-E.P.); bts@jbnu.ac.kr (T.-S.B.)
2. Dental Clinic of Ebarun, Suncheon-si 57999, Jeollanam-do, Korea; sumse1205@naver.com
* Correspondence: Yjang@jbnu.ac.kr (Y.-S.J.); mh@jbnu.ac.kr (M.-H.L.)
† These authors contributed equally to the paper.

Abstract: In this study, a hydrogel using single and double crosslinking was prepared using GelMA, a natural polymer, and the effect was evaluated when the double crosslinked hydrogel and tannic acid were treated. The resulting hydrogel was subjected to physicochemical property evaluation, biocompatibility evaluation, and animal testing. The free radicals generated through APS/TEMED have a scaffold form with a porous structure in the hydrogel, and have a more stable structure through photo crosslinking. The double crosslinked hydrogel had improved mechanical strength and better results in cell compatibility tests than the single crosslinked group. Moreover, in the hydrogel transplanted into the femur of a rat, the double crosslinked group showed an osteoinductive response due to the attachment of bone minerals after 4 and 8 weeks, but the single crosslinked group did not show an osteoinductive response due to rapid degradation. Treatment with a high concentration of tannic acid showed significantly improved mechanical strength through H-bonding. However, cell adhesion and proliferation were limited compared to the untreated group due to the limitation of water absorption capacity, and no osteoinduction reaction was observed. As a result, it was confirmed that the treatment of high-concentration tannic acid significantly improved mechanical strength, but it was not a suitable method for improving bone induction due to the limitation of water absorption.

Keywords: gelatin methacryloyl; osteoinduction; tannic acid; crosslinking; hydrogel; biodegradable

1. Introduction

Bone defects are health-threatening diseases and are caused by various factors such as trauma, genetics, and cancer. The number of patients increases with age. Although bones can be regenerated, the ability widely varies from person to person. Currently, the most common method to recover the damaged bone defects is the direct implantation of a bone-grafted material into the defective area. Bone grafting must include essential elements of bone regeneration, namely osteoinduction, osteoconduction, and osteogenesis, in conjunction with the final bonding between the bone and the graft material [1]. In bone tissue engineering, various complex processes involving cell adhesion, migration, proliferation, differentiation, and matrix formation are used while applying biomaterials to induce bone generation [2]. To recover functions, often, biomaterials containing bioactive substances are used [3]. Hydrogels made of natural and synthetic biomaterials that similarly mimic the structure and biological properties of the natural extracellular matrix have long been studied as candidates for cell delivery in medicine and dentistry [4].

Gelatin is a type of derived protein partially extracted from collagen by thermal or chemical denaturation. It is suitable for hydrogels due to its retentive ability for a motif of peptides that are degraded by matrix metalloproteinase (MMP) and arginine-glycine-aspartic acid (RGD) related to cell adhesion. Additionally, gelatin has an excellent biocompatibility and swelling ratio [5–8]. However, gelatin dissolves in water at a body temperature above 37 °C and does not offer structural stability. Its physical properties are improved by chemical crosslinking with glutaraldehyde or 1-ethyl-3-(3-dimethylaminopropyl)-carbodiimide) (EDC)/(N-hydroxysuccinimide) (NHS) [9,10]. However due to crosslinker toxicity, many restrictions were imposed as referred from previous studies [11–13].

The recently developed gelatin methacryloyl (GelMA) can be produced by chemical modification of the amine and hydroxyl groups of gelatins, through which the hydrogel can be covalently crosslinked in the presence of photoinitiators and light [14]. In addition, GelMA hydrogels irreversibly change some structures due to hydrolysis and chemical modification but retain some properties of collagen and gelatin, such as cell adhesion, heat sensitivity, and enzymatic degradation [6,14,15]. GelMA hydrogels support the formation of novel ECMs, are enzymatically degradable, can be produced at low cost, are readily crosslinked under physiological conditions, and show potential for tissue engineering [16].

Irgacure 2959 (2-hydroxy-4'-(2-hydroxyethoxy)-2-methylpropiophenone) is the most commonly used in tissue engineering applications [17–20]. However, its low water solubility and UV light (365 nm) exposure cause potentially harmful effects on cells and tissues. Prolonged exposure to UV light may damage the DNA and cellular functions [21–23]. However, visible light (VL) uses a longer wavelength (405 nm) and penetrates further into the tissue during treatments. Additionally, no heat is generated, and cell damage is minimal [24]. The biocompatibility of GelMA hydrogel in bone tissue engineering has been demonstrated by many studies [25–30]. However, hydrogels fabricated with GelMA have lower mechanical strength than other natural and synthetic polymeric hydrogels [31–33].

Recently, hydrogels made with double crosslinking (DC) are attracting a lot of attention due to their excellent mechanical performance [34]. In addition, methods have been proposed to adjust the mechanical properties of GelMA using several crosslinking steps. For example, Rizwan et. al. achieved double crosslinking by performing physical crosslinking and then photo crosslinking [35]. In another study, Zhou et. al. used enzyme crosslinking followed by photo crosslinking to improve the viscosity of GelMA for bioprinting [36].

Tannic acid (TA) is a natural polyphenol compound with biological antioxidant and antibiotic properties [37]. However, TA forms an amorphous structure with a complex coagulation behavior in hydrogels making it difficult to control these strong interactions [38,39]. Accordingly, macromolecules such as DNA proteins are used to balance the change [40], or multiple steps are applied on TA under controlled conditions [41]. Similar to polydopamine inspired by mussels, a high pyrogallol and catechol content of the TA molecular structure can improve compressive and tensile properties via bonding [42].

In this study, based on the excellent binding ability of the GelMA hydrogel fabricated using double crosslinking and TA (see Figure 1), it was evaluated through analysis whether the hydrogel network could be strengthened in a well-arranged manner by TA. In addition, changes in mechanical properties were observed when the double crosslinked hydrogel and TA were applied to the hydrogel, and the effect on biodegradation and osteoinduction when finally implanted in an animal model was observed. Based on this, it was attempted to confirm whether the manufacturing method using double crosslinking showed better effects than the manufacturing method using single crosslinking. In addition, we tried to determine whether the improvement of mechanical strength when TA was combined and whether TA had an effect on the improvement of bone induction.

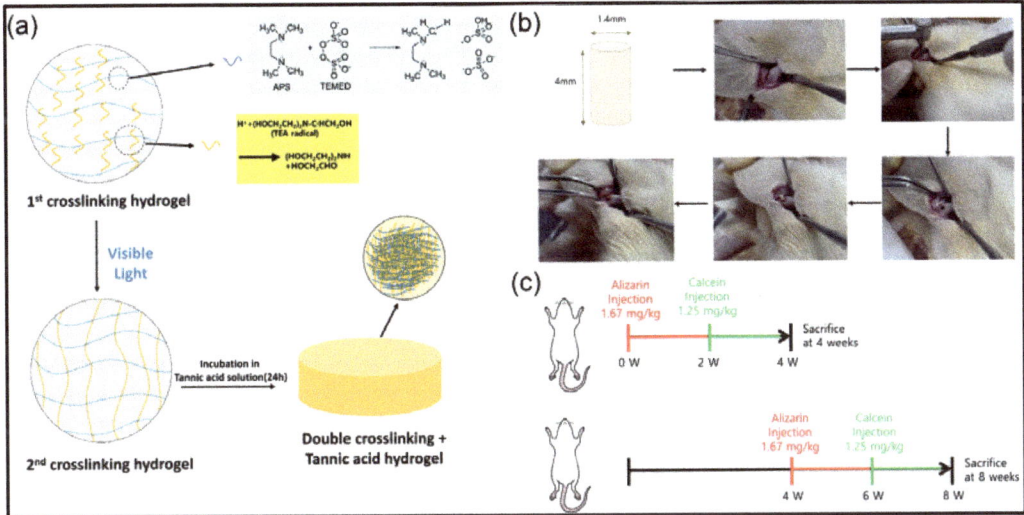

Figure 1. Schematic diagram of (**a**) Fabrication of double crosslinked hydrogel and tannic acid treatment, (**b**) Sample type and location for in vivo test (Sample placement in the femur), (**c**) Group division according to the schedule of staining reagent injection time for checking and analyzing bone formation behavior.

2. Materials and Methods

All materials used in this experiment, including type A gelatin, tetramethyl ethylene diamine (TEMED), ammonium persulfate (APS), methacrylic anhydride (MA), and tannic acid (TA), were purchased from Sigma Aldrich (Yongin, Korea). All chemicals were used without further purification.

2.1. Systhesis of Gelatin Methacryloyl

According to a previously reported study [43], methacryloyl-bonded GelMA macromonomer was synthesized. A total of 10% (w/v) of gelatin was completely dissolved in Dulbecco's phosphate-buffered saline (DPBS) at 60 °C and magnetically stirred. Then, 8 mL of MA were added to the gelatin solution and stirred at 50 °C for 2 h. The GelMA solution was then diluted in pre-made DPBS to increase the volume by 5 times and terminate the reaction. The GelMA solution was dialyzed against deionized water for 1 week at 40 °C in a 12–14 kDA cutoff tube. Subsequently, the solution was freeze-dried for 5 days, and the resulting GelMA was stored in a −20 °C freezer until further use.

2.2. Fabrication of Single and Double Crosslinking Hydrogels

The concentration of GelMA solution was fixed at 15% based on previous experimental results [44]. The prepared GelMA foam was dissolved in deionized water at 50 °C for 2 h. The hydrogel prepared by single crosslinking was prepared using low-temperature crosslinking and light crosslinking.

A hydrogel using low-temperature crosslinking was prepared by dissolving 14% (w/v) ammonium persulfate (APS) and 7% (w/v) tetramethylethylenediamine (TEMED). This prepolymer solution was pipetted into cylindrical (1.5 mm diameter, 1 mm thickness) polystyrene molds and placed in a freezer set to −20 °C. Low-temperature crosslinking was allowed to proceed for 18 h, and the resulting hydrogel was thawed and hydrated in dH$_2$O prior to use.

For photo crosslinking, 1.88 (v/v) triethanolamine (TEA), 1.25 (w/v) Vinylcaprolactam (VC), and 0.5 mM Eosin Y disodium salt were sequentially added to the prepared 15% (w/t) GelMA solution and mixed, and then the prepolymer solution was pipetted into a cylindrical (1.5 mm diameter, 1 mm thick) polystyrene mold and exposed to visible light for 120 s.

For double crosslinking, 14% (w/v) ammonium persulfate (APS) and 7% (w/v) tetramethylethylenediamine (TEMED) were sequentially added to the prepolymer solution used for photo crosslinking and mixed. Then, low-temperature crosslinking was per-formed in a freezer set at -20 °C. for 18 h, and then exposed to visible light for 120 s before thawing to form a double crosslinked hydrogel.

2.3. Fabrication of Hydrogel Applied with TA

The fabricated hydrogel was immersed in the tannic acid (TA) solution of previously pre-pared concentrations (10%, 50%, and 100% (w/v)), and shaken for 24 h on a shaker. Then, the hydrogel was washed three times with deionized water to remove excess TA. Hydrogels used for the experiment were fabricated with a 10 mm diameter and a 3 mm height and were referred to as GelMA-S (low-temperature crosslinking), GelMA-V (photo crosslinking), and GelMA-D (double crosslinking). Depending on the concentration of TA, additional indicators, T10, T50, and T100, were used.

2.4. Fourier Transform Infrared Spectroscopy (FT-IR) Characterization

FT-IR analysis was used to investigate the intermolecular interactions between TA, double crosslinking, and the GelMA Spectra that were obtained at room temperature using a FT-IR spectrometer (Perkin Elmer, Waltham, MA, USA). FT-IR analysis was performed within the wavelength 4000–500 cm^{-1} (KBr) using the attenuated total reflectance (ATR) method.

2.5. Mechanical Tests

Mechanical properties of the hydrogel were measured using a universal testing machine (Instron 5569, Instron, Norwood, MA, USA). All the first compression tests were performed on a cylindrical hydrogel at a 2 mm/min rate with up to 95% maximum load of a 50 N load cell. After the first compression test, damaged samples were eliminated, and the second compression test was conducted at a 0.5 mm/min rate and up to a 95% maximum load of a 500 N load cell. The second compression test data were automatically calculated using the Bluehill 2 software. The compressive modulus was calculated as the slope of the linear region (0–20%) of the stress-strain curve. All samples were hydrated during the test.

2.6. Swelling Ratio Tests

The prepared GelMA hydrogel was incubated at 37 °C for 24 h. After the sample was removed from the solution and the residual liquid separated using Kimwipe, the weight, W_s, was measured. The weight of the freeze-dried hydrogel was measured as W_d. The swelling ratio was calculated according to Equation (1).

$$\text{Swelling Ratio: SR} = (W_s - W_d)/W_d \tag{1}$$

2.7. FE-SEM Characterization

The freeze-dried hydrogel was placed on a wafer for platinum coating. The shape of the hydrogel was observed using a FE-SEM (Hitachi, Tokyo, Japan).

2.8. Evaluation of In-Vitro Cell Proliferation

Osteoblast cells were used in this study, MC3T3-E1 (ATCC; American Type Culture Collection), to evaluate their effect on bone formation. For the culture medium, a nutrient component, 10% fetal bovine serum (Gibco Co., Waltham, MA, USA) containing an antibiotic (penicillin), was added to an α-MEM (Gibco Co., Waltham, MA, USA) medium. The cell culture was conducted in an incubator (Thermo Electron Corporation, Waltham, MA, USA) in a 5% CO_2 atmosphere at 37 °C. A water soluble tetrazolium (WST) assay was used to evaluate cell proliferation by placing samples in a 48-well plate and incubating the MC3T3-E1 cells with a cell density of 1×10^4 cells mL^{-1} for 1, 3, and 7 days. Then, the medium was removed, replaced with 400 μL of CCK-8 (Enzo Life Science Inc., Farmingdale, NY, USA) reagent mixed with α-MEM medium, and stored in the incubator with 5% of CO_2. After 90 min, 100 μL were added to a 96-well plate, and the absorbance was measured at 450 nm using the ELISA reader (Molecular devices, Silicon Valley, CA, USA).

2.9. Evaluation of Bone Regeneration and Mineral Activity In Vivo

The effect of bone remodeling was compared to the rat femur defect model applying GelMA-P, GelMA-PT100, GelMA-VP, and GelMA-VPT100 groups (n = 3). Experiments in this study were conducted under the protocols approved by the Institutional Animal Care and Use Committee of the Chonbuk National University Laboratory Animal Center (CBNU 2020-094) following the declaration of Helsinki. The prepared freeze-dried hydrogel samples were implanted on the proximal femur from the outer side for each rat. Male Sprague–Dawley rats (n = 16), used in this experiment, were about 8 weeks old with an average weight of 280 g. The rats were purchased from Damul Science (Daejeon, Korea) and used after a week's adjusting period. For anesthesia, 0.6 mL/kg tiletamine and zolazepam (Zoletil 50, Virbac Laboratories, Carros, France) and 0.4 mL/kg xylazine hydrochloride (Rompun, Bayer Korea, Seoul, Korea) were intramuscularly injected into the leg of each rat. The surgical site of the anesthetized rat was shaved, and an approximately 1 cm incision was made on the femur after sterilization using povidone iodine. After raising the flap due to the incision, a contra-angle handpiece (X-smart Endodontic Motor, Dentsply Maillefer, Switzerland) equipped with a 1.6 mm pilot round head bur (H1.31-0.16, Lemgo, Germany) was used to create a hole in the cortical bone. After the surgery, an antibiotic (Amikacin, Samu Median Co., Ltd., Seoul, Korea) was subcutaneously injected (0.6 mL/kg). At 4 and 8 weeks after implanting the hydrogels, the rats were sacrificed to obtain the femoral bone blocks containing the sample.

Fluorescence Staining

Alizarin complexone (red) and calcein (green) fluorescent materials were used to observe the mineralized bones. To evaluate a bone-forming ability on the samples after 4 and 8 weeks, Alizarin complexone (red) solution (1.67 mL/ kg, body weight) was injected at 0 and 4 weeks, and calcein (1.25 mL/kg, body weight) was injected into the peritoneum at week 2 and week 6. After 4 and 8 weeks, the femoral bone blocks were obtained from the sacrificed rats and were fixed in 10% formaldehyde solution to fabricate a resin-embedded tissue slide. Then, all blocks were dehydrated using an increased concentration of ethanol, and methyl methacrylate (MMA, JUSEI Chemical Co. Ltd., Tokyo, Japan) was inserted into the bone. The bone blocks with penetrated MMA were embedded in an activated MMA resin. The embedded blocks in the resin were sectioned along the longitudinal axis of the embedded sample. The fluorescent-stained tissue slide of the sectioned sample was observed using a Super Resolution Confocal Laser Scanning Microscope (Carl Zeiss AG, Oberkochen, Germany). Histological images of bone staining with alizarin complexone (red) and calcein (green) were obtained at 543 nm and 488 nm, respectively.

2.10. Statistical Processing

To evaluate the difference among groups, SPSS ver 21.0 (SPSS Inc., Chicago, IL, USA) software was used. One-way ANOVA was used to assess three or more groups within one factor, and Tukey's postmortem analysis was used to evaluate the average. In all experiments, if the p-value was < 0.05, it would determine significant differences in the groups.

3. Results and Discussion

3.1. Formation of GelMA and TA Treatment of GelMA Hydrogel

We prepared a hydrogel synthesized by single crosslinking and double crosslinking according to the plan, and TA solutions with different concentrations were applied. As shown in Figure 2a, it can be seen that GelMA-D has changed color under the influence of Eosin Y. In addition, GelMA (15% w/v) hydrogel (cylindrical height 2 mm, diameter 10 mm) changed to opaque with a decrease in size after 24 h treatment of TA (10%, 50%, 100% w/v), which shows a real interaction between GelMA and TA.

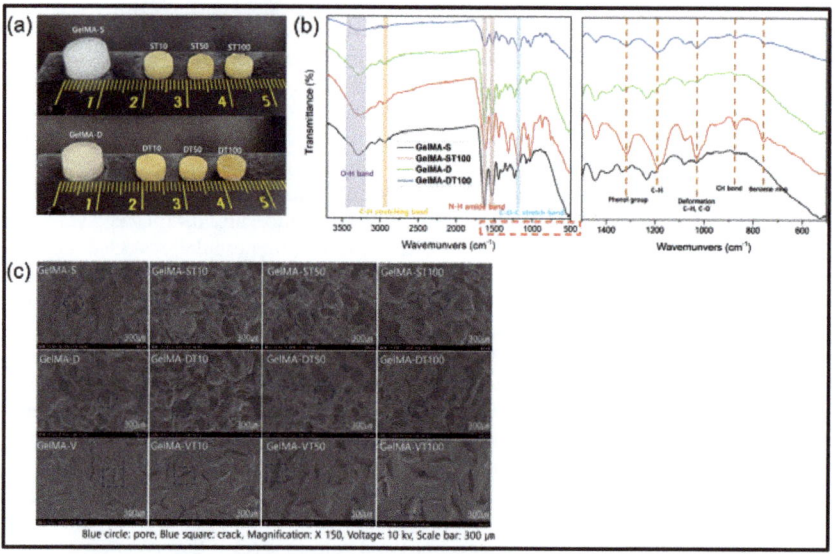

Figure 2. (a) The representative morphological changes from pristine GelMA hydrogels (left) to GelMA-TA (right) after a 24 h incubation in TA solution. Concentrations of GelMA and TA are 15% (w/v), and 10%, 50%, 100% (w/v), respectively. (b) FT-IR spectra of hydrogels. (c) FE-SEM of GelMA hydrogels, and GelMA-TA hydrogel 10%, 50%, 100% TA, 24 h treatment time. Red circles indicate pores, and squares indicate cracks. The image was measured using a voltage of 10 kv, the magnification was 150×, and the scale bar was 300 μm.

To confirm the interaction further, the Fourier Transform Infrared (FT-IR) study was performed (Figure 2b). Spectra from the remaining groups except for GelMA-S generally show that high transmittance at about 3100–3600 cm^{-1} is closely related to hydrogen bonding, which is due to the shift of O-H groups by additional bonding [45,46]. In addition, the peaks of the amide groups (I, II, III) characteristic of GelMA hydrogels were observed at around 1612 cm^{-1}, 1520 cm^{-1}, 1436 cm^{-1} without appreciable changes in intensity and frequency. These results show that covalent bonds are not present [47]. The peak at 1319 cm^{-1} is caused by the phenol group of TA. The peak at 1198 cm^{-1} is due to C-H, and the vibration peak at 1100–1000 cm^{-1} is due to C-O and C-H deformation. The peak in 550–900 cm^{-1}, which is characteristic of TA, is based on the C-H bond of the benzene ring [48].

Morphological analysis of the prepared GelMA and TA treatment of GelMA hydrogels was measured by FE-SEM (Figure 2c). For comparison, a hydrogel (GelMA-V) formed by photo crosslinking was additionally observed. GelMA-V showed agglomerated surfaces and partial cracks due to polymerization by radicals. The hydrogel formed by low-temperature crosslinking (GelMA-S) had a porous microstructure as the ice crystals formed by the APS/TEMED reaction were removed. The hydrogel formed by double crosslinking (GelMA-D) was photo crosslinked after low-temperature crosslinking and showed a similar shape to GelMA-S. Although an increase in pore size was observed with increasing TA concentration (10%, 50%, and 100%) in GelMA-V, the size did not increase noticeably after 50% concentration. In GelMA-S, it was confirmed that the wall of the porous structure was slightly thickened by the binding of TA, and in GelMA-D, more TA binding than GelMA-P was observed. However, looking at the overall trend, no significant difference in surface shape was observed between GelMA-S and GelMA-D groups.

3.2. Mechanical Properties of Htdrogel

Figure 3a shows a typical compressive stress-strain curve. In the case of the TA-treated group, it was confirmed that the fracture stress was significantly improved. In particular, the untreated and TA-treated hydrogel groups showed the same pattern and showed stronger fracture stress in the double crosslinked group than in the single crosslinked group.

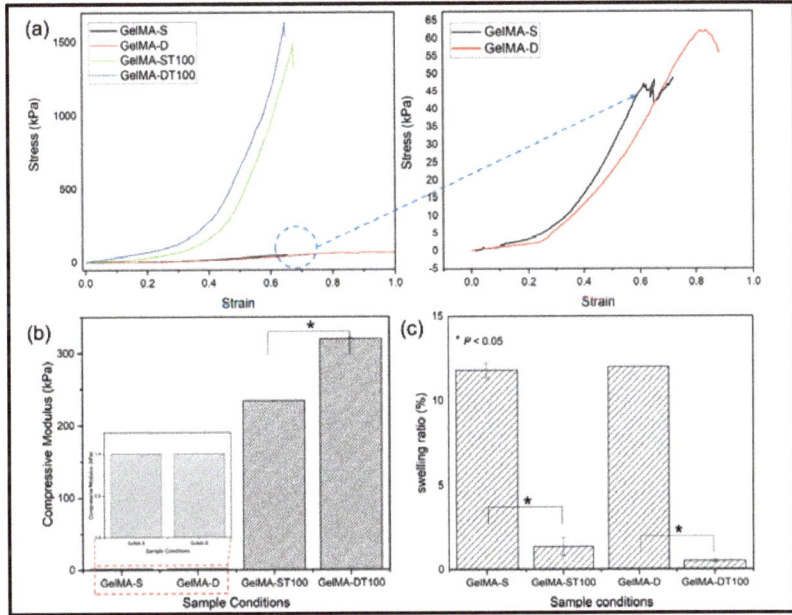

Figure 3. Mechanical properties of hydrogels (GelMA-S and GelMA-D) and after treated with the 100 w/v% of TA. The compression deformation of all hydrogels was performed up to 95%; (**a**) Stress-strain Curve for hydrogels, (**b**) Compressive Modulus through (0~20% slope), (**c**) Swelling ratio of GelMA and GelMA-TA hydrogels. (Data are presented as mean ± SD, $n = 3$, * $p < 0.05$).

These results showed the same trend in the compressive stress (Figure 3b). The compressive stresses of each group were GelMA-S: 46 kPa; GelMA-D: 62 kPa; GelMA-ST100: 234 kPa; GelMA-DT100 showed 319 kPa, and it was confirmed that the improvement was about 1.3 times between the single crosslinked and double crosslinked groups. In addition, it was confirmed that there was a difference of about five times between the untreated group and the TA group. TA bound to the hydrogel strengthened the bond with the GelMA network through hydrogen bonding and hydrophobic force, and showed improvement in mechanical strength [49]. Mechanical properties can be affected by the water content of the hydrogel. Therefore, the GelMA-S, GelMA-D, GelMA-ST100, and GelMA-DT100 groups were immersed in PBS and then cultured in an incubator at 37 °C for 24 h.

The smoothing behavior of the hydrogel was then characterized and shown in Figure 3c. The swelling behavior of GelMA-S and GelMA-D showed similar results, and no significant difference was observed. Similarly, no significant difference was observed in the swelling behavior of GelMA-ST100 and GelMA-DT100. However, a significant difference was observed when comparing the untreated group and the TA treated group. It could be confirmed that the TA-treated group was significantly restricted in swelling behavior compared to the untreated group. The low swelling behavior of the TA-treated group showed that the interaction of H-bonds between TA and GelMA could compress the structure of the hydrogel and limit water absorption [43]. As a result, it was confirmed that the low swelling ratio of the TA-treated hydrogel group indicates a higher mechanical strength.

3.3. Cytocompatibility of Hydrogel

We investigated the cellular compatibility of GelMA and GelMA-TA hydrogels. Pro-osteoblasts (MC3T3-E1) were cultured in flat 48-well plates with or without hydrogels. According to previously known research results, the differentiation of MC3T3-E1 cells showed better cell activity in soft materials than in hard materials [50]. The single and double crosslinked hydrogels used in the study have soft matrix properties, whereas the TA-treated GelMA hydrogels have more rigid matrix properties. Our experiments also showed similar results. Cell viability was confirmed using quantitative detection of cck8 at 1, 3, and 7 days after seeding (Figure 4). UV absorbance increased with increasing incubation time for days 1–3 due to cell proliferation, but there was no statistical difference in cell viability in all groups. However, on day 7, there was a change between each group. GelMA-D showed better UV absorbance, and a significant difference occurred compared to other groups.

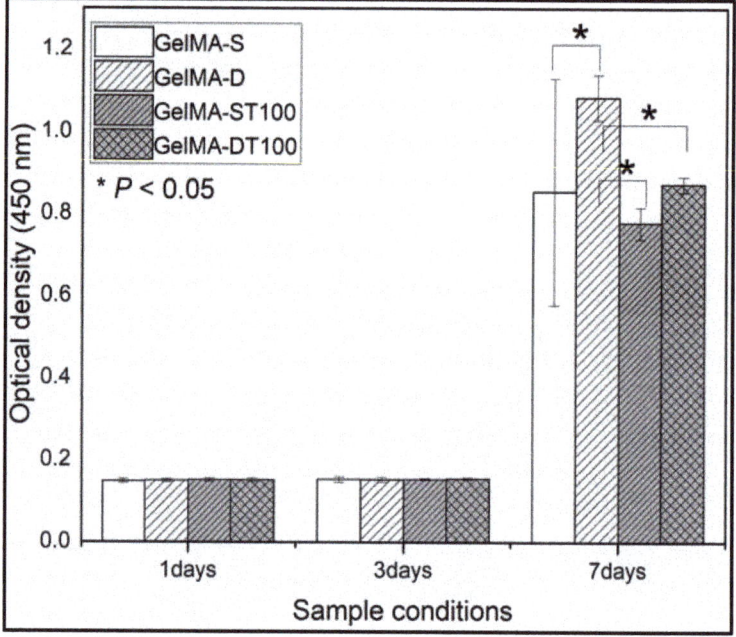

Figure 4. The proliferation of cells measured by CCK8 assays (WST) on days 1, 3, and 7. (Data are presented as mean ± SD, $n = 3$, * $p < 0.05$).

The group treated with TA showed lower absorbance than the existing GelMA group, but there was no significant difference except for the GelMA-D group. These results also suggested that the hydrogel can be applied to in vivo use. A high concentration of TA treatment continuously released an excess of TA in the hydrogel, and it was confirmed that the released TA affects cell activity. Although the exact mechanism of the effect of TA on cellular activity is unknown, it may be due to the presence of galloyl groups in TA. A galloyl group of tannic acid interacts with proteins through hydrophobic bonding and electrostatic and hydrophobic interactions [51]. The electrostatic interaction induced by tannic acid at 7.4 pH is relatively weaker than the hydrogen bonding and fails to stabilize proteins in their native form [52]. This interaction of tannic acid suggests the adhesion prevention of osteoblasts. Nevertheless, these results suggest that the hydrogel may be applicable for in vivo use.

3.4. In Vivo Osteoinduction

In vivo experiments were conducted with GelMA-S, GelMA-D, GelMA-ST100, and Gel-MA-DT100 groups. Osteoinduction and biodegradation were observed by implanting a hydrogel sample prepared in the form of a rod into the femur of a rat. After 4 and 8 weeks, mice were sacrificed and the results of new bone formation according to osteoinduction in the hydrogel are shown in Figure 5 through fluorescence images. Alizarin complexon (red) and calcein (green) staining was performed every 2 weeks to observe bone formation. After 4 weeks of sample transplantation, most of the staining intervals of alizarin red and calcein green were consistent, and there was no significant difference in the rate of new bone formation.

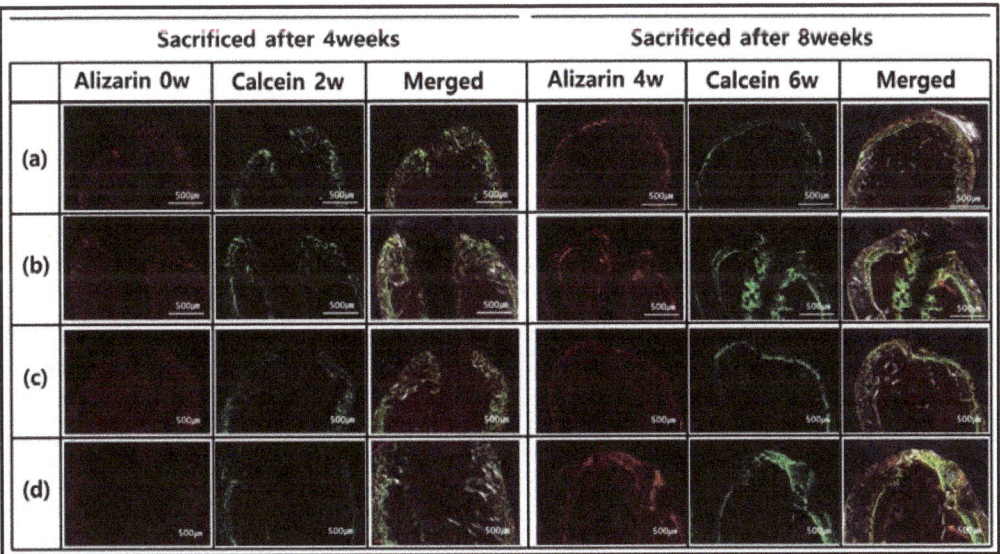

Figure 5. Fluorescence microscopy imaging after 4 and 8 weeks after implantation (red and green represent alizarin and calcein labeling, respectively): (**a**) GelMA-S, (**b**) GelMA-D, (**c**) GelMA-ST100, and (**d**) GelMA-DT100.

However, 8 weeks after implantation, different types were observed for each group. In GelMA-S and GelMA-ST100, the hydrogel morphology and osteoinductive reaction were not confirmed. On the other hand, it was confirmed that the formation of new bone and the concentration of calcein green increased in GelMA-D and GelMA-DT100. In particular, in GelMA-D, it was confirmed that many of these minerals were generated inside the hydrogel and maintained its shape. In the process of bone remodeling, bones are synthesized by osteoblasts, and small mineral crystals are deposited between collagen molecules. Over the next few days, these crystals fill the space occupied by water, resulting in bone mineralization [53]. For hydrogel samples, it is important to have adequate strength to retain their shape and the ability to absorb moisture. GelMA-D showed that many of the main minerals were settled in the hydrogel before the hydrogel was decomposed and replaced with the degradation (Figure 5b). However, in the hydrogel treated with TA, water absorption was inhibited, and it was confirmed that the main mineral was deposited only on the outside (Figure 5d). These morphologies tend to be consistent with Villanueva Osteochrome Staining (Figure 6).

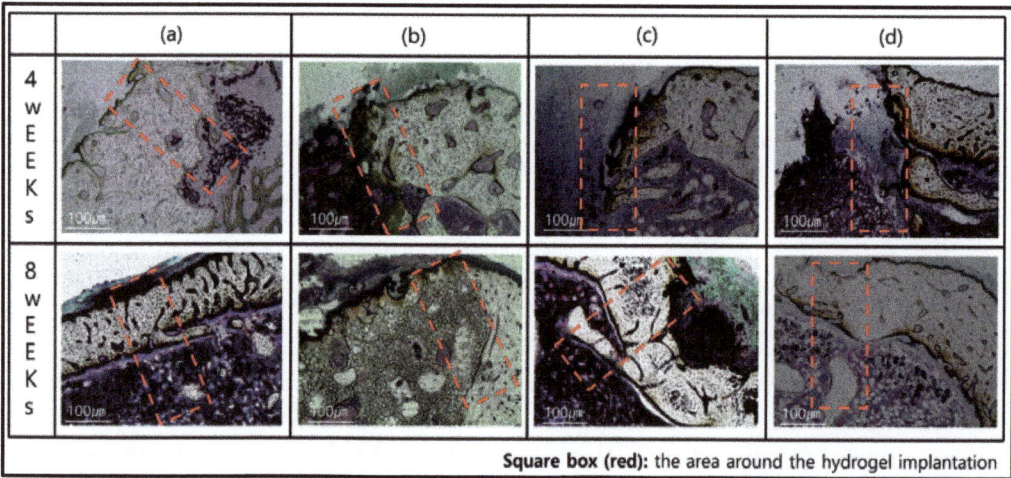

Square box (red): the area around the hydrogel implantation

Figure 6. Histological images of Villanueva Osteochrome Staining for bone tissues after 4 and 8 weeks: (**a**) GelMA-S, (**b**) GelMA-D, (**c**) GelMA-ST100, and (**d**) GelMA-DT100.

GelMA-S did not adhere to the bone 4 weeks after transplantation, and it was confirmed that rapid degradation occurred (Figure 6a). On the other hand, in GelMA-D, osseointegration was successfully achieved 4 weeks after transplantation, and it was confirmed that the external bone mineral penetrated into the sample. In addition, it was observed that the bone mineral formed in the sample was gradually converted into the new bone after 8 weeks. It is judged that the induction reaction for bone remodeling occurred smoothly only in GelMA-D (Figure 6b). After 4 weeks of implantation of GelMA-ST100 and GelMA-DT100 treated with TA, the shape of the sample was maintained, but it did not adhere to the bone and there was no osteoinductive reaction (Figure 6c,d). The location where mineral crystals are deposited between collagen fibers, one of the components of bone, and where mineralization begins, is called the hole zone [52]. This region is the first to deposit calcium and phosphorus in the inter-molecular space created when collagen molecules move out of position when forming fibers [54].

GelMA-S can be exchanged smoothly in body fluids through its porous structure, but it has a disadvantage in that rapid degradation can occur because the reaction range by the enzyme is large. Double crosslinking solved these shortcomings. The improved strength provided a favorable environment for calcium and phosphorus deposition and new bone formation before being degraded by enzymes. On the other hand, TA treatment gave high mechanical strength but did not provide a suitable environment for the formation of new bone by osteoinduction. In addition, although the exact mechanism of TA has not been elucidated, it cannot be ruled out that the adhesion of progenitor cells is hindered by the nature of TA, which does not stabilize the natural protein in the environment of pH 7.4 [55,56].

4. Conclusions

In this study, it was evaluated whether a hydrogel produced by double crosslinking using GelMA, a natural polymer, was suitable for improving bone induction and showed better results when additionally combined with TA.

As a result, the double crosslinked group showed improved mechanical strength and better cell compatibility than the single crosslinked group. In addition, the single crosslinked group transplanted into the rat femur showed no osteoinduction response due to rapid degradation after 4 and 8 weeks, but the double crosslinked group showed bone mineral binding and osteoinduction. The hydrogel group treated with a high concentration of TA showed a significant improvement in mechanical strength through H-bonding. However, cell adhesion and proliferation were limited compared to the untreated group, and osteoinduction was not observed in the TA-treated single and double crosslinked groups.

In conclusion, it was found that double crosslinking is a suitable method for improving the strength and bone induction of hydrogels compared to single crosslinking. Conversely, it was confirmed that the binding of high-concentration TA significantly improved mechanical strength, but delayed bone remodeling by limiting water absorption. In future research, it is expected that the development of functional hydrogels that can be used for bone tissue engineering by supporting bioactive substances or drugs on the double crosslinked hydrogel is expected.

Author Contributions: Conceptualization, methodology, and writing—original draft preparation, J.-B.C.; writing—review and editing, S.-M.B.; validation and formal analysis, J.-E.P.; investigation and resources, T.-S.B.; project administration, Y.-K.K.; supervision and funding acquisition, M.-H.L. and Y.-S.J. All authors have read and agreed to the published version of the manuscript.

Funding: This research was funded by a National Research Foundation of Korea (NRF) grant funded by the Korea Government (MSIP) No. 2019R1C1C1003784.

Institutional Review Board Statement: The study was conducted according to the guidelines of the Declaration of Helsinki and approved by the Institutional Animal Care and Use Committee of the Chonbuk National University Laboratory Animal Center (CBNU 2020-094).

Informed Consent Statement: Not applicable.

Acknowledgments: This study has reconstructed the data of Ji-Bong Choi's doctoral dissertation submitted in 2021 (title: "Fabrication and characterization of biodegradable gelatin-methacrylate hydrogel for tissue engineering").

Conflicts of Interest: The authors declare no conflict of interest.

References

1. Giannoudis, P.V.; Dinopoulos, H.; Tsiridis, E. Bone substitutes: An update. *Injury* 2005, *36*, S20–S27. [CrossRef] [PubMed]
2. Salgado, A.J.; Coutinho, O.P.; Reis, R.L. Bone tissue engineering: State of the art and future trends. *Macromol. Biosci.* 2004, *4*, 743–765. [CrossRef] [PubMed]
3. Kim, S.; Von Recum, L. Endothelial stem cells and precursors for tissue engineering: Cell source, differentiation, selection, and application. *Tissue Eng. Part B Rev.* 2008, *14*, 133–147. [CrossRef] [PubMed]
4. Annabi, N.; Tamayol, A.; Uquillas, J.A.; Akbari, M.; Bertassoni, L.E.; Cha, C.; Camci-Unal, G.; Dokmeci, M.R.; Peppas, N.A.; Khademhosseini, A. 25th anniversary article: Rational design and applications of hydrogels in regenerative medicine. *Adv. Mater.* 2014, *26*, 85–124. [CrossRef]
5. Chen, Y.C.; Lin, R.Z.; Qi, H.; Yang, Y.; Bae, H.; Melero-Martin, J.M.; Khademhosseini, A. Functional human vascular network generated in photocrosslinkable gelatin methacrylate hydrogels. *Adv. Funct. Mater.* 2012, *22*, 2027–2039. [CrossRef]
6. Nichol, J.W.; Koshy, S.T.; Bae, H.; Hwang, C.M.; Yamanlar, S.; Khademhosseini, A. Cell-laden microengineered gelatin methacrylate hydrogels. *Biomaterials* 2010, *31*, 5536–5544. [CrossRef]
7. Lin, R.-Z.; Chen, Y.-C.; Moreno-Luna, R.; Khademhosseini, A.; Melero-Martin, J.M. Transdermal regulation of vascular network bioengineering using a photopolymerizable methacrylated gelatin hydrogel. *Biomaterials* 2013, *34*, 6785–6796. [CrossRef]
8. Hosseini, V.; Ahadian, S.; Ostrovidov, S.; Camci-Unal, G.; Chen, S.; Kaji, H.; Ramalingam, M.; Khademhosseini, A. Engineered contractile skeletal muscle tissue on a microgrooved methacrylated gelatin substrate. *Tissue Eng. Part A* 2012, *18*, 2453–2465. [CrossRef]

9. Park, C.; Vo, C.L.-N.; Kang, T.; Oh, E.; Lee, B.-J. New method and characterization of self-assembled gelatin–oleic nanoparticles using a desolvation method via carbodiimide/N-hydroxysuccinimide (EDC/NHS) reaction. *Eur. J. Pharm. Biopharm.* **2015**, *89*, 365–373. [CrossRef]
10. Yu, T.; Wang, W.; Nassiri, S.; Kwan, T.; Dang, C.; Liu, W.; Spiller, K.L. Temporal and spatial distribution of macrophage phenotype markers in the foreign body response to glutaraldehyde-crosslinked gelatin hydrogels. *J. Biomater. Sci. Polym. Ed.* **2016**, *27*, 721–742. [CrossRef]
11. Montemurro, F.; De Maria, C.; Orsi, G.; Ghezzi, L.; Tinè, M.R.; Vozzi, G. Genipin diffusion and reaction into a gelatin matrix for tissue engineering applications. *J. Biomed. Mater. Res. Part B Appl. Biomater.* **2017**, *105*, 473–480. [CrossRef]
12. Focaroli, S.; Teti, G.; Salvatore, V.; Durante, S.; Belmonte, M.M.; Giardino, R.; Mazzotti, A.; Bigi, A.; Falconi, M. Chondrogenic differentiation of human adipose mesenchimal stem cells: Influence of a biomimetic gelatin genipin crosslinked porous scaffold. *Microsc. Res. Tech.* **2014**, *77*, 928–934. [CrossRef]
13. Xie, X.; Tian, J.-K.; Lv, F.-Q.; Wu, R.; Tang, W.-B.; Luo, Y.-K.; Huang, Y.-Q.; Tang, J. A novel hemostatic sealant composed of gelatin, transglutaminase and thrombin effectively controls liver trauma-induced bleeding in dogs. *Acta Pharmacol. Sin.* **2013**, *34*, 983–988. [CrossRef]
14. Van Den Bulcke, A.I.; Bogdanov, B.; De Rooze, N.; Schacht, E.H.; Cornelissen, M.; Berghmans, H. Structural and rheological properties of methacrylamide modified gelatin hydrogels. *Biomacromolecules* **2000**, *1*, 31–38. [CrossRef]
15. Benton, J.A.; De Forest, C.A.; Vivekanandan, V.; Anseth, K.S. Photocrosslinking of gelatin macromers to synthesize porous hydrogels that promote valvular interstitial cell function. *Tissue Eng. Part A* **2009**, *15*, 3221–3230. [CrossRef]
16. Schuurman, W.; Levett, P.A.; Pot, M.W.; van Weeren, P.R.; Dhert, W.J.; Hutmacher, D.W.; Melchels, F.P.; Klein, T.J.; Malda, J. Gelatin-methacrylamide hydrogels as potential biomaterials for fabrication of tissue-engineered cartilage constructs. *Macromol. Biosci.* **2013**, *13*, 551–561. [CrossRef]
17. Liu, W.; Heinrich, M.A.; Zhou, Y.; Akpek, A.; Hu, N.; Liu, X.; Guan, X.; Zhong, Z.; Jin, X.; Khademhosseini, A. Extrusion bioprinting of shear-thinning gelatin methacryloyl bioinks. *Adv. Healthc. Mater.* **2017**, *6*, 1601451. [CrossRef]
18. Zhao, X.; Sun, X.; Yildirimer, L.; Lang, Q.; Lin, Z.Y.W.; Zheng, R.; Zhang, Y.; Cui, W.; Annabi, N.; Khademhosseini, A. Cell infiltrative hydrogel fibrous scaffolds for accelerated wound healing. *Acta Biomater.* **2017**, *49*, 66–77. [CrossRef]
19. Hu, J.; Hou, Y.; Park, H.; Choi, B.; Hou, S.; Chung, A.; Lee, M. Visible light crosslinkable chitosan hydrogels for tissue engineering. *Acta Biomater.* **2012**, *8*, 1730–1738. [CrossRef]
20. Lee, B.H.; Lum, N.; Seow, L.Y.; Lim, P.Q.; Tan, L.P. Synthesis and characterization of types A and B gelatin methacryloyl for bioink applications. *Materials* **2016**, *9*, 797. [CrossRef]
21. Noshadi, I.; Hong, S.; Sullivan, K.E.; Sani, E.S.; Portillo-Lara, R.; Tamayol, A.; Shin, S.R.; Gao, A.E.; Stoppel, W.L.; Black, L.D., III. In vitro and in vivo analysis of visible light crosslinkable gelatin methacryloyl (GelMA) hydrogels. *Biomater. Sci.* **2017**, *5*, 2093–2105. [CrossRef]
22. Fedorovich, N.E.; Oudshoorn, M.H.; van Geemen, D.; Hennink, W.E.; Alblas, J.; Dhert, W.J. The effect of photopolymerization on stem cells embedded in hydrogels. *Biomaterials* **2009**, *30*, 344–353. [CrossRef]
23. Kappes, U.P.; Luo, D.; Potter, M.; Schulmeister, K.; Rünger, T.M. Short-and long-wave UV light (UVB and UVA) induce similar mutations in human skin cells. *J. Investig. Dermatol.* **2006**, *126*, 667–675. [CrossRef]
24. Jones, C.A.; Huberman, E.; Cunningham, M.L.; Peak, M.J. Mutagenesis and cytotoxicity in human epithelial cells by far-and near-ultraviolet radiations: Action spectra. *Radiat. Res.* **1987**, *110*, 244–254. [CrossRef]
25. Kielbassa, C.; Roza, L.; Epe, B. Wavelength dependence of oxidative DNA damage induced by UV and visible light. *Carcinogenesis* **1997**, *18*, 811–816. [CrossRef]
26. Elisseeff, J.; Anseth, K.; Sims, D.; McIntosh, W.; Randolph, M.; Langer, R. Transdermal photopolymerization for minimally invasive implantation. *Proc. Natl. Acad. Sci. USA* **1999**, *96*, 3104–3107. [CrossRef]
27. Sadat-Shojai, M.; Khorasani, M.-T.; Jamshidi, A. 3-dimensional cell-laden nano-hydroxyapatite/protein hydrogels for bone regeneration applications. *Mater. Sci. Eng. C* **2015**, *49*, 835–843. [CrossRef]
28. Zuo, Y.; Liu, X.; Wei, D.; Sun, J.; Xiao, W.; Zhao, H.; Guo, L.; Wei, Q.; Fan, H.; Zhang, X. Photo-cross-linkable methacrylated gelatin and hydroxyapatite hybrid hydrogel for modularly engineering biomimetic osteon. *ACS Appl. Mater. Interfaces* **2015**, *7*, 10386–10394. [CrossRef]
29. Annabi, N.; Rana, D.; Sani, E.S.; Portillo-Lara, R.; Gifford, J.L.; Fares, M.M.; Mithieux, S.M.; Weiss, A.S. Engineering a sprayable and elastic hydrogel adhesive with antimicrobial properties for wound healing. *Biomaterials* **2017**, *139*, 229–243. [CrossRef]
30. Rothrauff, B.B.; Yang, G.; Tuan, R.S. Tissue-specific bioactivity of soluble tendon-derived and cartilage-derived extracellular matrices on adult mesenchymal stem cells. *Stem Cell Res. Ther.* **2017**, *8*, 133. [CrossRef]
31. Chen, X.; Bai, S.; Li, B.; Liu, H.; Wu, G.; Liu, S.; Zhao, Y. Fabrication of gelatin methacrylate/nanohydroxyapatite microgel arrays for periodontal tissue regeneration. *Int. J. Nanomed.* **2016**, *11*, 4707. [CrossRef]
32. Bartnikowski, M.; Akkineni, A.R.; Gelinsky, M.; Woodruff, M.A.; Klein, T.J. A hydrogel model incorporating 3D-plotted hydroxyapatite for osteochondral tissue engineering. *Materials* **2016**, *9*, 285. [CrossRef] [PubMed]
33. Eslami, M.; Vrana, N.E.; Zorlutuna, P.; Sant, S.; Jung, S.; Masoumi, N.; Khavari-Nejad, R.A.; Javadi, G.; Khademhosseini, A. Fiber-reinforced hydrogel scaffolds for heart valve tissue engineering. *J. Biomater. Appl.* **2014**, *29*, 399–410. [CrossRef] [PubMed]

34. Ramón-Azcón, J.; Ahadian, S.; Estili, M.; Liang, X.; Ostrovidov, S.; Kaji, H.; Shiku, H.; Ramalingam, M.; Nakajima, K.; Sakka, Y. Dielectrophoretically aligned carbon nanotubes to control electrical and mechanical properties of hydrogels to fabricate contractile muscle myofibers. *Adv. Mater.* **2013**, *25*, 4028–4034. [CrossRef]
35. Cha, C.; Shin, S.R.; Gao, X.; Annabi, N.; Dokmeci, M.R.; Tang, X.; Khademhosseini, A. Controlling mechanical properties of cell-laden hydrogels by covalent incorporation of graphene oxide. *Small* **2014**, *10*, 514–523. [CrossRef]
36. Zawko, S.A.; Suri, S.; Truong, Q.; Schmidt, C.E. Photopatterned anisotropic swelling of dual-crosslinked hyaluronic acid hydrogels. *Acta Biomater.* **2009**, *5*, 14–22. [CrossRef]
37. Rizwan, M.; Peh, G.S.; Ang, H.-P.; Lwin, N.C.; Adnan, K.; Mehta, J.S.; Tan, W.S.; Yim, E.K. Sequentially-crosslinked bioactive hydrogels as nano-patterned substrates with customizable stiffness and degradation for corneal tissue engineering applications. *Biomaterials* **2017**, *120*, 139–154. [CrossRef]
38. Zhou, M.; Lee, B.H.; Tan, L.P. A dual crosslinking strategy to tailor rheological properties of gelatin methacryloyl. *Int. J. Bioprint.* **2017**. [CrossRef]
39. Sahiner, N.; Sagbas, S.; Sahiner, M.; Silan, C.; Aktas, N.; Turk, M. Biocompatible and biodegradable poly (Tannic Acid) hydrogel with antimicrobial and antioxidant properties. *Int. J. Biol. Macromol.* **2016**, *82*, 150–159. [CrossRef]
40. Le Bourvellec, C.; Renard, C. Interactions between polyphenols and macromolecules: Quantification methods and mechanisms. *Crit. Rev. Food Sci. Nutr.* **2012**, *52*, 213–248. [CrossRef]
41. Chen, Y.-N.; Peng, L.; Liu, T.; Wang, Y.; Shi, S.; Wang, H. Poly (vinyl alcohol)–tannic acid hydrogels with excellent mechanical properties and shape memory behaviors. *ACS Appl. Mater. Interfaces* **2016**, *8*, 27199–27206. [CrossRef]
42. Shin, M.; Ryu, J.H.; Park, J.P.; Kim, K.; Yang, J.W.; Lee, H. DNA/tannic acid hybrid gel exhibiting biodegradability, extensibility, tissue adhesiveness, and hemostatic ability. *Adv. Funct. Mater.* **2015**, *25*, 1270–1278. [CrossRef]
43. Choi, J.-B.; Kim, Y.-K.; Byeon, S.-M.; Park, J.-E.; Bae, T.-S.; Jang, Y.-S.; Lee, M.-H. Fabrication and characterization of biodegradable gelatin methacrylate/biphasic calcium phosphate composite hydrogel for bone tissue engineering. *Nanomaterials* **2021**, *11*, 617. [CrossRef]
44. Fan, H.; Wang, L.; Feng, X.; Bu, Y.; Wu, D.; Jin, Z. Supramolecular hydrogel formation based on tannic acid. *Macromolecules* **2017**, *50*, 666–676. [CrossRef]
45. Liu, B.; Wang, Y.; Miao, Y.; Zhang, X.; Fan, Z.; Singh, G.; Zhang, X.; Xu, K.; Li, B.; Hu, Z. Hydrogen bonds autonomously powered gelatin methacrylate hydrogels with super-elasticity, self-heal and underwater self-adhesion for sutureless skin and stomach surgery and E-skin. *Biomaterials* **2018**, *171*, 83–96. [CrossRef] [PubMed]
46. Zhao, X.; Lang, Q.; Yildirimer, L.; Lin, Z.Y.; Cui, W.; Annabi, N.; Ng, K.W.; Dokmeci, M.R.; Ghaemmaghami, A.M.; Khademhosseini, A. Photocrosslinkable gelatin hydrogel for epidermal tissue engineering. *Adv. Healthc. Mater.* **2016**, *5*, 108–118. [CrossRef]
47. Kozlovskaya, V.; Kharlampieva, E.; Drachuk, I.; Cheng, D.; Tsukruk, V.V. Responsive microcapsule reactors based on hydrogen-bonded tannic acid layer-by-layer assemblies. *Soft Matter* **2010**, *6*, 3596–3608. [CrossRef]
48. Mao, A.S.; Shin, J.-W.; Mooney, D.J. Effects of substrate stiffness and cell-cell contact on mesenchymal stem cell differentiation. *Biomaterials* **2016**, *98*, 184–191. [CrossRef]
49. Coleman, M.M.; Lee, K.H.; Skrovanek, D.J.; Painter, P.C. Hydrogen bonding in polymers. 4. Infrared temperature studies of a simple polyurethane. *Macromolecules* **1986**, *19*, 2149–2157. [CrossRef]
50. Ma, M.; Zhong, Y.; Jiang, X. Thermosensitive and pH-responsive tannin-containing hydroxypropyl chitin hydrogel with long-lasting antibacterial activity for wound healing. *Carbohydr. Polym.* **2020**. [CrossRef]
51. Topkaya, S.N. Gelatin methacrylate (GelMA) mediated electrochemical DNA biosensor for DNA hybridization. *Biosens. Bioelectron.* **2015**, *64*, 456–461. [CrossRef] [PubMed]
52. Muhoza, B.; Xia, S.; Zhang, X. Gelatin and high methyl pectin coacervates crosslinked with tannic acid: The characterization, rheological properties, and application for peppermint oil microencapsulation. *Food Hydrocoll.* **2019**, *97*, 105174. [CrossRef]
53. Glimcher, M.J. Mechanism of calcification: Role of collagen fibrils and collagen-phosphoprotein complexes in vitro and in vivo. *Anat. Rec.* **1989**, *224*, 139–153. [CrossRef] [PubMed]
54. Shin, C.S.; Cho, J.Y. Bone remodeling and mineralization. *J. Korean Endocrinol. Soc.* **2005**, *20*, doi. [CrossRef]
55. Yu, X.; Biedrzycki, A.H.; Khalil, A.S.; Hess, D.; Umhoefer, J.M.; Markel, M.D.; Murphy, W.L. Nanostructured mineral coatings stabilize proteins for therapeutic delivery. *Adv. Mater.* **2017**, *29*, 1701255. [CrossRef]
56. Minteer, S.D. *Enzyme Stabilization and Immobilization*; Springer: Berlin/Heidelberg, Germany, 2017; pp. 94–104.

Article

Mechanical Properties and Thermal Conductivity of Thermal Insulation Board Containing Recycled Thermosetting Polyurethane and Thermoplastic

Ping He *, Haoda Ruan, Congyang Wang and Hao Lu

College of Mechanical and Electrical Engineering, Anhui Jianzhu University, Hefei 230601, China; rhd@stu.ahjzu.edu.cn (H.R.); wangcongyang@stu.ahjzu.edu.cn (C.W.); luhao6083@sina.cn (H.L.)
* Correspondence: heping@ahjzu.edu.cn; Tel.: +86-177-0560-8398

Abstract: This study used a mechanochemical method to analyze the recycling mechanism of polyurethane foam and optimize the recycling process. The use of mechanochemical methods to regenerate the polyurethane foam powder breaks the C–O bond of the polyurethane foam and greatly enhances the activity of the powder. Based on orthogonal test design, the mesh, proportion, temperature, and time were selected to produce nine recycled boards by heat pressing. Then, the influence of four factors on the thermal conductivity and tensile strength of the recycled board was analyzed. The results show that 120 mesh polyurethane foam powder has strong activity, and the tensile strength can reach 9.913 Mpa when it is formed at 205 °C and 40 min with 50% PP powder. With the help of the low thermal conductivity of the polyurethane foam, the thermal conductivity of the recycled board can reach 0.037 W/m·K at the parameter of 40 mesh, 80%, 185 °C, 30 min. This research provides an effective method for the recycling of polyurethane foam.

Keywords: mechanochemical method; recycled polyurethane foam; orthogonal test; tensile strength; thermal conductivity

Citation: He, P.; Ruan, H.; Wang, C.; Lu, H. Mechanical Properties and Thermal Conductivity of Thermal Insulation Board Containing Recycled Thermosetting Polyurethane and Thermoplastic. *Polymers* **2021**, *13*, 4411. https://doi.org/10.3390/polym13244411

Academic Editors: Alexey Iordanskii and Vetcher Alexandre

Received: 8 November 2021
Accepted: 13 December 2021
Published: 16 December 2021

Publisher's Note: MDPI stays neutral with regard to jurisdictional claims in published maps and institutional affiliations.

Copyright: © 2021 by the authors. Licensee MDPI, Basel, Switzerland. This article is an open access article distributed under the terms and conditions of the Creative Commons Attribution (CC BY) license (https://creativecommons.org/licenses/by/4.0/).

1. Introduction

Polyurethane is widely used in the construction industry, automobile industry, coatings, and clothing applications, because of its good stability, corrosion resistance, low density, and thermal conductivity [1]. Therefore, the production of polyurethane is also increasing. At present, the annual output of polyurethane is close to 30 million tons, accounting for 7.9% of the total output of plastics. It is the fifth most used polymer in the world [2]. Polyurethanes are generally divided into the following categories: flexible foams, rigid foams, and shells (coatings, adhesives, sealants, elastomers), which are used for the different applications shown in Table 1 [3–6].

In the process of production and consumption, a large number of polyurethane foam wastes have appeared. Due to the small pile-up density (about 30 kg/m³) and difficulty in natural degradation, polyurethane foam has caused serious environmental problems [7]. Many countries are researching biodegradable polyurethane foam, but the high price makes the traditional polyurethane foams cannot be replaced in a short time [8,9] Therefore, how to properly handle polyurethane foam waste is worth studying.

The treatment methods of polyurethane foam waste are landfill, incineration, and recovery [10,11]. The proportion of landfill waste can be as high as 50%. Because of the damage to the ecology and the environment, and the continuous depletion of oil reserves, many countries restrict or even prohibit the landfill of polymer waste [2]. Incineration, as another treatment method of polyurethane foam, occupies an important position. Incineration uses polyurethane waste as fuel to recover energy. In fact, polyurethane combustion can provide the same amount of heat as coal by weight [7]. However, flame retardants are added to many polyurethane foams, which greatly hinders the combustion of polyurethane.

The incomplete combustion of polyurethane will produce toxic gases (such as CO, NOx) and pollute the atmosphere. Therefore, recycling will become the best way to deal with polyurethane foam.

Table 1. Categories of PU applications.

Categories	Applications	Production
Flexible foams	Furniture, carpet, bedding, matrasses	36%
Rigid foams	Commercial refrigerators, insulation board, packaging	32%
Elastomers	Implants, medical devices, shoe soles	8%
Adhesives and sealant	Casting, sealants	6%
Coatings	Aircraft, vehicles (bumpers, side panels)	14%
Binders	Assembling of wood boards, rubber, elastomeric flooring surfaces	4%

After the polyurethane foam is cured, it cannot be reshaped by heating it again. The good performance of polyurethane foam makes recycling more difficult. At present, there are two methods to recycle waste polyurethane foam: physical recycling and chemical recycling [12,13].

The physical recycling method does not change the chemical structure. The polyurethane foam is broken into particles or powders, which can be directly used as filler or reshaped with adhesives [14]. Nowadays, the physical recycling method of polyurethane foam has been widely used. Yang et al. [15] crushed rigid polyurethane foam into particles to enhance the mechanical properties of rigid polyurethane foam (PUF) and phenolic foam (PF). The results show that when the particle polyurethane foam (PPU) content is 5 wt%, the compressive strength of PUF and PF has an increase of nearly 20%. Gama et al. [16] reported that PUF waste particles can be mixed with MDI and then molded at 100–200 °C and 30–200 bar pressure. The product of this method has been useful as insulation panels, carpets, and furniture. Moon et al. [17] use low-temperature pulverization to pulverize flexible polyurethane foam into powder. The polyurethane foam powder is treated by ultrasonic, and the original polyurethane foam is added to prepare mixed PUF. The results show that the car seat cushion made of mixed PUF has higher comfort than pure PUF and reduces the hardness and hysteresis loss. The physical recycling method is simple in operation and low cost, but its application range is limited, and its potential has not been extensively developed.

Chemical recycling methods, also known as raw material recovery, include alcoholysis, hydrolysis, glycolysis, acidolysis, etc. [18,19], which degrade polyurethane foam into oligomers and smaller molecules. The raw materials recovered by the chemical method can be used in new polyurethane foam or other products. Valle et al. [20] used castor oil to successfully decompose flexible polyurethane foam waste. The results show that increasing the concentration of Decomposed polyurethane (DP) will increase the elongation at break, reduce the tensile strength and the cell size. Heiran et al. [21] used different glycols and catalysts for the glycolysis of waste polyurethane foam. Parameters such as temperature and material ratio are determined. The recovered raw materials can be used to prepare new polyurethanes and be used in boiler insulation and protective coatings. Gama et al. [22] depolymerized flexible polyurethane foam with succinic acid to obtain recycled polyol. The recycled polyol will replace part of the original polyol to produce polyurethane foam. The results show that 30% recycled polyol has no obvious effect on the morphology and density of the polyurethane foam. The chemical recycling method follows the principle of degradation and is the best method for recycling polyurethane foam in theory. However, the process is complicated, and the separation and purification process are very expensive, which is difficult for industrial application.

Mechanochemistry is based on the physical method and accumulates mechanical energy and thermal energy, through long-term mechanical force action to make solid reactants react chemically without solvent and change the chemical structure of substances [23,24]. Although the thermosetting plastics cannot be reduced to raw materials by using the mechanochemical method, such as the chemical recovery method, it can interrupt the network crosslinking structure of thermosetting plastics, reducing the crosslinking degree and improving the activity of recycled powder. Hu et al. [25] used the mechanochemical method to recover thermosetting phenolic resin, and the tensile strength of the recycled material could reach 8.13 Mpa.

In summary, the mechanical method is feasible for recycling thermosetting plastics, but it is mainly focused on the mechanical properties of recycled materials, which is undoubtedly a waste for polyurethane foam with high thermal insulation capacity. This research is an attempt to recycle polyurethane foam and make it into an insulation material that can be used in buildings. Mechanochemical method was used to recover polyurethane foam as filler, recycled polypropylene as the matrix, without adding any other adhesive, only change the polyurethane particle size, proportion, and heat pressing parameters, and the thermal conductivity and tensile strength of the product were evaluated. The recovery process of polyurethane foam by the mechanochemical method is shown in Figure 1.

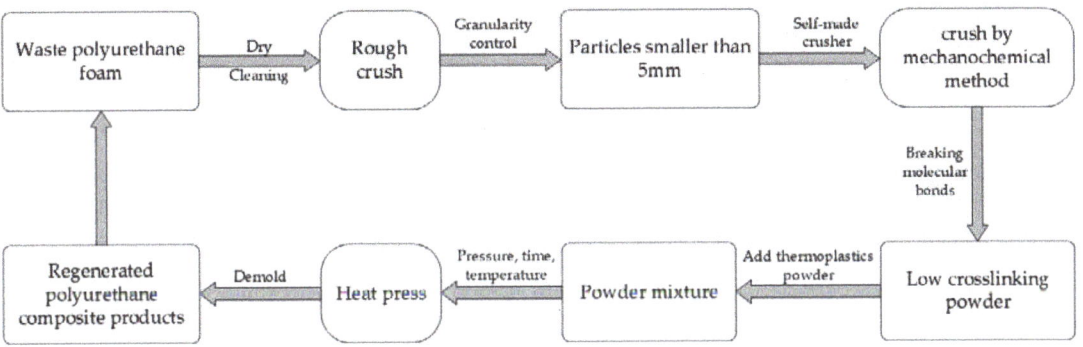

Figure 1. Recovery process of polyurethane foam by mechanochemical method.

2. Materials and Methods

2.1. Materials

The waste polyurethane foam used in this study is rigid polyurethane insulation board (Aoyang Insulation Material Corp., Langfang, China). As shown in Figure 2, the outer side of this board is a fireproof layer composed of non-woven fabric and inorganic paste, and the middle is rigid polyurethane foam. This board is the most commonly used type of building insulation material in China. The matrix material is recycled polypropylene (ZhongLian Plastic Corp., Dongguan, China).

2.2. Experiment Process

2.2.1. PUF Crushing Process

The crushing of the waste polyurethane foam is carried out in a self-made crusher specially designed for long-term crushing in the laboratory. To fit the actual recovery conditions, the fireproof layer was retained. It can enhance the strength of the recycled broad and improve the economic benefit. First, the polyurethane foam was manually cut into small pieces of 2 square centimeters, and then rough crushed into particles smaller than 5 mm, last, crushed into low crosslinking powder with the self-made crusher. As shown in Figure 2d, the self-made crusher is equipped with three sets of cutter teeth and two grinding discs. There are shear force, grinding force, extrusion force, and other mechanical forces in the grinding process. As the material is pulverized and heat energy accumulates, the network structure of polyurethane foam is broken and active groups are formed. The speed of the crusher is set to 1500 r/min, and the crushing time is 40 min. This crushing condition was obtained by previous studies in the laboratory [26]. Faster speed and longer time will enhance the crushing effect, but the mechanical energy consumption is greatly increased, and the efficiency is lower. At 1500 r/min and 40 min, polyurethane foam can be effectively degraded and has the highest cost performance. Polyurethane foam powder is shown in Figure 2c.

Figure 2. Appearance of (**a**) Waste PUF board, (**b**) PUF pieces, (**c**) PUF powder, (**d**) self-made crusher, and (**e**) microscopic morphology of PUF powder.

2.2.2. Heat Press Process

The polyurethane foam powder was molded with the flat vulcanizing machine XLB350X (Qicai Hydraulic Machinery Corp., Shanghai, China). To facilitate demolding, a layer of PET film is laid on the bottom of the mold. The melting point of PET is above 250 °C, which can prevent the melt from bonding with the mold. Mix the waste polyurethane foam powder and recycled PP material evenly, lay the mixture in the mold, put another layer of PET film on the powder, cover the press mold. Preheat mold at 175 °C for 10 min before each experiment. Then the heat pressing is carried out at the temperature and time in the table. After the heat pressing, the exhaust is released for 10 min, and then kept warm for 10 min. Finally, remove the mold and cool it to room temperature before demolding. The size of the board is $150 \times 150 \times 5$ mm^3, as shown in Figure 3.

Figure 3. Recycled boards formed by heat press.

2.3. Performance Testing

2.3.1. PUF Powder Testing

The distribution of polyurethane foam powder was determined by laser particle size analyzer BT-9300ST (Bettersize Instruments Crop., Dandong, China). Distilled water and sodium pyrophosphate were added to the powder to make a suspension, and the powder was dispersed by ultrasonic for 3 min. The cycle speed during the test was 1600 rpm, and the average value was taken for 6 tests.

The Fourier transform infrared spectrometer FTIR-850 (GangDong Sci&Tech Ltd., Tianjin, China) was used to study the molecular structure changes of polyurethane foam powders. Three meshes of powders (40, 80, 120) were added into potassium bromide and made into press sheets, which were determined by 32 scanning times.

2.3.2. Scanning Electron Microscope (SEM)

Scanning electron microscope EVO-18 (Carl Zeiss AG, Oberkochen, Baden-Württemberg, Germany) was used to observe the microstructure of polyurethane powder and recycled board. Considering the low electrical conductivity of polyurethane, the recycled boards were cut into $5 \times 5 \times 5$ mm^3 samples and coated with gold under vacuum. The acceleration voltage is selected as 20 kV, which can satisfy the analysis of most elements. Compared with the lower acceleration voltage, 20 kV can obtain a higher resolution and help us observe the composition information inside the sample.

2.3.3. Thermal Conductivity Testing

Thermal conductivity has always been considered as the main parameter related to the practical application of polyurethane foam. Heat flow meter apparatus DRPL-III (XiangYi, Instrument Co., Ltd., Xiangtan, China) was used to detect the thermal conductivity of nine boards. Based on the ISO 8301 standard, select Two heat flow meter configurations, set the cold surface to 25 °C, the hot surface to 40 °C, and the pressure to 80 N. Thermal conductivity is calculated according to Formula (1). The experiments were repeated three times for each sample to obtain an average value. The samples measured for thermal conductivity were polished and refined to reduce thermal contact resistance.

$$\lambda = 0.5(f_1 e_1 + f_2 e_2)\frac{d}{\Delta T} \quad (1)$$

where:

λ = thermal conductivity (W·m^{-1}·K^{-1})
f = calibration factor (W·m^{-2}·V^{-1})
e = heat flow meter output (V)
d = average specimen thickness (m)

2.3.4. Tensile Strength Testing

Materials testing System AGS-X (Shimadzu Corp., Kyoto, Japan) was used to test the tensile strength of recycled boards, based on the International Organization for Standardization (ISO) 527 and ASTM D638. the recycled boards were made into standard-size samples (width 10 mm, gage length 50 mm), the test speed was set to 1 mm/min, and the maximum load that the sample could bear was measured. The experiment was repeated three times to determine the tensile strength.

3. Results and Discussion

The main purpose of this work is to develop a kind of recycled sheet with better mechanical properties and lower thermal conductivity. Thus, this can be viewed as an optimization problem with two objectives. The optimization process is mainly divided into the following parts:

(1) Analyze the crushing effect of polyurethane foam powder;

(2) Design and complete the orthogonal test;
(3) Take the thermal conductivity and tensile strength as the response values, analyze the influence of factors on them;
(4) Multi-objective optimization selection of recycled board.

3.1. Analysis of Crushing Effect

3.1.1. Particle Size Distribution of PUF Powder

Figure 4 showed the size distribution of the pulverized PUF powder. The average size of PUF powder is 245 μm, which will not greatly affect the board forming and can retain certain thermal insulation performance. The particle size distribution is: 52.89% = 177–420 μm; 22.83% = 180–125 μm; 8.74% = 75–125 μm; 4.21% < 75 μm.

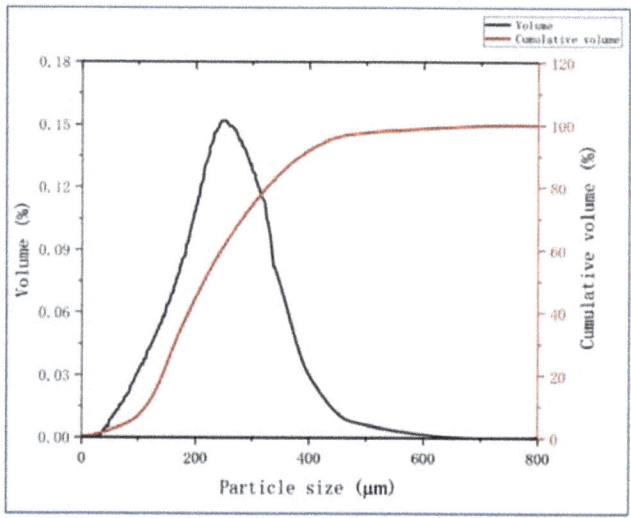

Figure 4. The particle size distribution of PUF powder.

3.1.2. FTIR (Fourier Transform Infrared Spectroscopy) Analysis

FTIR is used to analyze the molecular structure and group changes in the degradation process, as shown in Figure 5. The characteristic peak of amino (–NH–) is 3317.6 cm^{-1} in 40 mesh. The characteristic peak of the pulverized particle size is widened in the 200 mesh, and a large amount of hydroxyl (–OH–) appeared and the characteristic peak of the large concentration of amino (–NH–) is formed by coincidence. This is the result of the C–O bond breaking to form the hydroxyl (–OH–) group. With different mesh numbers, cyanate group characteristic peaks appeared in wave numbers 3317.6–3369.4 cm^{-1}, indicating the rupture of the carbamate group at the C–O bond and the emergence of a new isocyanate group.

At wavenumber 1226.5 cm^{-1}, the stretching vibration peak of carbamate group C–O changed significantly, indicating that the carbamate group on the main chain of the polymerization is gradually reduced and the cross-linking structure is gradually destroyed.

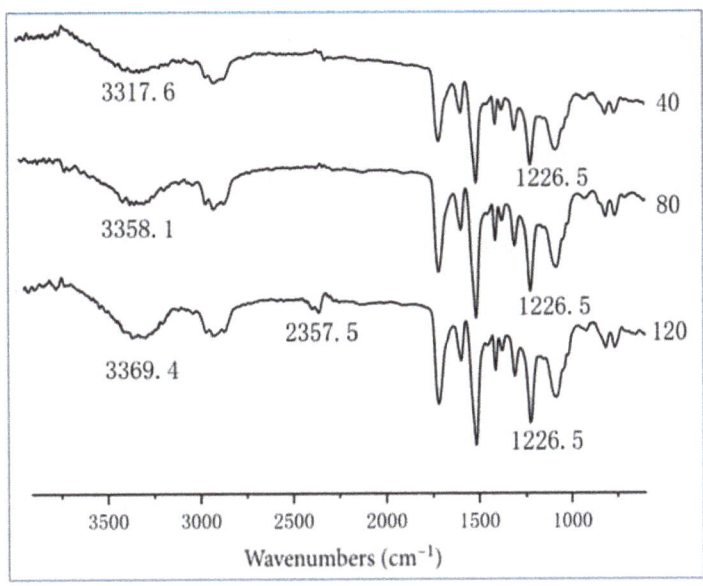

Figure 5. ATR-FTIR spectra of polyurethane powders with different mesh numbers.

3.1.3. Microstructure of PUF Powder

The micrograph of PUF powder is shown in Figure 6a (40 mesh, loading voltage is 20 kV, magnification is 50 times). It can be seen that there are a lot of fibers in PUF powder, which come from the fireproof layer mentioned above. In this study, choosing to retain these fibers can not only reduce the pre-treatment cost but also effectively improve the mechanical properties of plastics by adding fibers into plastics. Micrographs of 40 mesh, 80 mesh, and 120 mesh powders are shown in Figure 7 (loading voltage 20 kV, magnification is 300 times). It can be seen that the shapes of the three powders are similar, but the 40-mesh powder retains more of the pore structure of polyurethane foam, which can be seen more clearly in Figure 6b (40 mesh, loading voltage 20 kV, magnification 200 times). This explains the influence of mesh number on thermal conductivity well. The melted polypropylene covers the surface of polyurethane foam powder and generates bubbles again, which can greatly reduce the thermal conductivity. However, the existence of a large number of bubble structures also causes it to become a weak point in the tensile test.

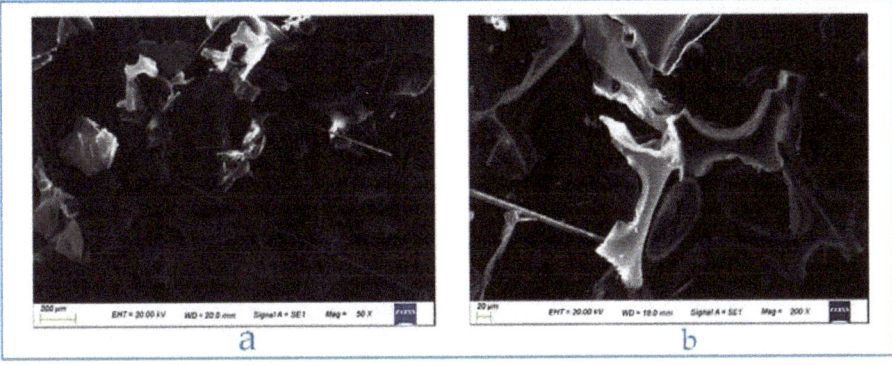

Figure 6. Microscopic morphology of PUF powder with different magnification. (**a**) 50 times (**b**) 200 times.

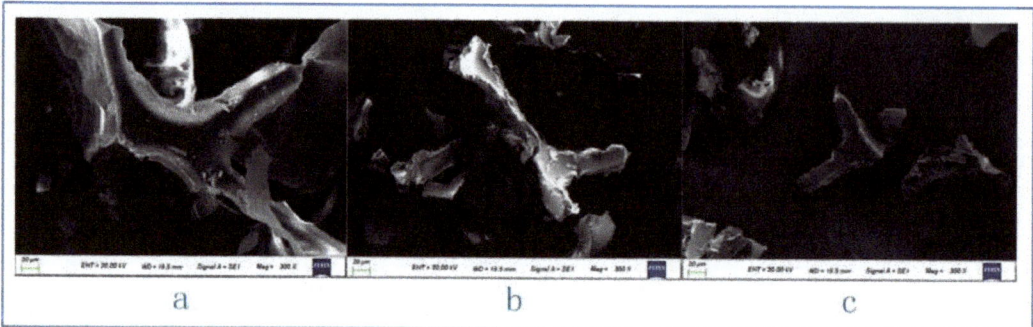

Figure 7. Microscopic morphology of PUF powders with different mesh numbers. (**a**) 40 mesh. (**b**) 80 mesh. (**c**) 120 mesh.

3.2. Orthogonal Test Analysis

3.2.1. Orthogonal Test Design

In orthogonal test design, the choice of factors and levels is very important. Based on the results of particle size analysis and infrared spectrum analysis, and the previous studies [26]. Four factors were determined, namely, mesh number (A), PUF proportion (B), temperature (C), and time (D).

(A) Mesh: After a single factor test on the mesh level, the mesh level is set to 40 mesh, 80 mesh, and 120 mesh. The larger the particle size of the powder, the more it can preserve the thermal insulation properties of the polyurethane foam itself, but too large particles will reduce the bond strength of the polyurethane foam powder and polypropylene, and the mechanical properties of the recycled board are poor. To improve the mechanical properties, a large amount of polypropylene powder is added, which makes the recycling of polyurethane foam secondary and goes against the goal. At the same time, the smaller the particle size of the powder, the better its mechanical properties. When polyurethane powder finer than 200 mesh is used and the addition amount exceeds 50%, the tensile strength of the finished product can be close to 20 Mpa. However, it has to be considered that the limited output of ultra-fine powder will greatly increase the cost of pulverization. Therefore, under the current conditions, the selection of 40 mesh, 80 mesh, and 120 mesh is more reasonable.

(B) PUF proportion: Due to the decision to recycle polyurethane foam as the main body, the proportion of polyurethane foam powder small addition was set to 50%. Considering the maximum proportion, although 100% polyurethane foam powder can be molded under high temperature and high pressure (180 °C, 35 Mpa), the molding effect is poor, and the mechanical properties are not ideal. Therefore, the maximum addition proportion of polyurethane foam is set to 80%.

(C) Temperature: The temperature is selected to make the polypropylene powder obtain fluidity. In fact, the polyurethane foam will also have a certain degree of plasticity at a certain temperature, which will help the molding of the product. Considering the melting temperature of polypropylene, a series of tests were carried out in the range of 165–215 °C. When the temperature is lower than 185 °C, the polypropylene powder has begun to flow, but the molding effect is not satisfactory. The bonding strength of the polyurethane foam powder and polypropylene is very poor, and even the powder fell off when the final product was taken. When the heating time is extended, the effect will be improved, but the mechanical properties are still not ideal and not economical enough. Therefore, it was decided to set the temperature range to 185 °C, 195 °C, and 205 °C. Under these conditions, the molding effect is the best.

(D) Time: As mentioned before, polypropylene takes time to melt and combine with the polyurethane foam powder. A series of tests were conducted in the range of 10–60 min, and it was found that 30–50 min is the most reasonable range. Too short heating time will affect the mechanical properties, and the longer time is meaningless.

Table 2 lists the details of the factors and their levels. Based on orthogonal test design table L9 (3^4), a total of 9 groups of tests were conducted, as shown in Table 3, with each row representing one test.

Table 2. The parameters and levels for processing.

Levels	A (Mesh)	B (Proportion)	C (Temperature)	D (Time)
1	40	50%	185	30
2	80	65%	195	40
3	120	80%	205	50

Table 3. DOE for final experimentation.

Exp No.	Mesh	Proportion	Temperature (°C)	Time (min)
1	40	50%	185	30
2	40	65%	195	40
3	40	80%	205	50
4	80	50%	195	50
5	80	65%	205	30
6	80	80%	185	40
7	120	50%	205	40
8	120	65%	185	50
9	120	80%	195	30

3.2.2. Results of Orthogonal Test

In this paper, the indexes of the orthogonal test are set as two: thermal conductivity, and tensile strength. Each set of experiments was run three times and the results were averaged. Experimental data are shown in Table 4.

Table 4. Orthogonal scheme and its results.

Test	A	B	C	D	Thermal Conductivity (W/m·K)	Tensile Strength (Mpa)
1	1	1	1	1	0.0711	0.9031
2	1	2	2	2	0.0645	0.4182
3	1	3	3	3	0.0555	0.4143
4	2	1	2	3	0.0982	3.5275
5	2	2	3	1	0.0726	0.9451
6	2	3	1	2	0.0596	0.2177
7	3	1	3	2	0.1263	9.9129
8	3	2	1	3	0.0803	1.4642
9	3	3	2	1	0.0614	1.1847

3.3. Performance Analysis of Board

3.3.1. Thermal Conductivity Analysis

Table 5 shows the results of the range analysis of thermal conductivity. K_i represents the average value of thermal conductivity under a certain factor. The mesh size (A) is positively correlated with the thermal conductivity. The larger the mesh size of the powder is, the larger the thermal conductivity is, that is to say, the worse the thermal insulation performance is (Figure 8A). Temperature (C) showed similar results to mesh (A) (Figure 8C). On the contrary, the higher the proportion of polyurethane powder (B), the lower the thermal conductivity (Figure 8B). When the value of factor time (D) increases, the corresponding value of Ki increases first and then decreases (Figure 8D). According to R-value, the factors can be arranged as B > A > D > C, indicating that the proportion of polyurethane has the greatest influence on thermal conductivity, followed by the particle size of polyurethane powder, temperature and time have less influence.

Table 5. Range analysis of thermal conductivity.

Elements	A	B	C	D
K_1	0.0637	0.0985	0.0703	0.0684
K_2	0.0768	0.0725	0.0747	0.0835
K_3	0.0893	0.0588	0.0848	0.078
R	0.0256	0.0397	0.0145	0.0151

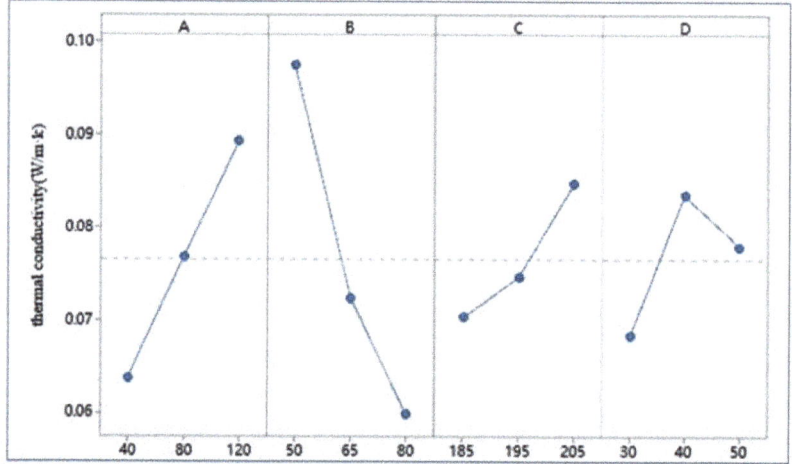

Figure 8. Effect of factors at different levels on thermal conductivity: (A) Mesh. (B) Proportion. (C) Temperature. (D) Time.

According to the results in Table 4, ANOVA is conducted for thermal conductivity, and statistical results were listed in Table 6. SS is the sum of squares of variables; DOF represents degrees of freedom; MS is mean square, that is, the ratio of SS to DOF; F and P are the values that determine whether the variable is significant. The high value of F and the low value of P indicate that the variable is more significant [27]. As can be seen from Table 6, the results of ANOVA are consistent with the previous range analysis, and the proportion of polyurethane powder has an important influence on the thermal conductivity. Because polyurethane foam powders do not melt when heated, there are many gaps between the powders, and the melted polypropylene seeps into these gaps and joins the powders. As the proportion of polyurethane foam powder increases, the proportion of polypropylene decreases, and polypropylene cannot be filled into the gap, resulting in a large number of bubbles. This is also the reason why the mesh size will affect the thermal conductivity. The smaller the mesh of the powder, the larger the particle size, and the gap is also larger. This can also be seen in the micrograph. To more intuitively express the influence of factors on thermal conductivity, the proportion and mesh number of polyurethane foams are selected as conditions to draw the Surface projection, as shown in Figure 9. Minitab software is used to analyze the linear regression equation of thermal conductivity coefficient $Υ_1$:

$$Υ_1 = -0.0233 + 0.000320A - 0.001323B + 0.0023C + 0.000482D \tag{2}$$

Table 6. ANOVA result for the thermal conductivity.

Variable	SS	DOF	MS	F	p	Contribution
A	0.002957	2	0.001479	25.64	<0.001	23.99%
B	0.007324	2	0.003662	63.51	<0.001	59.42%
C	0.000991	2	0.000496	8.59	0.002	8.04%
D	0.001052	2	0.000526	9.12	0.002	8.54%
Error	0.001038	18	0.000058			
Total	0.013363	26				100%

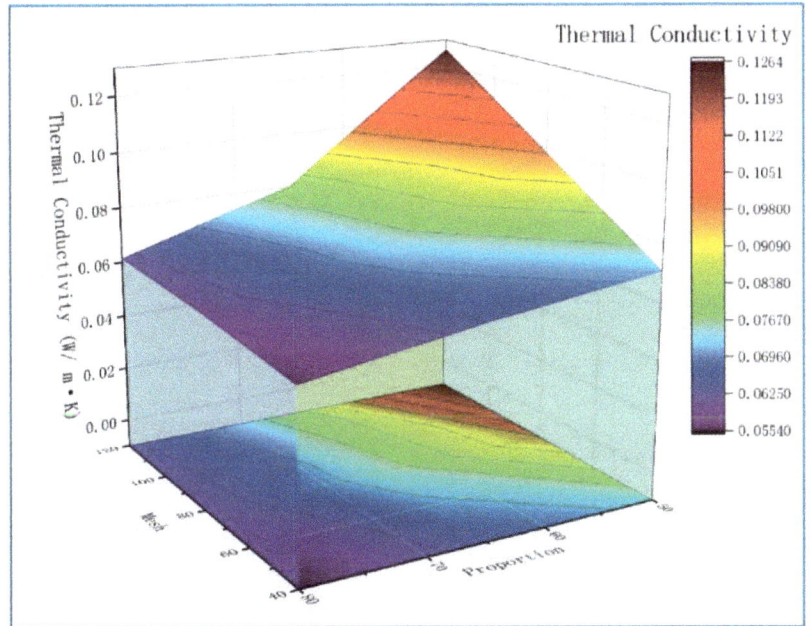

Figure 9. The surface projection of Mesh and Proportion affects thermal conductivity.

3.3.2. Tensile Strength Analysis of Board

Table 7 shows the results of the range analysis of tensile strength. According to R-value, the order of factor influence is: B > A > C > D. It can be seen that the influence trend of different factors on tensile properties is consistent with the thermal conductivity, as shown in Figure 10.

Table 7. Range analysis of tensile strength.

Elements	A	B	C	D
K_1	0.5544	4.7811	0.8616	1.0109
K_2	1.5634	0.9183	1.6860	3.4921
K_3	4.1872	0.6055	3.7574	1.8020
R	3.6328	4.1755	2.8958	2.4811

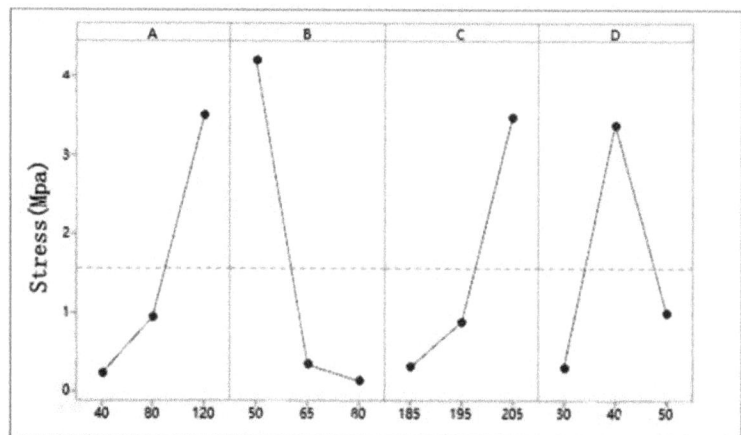

Figure 10. Effect of factors at different levels on stress: (**A**) Mesh. (**B**) Proportion. (**C**) Temperature. (**D**) Time.

ANOVA is conducted for tensile strength, and the results were shown in Table 8. Mesh, proportion, temperature, and time have significant effects on the tensile strength, and the ranking of their contribution degree is consistent with the range analysis results. To observe the tensile properties more intuitively, the stress-strain data are drawn, as shown in Figure 11. It can be seen that the stress of the No.7 broad is much higher than that of other broads. From its processing parameters (120 mesh, 50%, 205 °C, 40 min), this result is inevitable. 120 mesh PUF powder has stronger surface activity and can be better combined with PP powder. The high proportion of PP powder provides a higher tensile strength for the No.7 broad. The temperature of 205 °C, and the time of 40 min, provide the possibility of a good combination of PUF and PP. The second highest stress is no.4 broad, whose machining parameters are (80 mesh, 50%, 195 °C, 50 min). Comparing the thermal conductivity and tensile strength of these two boards, it can be seen that they have the highest values of both. High tensile strength means better powder bonding and lower porosity. This also makes the thermal conductivity higher.

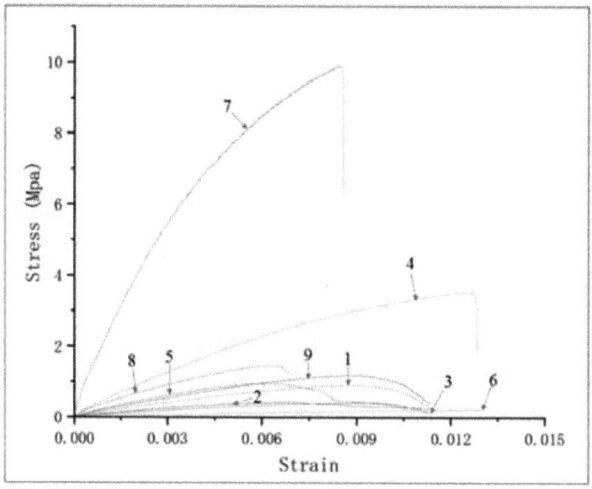

Figure 11. Stress versus strain curve (The numbers in the figure represent different experiment numbers).

Table 8. ANOVA result for the tensile strength.

Variable	SS	DOF	MS	F	p	Contribution
A	63.148	2	31.5739	66.68	<0.001	27.43%
B	96.860	2	48.4300	102.28	<0.001	42.08%
C	40.309	2	20.1545	42.56	<0.001	17.51%
D	29.870	2	14.9349	31.54	<0.001	12.98%
Error	8.523	18	0.4735			
Total	238.710	26				100%

Similarly, the proportion and mesh number of polyurethane foams are selected as conditions to draw the surface projection of its influence on tensile strength, as shown in Figure 12. The linear regression equation of tensile strength Y_2 is:

$$Y_2 = -22.37 + 0.04524A - 0.1392B + 0.1392C + 0.0390D \qquad (3)$$

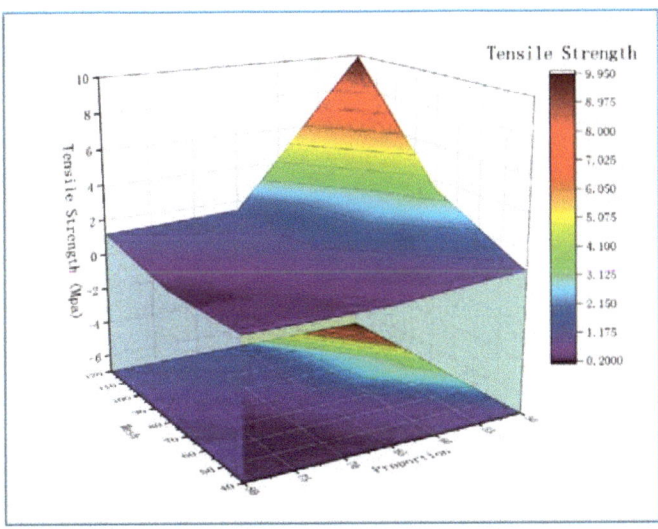

Figure 12. The surface projection that Mesh and Proportion affect tensile strength.

3.3.3. Microstructure of Recycled Board

Micrographs of nine boards are shown in Figure 13. The fireproof layer fibers are evident in Figure 13c,f. Note that each row has the same number of mesh and each column has the same proportion. Therefore, it is also easy to compare the effects of mesh and proportion on the board. The lowest thermal conductivity is board No. 3, which can also be seen by comparing 13c with other horizontal and longitudinal pictures. A large number of polyurethane foam powders provides the possibility of low thermal conductivity. A small amount of polypropylene cannot completely wrap the polyurethane foam powder and only plays a role of connection, which is also the reason why the tensile strength of No. 3 board is only 0.4143 Mpa. The opposite is board No. 7, whose micrograph is shown in 13 g. Board No. 7 contains 50%, 120 mesh polyurethane foam powder, which is completely coated with equal weight polypropylene. It is not available on other boards. It also brings excellent tensile strength to the No. 7 plate. It is worth mentioning that the thermal conductivity of No. 7 board is 0.1263 W/m·K, which is 54% of the thermal conductivity of pure polypropylene board (about 0.23 W/m·K). It is also proved that 120 mesh polyurethane powder has a great influence on reducing thermal conductivity.

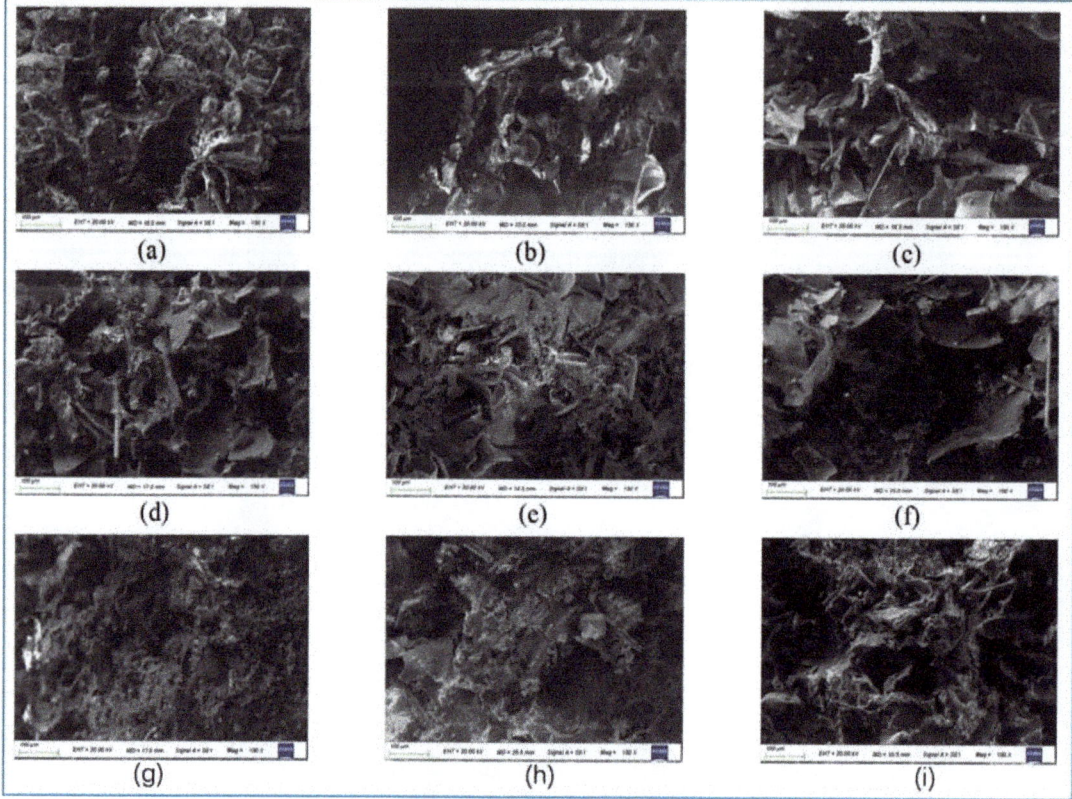

Figure 13. Photomicrographs for samples 1–9 (×150 magnification; (**a**–**i**) corresponds to experiments no. 1–9).

3.3.4. Parameter Selection of Recycled Board

Based on the above analysis, we can see that the thermal conductivity is positively correlated with the tensile strength, which makes it difficult to obtain an optimal solution and needs to be adjusted according to actual requirements. This study provides three parameters for reference: the lowest thermal conductivity, the maximum tensile strength, and the equilibrium selection.

The lowest thermal conductivity is selected as $A_1B_3C_1D_1$. Under this condition, the thermal conductivity is 0.037 W/m·K, the tensile strength is 0.133 Mpa. The low strength makes it difficult to use as a stand-alone material. However, it can be used as the building insulation boards, surrounded by brick, concrete, reinforced concrete, and other heavy materials. Or as the sandwich of steel board, to provide better insulation ability.

The maximum tensile strength was selected as $A_3B_1C_3D_2$. Under this condition, the thermal conductivity was 0.1253 W/m·K, the tensile strength was 9.913 Mpa. Its thermal insulation performance is general, but good mechanical properties can be used for room decoration panels, pipes, bumpers, gaskets.

In addition, according to range analysis, the influence of proportion (B) on the thermal conductivity is much higher than the other three factors. Although the proportion (B) has the highest influence on tensile strength, the difference between it and the other three factors is small. Therefore, B_3 is selected to obtain better thermal insulation performance, and $A_3C_3D_2$ is selected to obtain better tensile strength. Under the condition of $A_3B_3C_3D_2$, the thermal conductivity is 0.086 W/m·K, the tensile strength is 5.737 Mpa. Balanced performance can be applied to a wide range of uses, such as replacing lightweight aggregate concrete for the interior and exterior walls of buildings, roofs, and floors.

For comparison, the above results and the properties of the original material are listed in Table 9. It can be seen that the thermal conductivity is as low as 0.037 W/m·K, which is very close to the thermal conductivity of polyurethane foam. In the past research on thermosetting plastics, researchers focused on the mechanical properties of recycled plates. For example, Prestes et al. [14] Added 40% high-pressure Laminate powders into polypropylene, extruded by the extruder model and the tensile strength was 11.58 Mpa. Quadrini et al. [28] formed pure polyurethane foam powders by hot pressing. The tensile strength and compressive strength were 2.4 Mpa and 22 Mpa respectively. Considering only the mechanical properties, it is undoubtedly a waste of polyurethane foam with high insulation capacity. In terms of the highest thermal conductivity, the value of 0.1253 W/m·K is 54% of the thermal conductivity of pure polypropylene board, which has also met the requirements of China for thermal insulation materials. Its 9.913 Mpa tensile strength far exceeds that of polyurethane foam. Taking into account that under $A_3B_1C_3D_2$ condition, the proportion of polyurethane foam powder is 50%, the mechanical properties of this recycled board are not weaker than the study by Prestes et al.

Table 9. Performance and application of the board.

Parameter	Thermal Conductivity (W/m·k)	Tensile Strength (Mpa)	Applications
$A_1B_3C_1D_1$	0.037	0.133	Insulation board, the sandwich of steel broad
$A_3B_1C_3D_2$	0.1253	9.913	Room trim panels, pipes, bumpers, gaskets,
$A_3B_3C_3D_2$	0.086	5.737	Walls, roofs, floors
Polyurethane foam	0.022~0.030	0.3	
Polypropylene	0.23	29	

4. Conclusions

Following are the conclusions from this study:

The effect of mechanochemical pulverization of waste polyurethane foam on the appearance and molecular structure of PUF is studied. As a result, the mechanical and thermal energy is accumulated during a long period of crushing. Under the combined action, the C–O bond of PUF is broken, the network crosslinking structure is destroyed, and the activity of PUF powder is significantly improved.

Polyurethane foam powder and PP can be remolded into composite materials by heat pressing. Taking mesh, proportion, temperature, and time as factors and thermal conductivity, tensile strength, and density as indexes, the orthogonal test design method is established. Based on range analysis and variance analysis, the influence of each factor on the index is studied.

The results show that the proportion of polyurethane foam powder has the greatest influence on thermal conductivity and tensile strength, the second is mesh size, and the temperature and time have less influence. When the mesh number is 40 and the proportion is 80%, the lowest thermal conductivity 0.037 W/m·K and the tensile strength is 0.133 Mpa are obtained. The polyurethane foam powder at the age of 40 mesh retains a relatively complete bubble structure, but the melting of 20% polypropylene is not enough to fill it but will form new and smaller bubbles. When the mesh number is 120 and the proportion is 50%, the maximum tensile strength is 9.913 Mpa and the thermal conductivity is 0.1253 W/m·K. However, the value of 0.1253 W/m·K, which is 54% of the thermal conductivity of pure polypropylene board (about 0.23 W/m·K), has also reached the requirements of China for thermal insulation materials.

This study provides an effective method for the recovery of polyurethane foam, and as an example of applications can be expected: Insulation, roofs, bumpers, gaskets, etc. More research is needed to improve the properties of recycled polyurethane foam. The performance of the recycled boards was slightly worse than that of the original material, which is to be expected considering that no additives were added in this test. In the next test, additives will be selected and different thermoplastics will be tried.

Author Contributions: Conceptualization, P.H.; methodology, H.R.; software, H.R.; validation, P.H., H.R. and C.W.; investigation, H.L.; resources, P.H.; data curation, H.R.; writing—original draft preparation, H.R.; writing—review and editing, H.R.; visualization, C.W.; supervision, P.H.; project administration, P.H.; funding acquisition, P.H. All authors have read and agreed to the published version of the manuscript.

Funding: This research was funded by National Natural Science Foundation of China, grant number 51877001; Central Government to Guide Local Scientific and Technological Development, grant number 202107d06020004; Key Project of Anhui Provincial Department of Education College Excellent Talents Support Program, grant number gxyqZD2020034, and Doctor Anhui Jianzhu University Started Funding Project, grant number 2018QD14.

Institutional Review Board Statement: Not applicable.

Informed Consent Statement: Not applicable.

Data Availability Statement: Data generated or analyzed during this study are included in this published article.

Conflicts of Interest: The authors declare no conflict of interest.

References

1. Wang, M.; Zhang, X.; Zhang, W.; Lu, C.; Yuan, G. From Thermosetting to Thermoplastic: A Novel One-Pot Approach to Recycle Polyurethane Wastes via Reactive Compounding with Diethanolamine. *Prog. Rubber Plast. Recycl. Technol.* **2014**, *30*, 221–236. [CrossRef]
2. Plastics Europe Association of Plastics Manufacturers Plastics—The Facts 2019 an Analysis of European Plastics Production, Demand and Waste Data. Available online: https://www.plasticseurope.org/en/resources/market-data (accessed on 15 July 2021).
3. Deng, Y.; Dewil, R.; Appels, L.; Ansart, R.; Baeyens, J.; Kang, Q. Reviewing the thermo-chemical recycling of waste polyurethane foam. *J. Environ. Manag.* **2021**, *278*, 111527. [CrossRef]
4. Stachak, P.; Lukaszewska, I.; Hebda, E.; Pielichowski, K. Recent Developments in Polyurethane-Based Materials for Bone Tissue Engineering. *Polymers* **2021**, *13*, 946.
5. Magnin, A.; Pollet, E.; Phalip, V.; Averous, L. Evaluation of biological degradation of polyurethanes. *Biotechnol. Adv.* **2020**, *39*, 107457. [CrossRef] [PubMed]
6. Yang, H.; Yu, B.; Song, P.; Maluk, C.; Wang, H. Surface-coating engineering for flame retardant flexible polyurethane foams: A critical review. *Compos. Part B Eng.* **2019**, *176*, 107185. [CrossRef]
7. Yang, W.; Dong, Q.; Liu, S.; Xie, H.; Liu, L.; Li, J. Recycling and Disposal Methods for Polyurethane Foam Wastes. *Procedia Environ. Sci.* **2012**, *16*, 167–175. [CrossRef]
8. Wang, B.; Ma, S.; Li, Q.; Zhang, H.; Liu, J.; Wang, R.; Chen, Z.; Xu, X.; Wang, S.; Lu, N.; et al. Facile synthesis of "digestible", rigid-and-flexible, bio-based building block for high-performance degradable thermosetting plastics. *Green Chem.* **2020**, *22*, 1275–1290. [CrossRef]
9. Yuan, Y.; Sun, Y.; Yan, S.; Zhao, J.; Liu, S.; Zhang, M.; Zheng, X.; Jia, L. Multiply fully recyclable carbon fibre reinforced heat-resistant covalent thermosetting advanced composites. *Nat. Commun.* **2017**, *8*, 14657. [CrossRef]
10. Tantisattayakul, T.; Kanchanapiya, P.; Methacanon, P. Comparative waste management options for rigid polyurethane foam waste in Thailand. *J. Clean. Prod.* **2018**, *196*, 1576–1586. [CrossRef]
11. Kemona, A.; Piotrowska, M. Polyurethane Recycling and Disposal: Methods and Prospects. *Polymers* **2020**, *12*, 1752. [CrossRef]
12. Gharde, S.; Kandasubramanian, B. Mechanothermal and chemical recycling methodologies for the Fibre Reinforced Plastic (FRP). *Environ. Technol. Innov.* **2019**, *14*, 100311. [CrossRef]
13. Singh, R.; Singh, I.; Kumar, R.; Brar, G. Waste thermosetting polymer and ceramic as reinforcement in thermoplastic matrix for sustainability: Thermomechanical investigations. *J. Thermoplast. Compos. Mater.* **2019**, *34*, 523–535. [CrossRef]
14. Prestes, P.; Domingos, M.; Faulstich, D. Effect of high pressure laminate residue on the mechanical properties of recycled polypropylene blends. *Polym. Test.* **2019**, *80*, 106104. [CrossRef]
15. Yang, C.; Zhuang, Z.; Yang, Z. Pulverized polyurethane foam particles reinforced rigid polyurethane foam and phenolic foam. *J. Appl. Polym. Sci.* **2014**, *13*, 39734. [CrossRef]
16. Gama, N.V.; Ferreira, A.; Barros-Timmons, A. Polyurethane Foams: Past, Present, and Future. *Materials* **2018**, *11*, 1841. [CrossRef]
17. Moon, J.; Kwak, S.; Lee, J.; Kim, D.; Ha, J.; Oh, J. Synthesis of polyurethane foam from ultrasonically decrosslinked automotive seat cushions. *Waste Manag.* **2019**, *85*, 557–562. [CrossRef] [PubMed]
18. Gama, N.; Godinho, B.; Marques, G.; Silva, R.; Barros-Timmons, A.; Ferreira, A. Recycling of polyurethane by acidolysis: The effect of reaction conditions on the properties of the recovered polyol. *Polymers* **2021**, *219*, 123561. [CrossRef]

19. Godinho, B.; Gama, N.; Barros-Timmons, A.; Ferreira, A. Recycling of different types of polyurethane foam wastes via acidolysis to produce polyurethane coatings. *Sustain. Mater. Technol.* **2021**, *29*, e00330. [CrossRef]
20. Valle, V.; Aguirre, C.; Aldás, M.; Pazmiño, M.; Almeida-Naranjo, C. Recycled-based thermosetting material obtained from the decomposition of polyurethane foam wastes with castor oil. *J. Mater. Cycles Waste Manag.* **2020**, *22*, 1793–1800. [CrossRef]
21. Heiran, R.; Ghaderian, A.; Reghunadhan, A.; Sedaghati, F.; Thomas, S.; Haghighi, A. Glycolysis: An efficient route for recycling of end of life polyurethane foams. *J. Polym. Res.* **2021**, *28*, 22. [CrossRef]
22. Gama, N.; Godinho, B.; Marques, G.; Silva, R.; Barros-Timmons, A.; Ferreira, A. Recycling of polyurethane scraps via acidolysis. *Chem. Eng. J.* **2020**, *395*, 125102. [CrossRef]
23. James, S.; Adams, C.; Bolm, C.; Braga, D.; Collier, P.; Friscic, T.; Grepioni, F.; Harris, K.; Hyett, G.; Jones, W.; et al. Mechanochemistry: Opportunities for new and cleaner synthesis. *Chem. Soc. Rev.* **2012**, *41*, 413–447. [CrossRef]
24. Zhang, C.; Zhuang, L.; Yuan, W.; Wang, J.; Bai, J. Extraction of lead from spent leaded glass in alkaline solution by mechanochemical reduction. *Hydrometallurgy* **2016**, *165*, 312–317. [CrossRef]
25. Hu, J.; Dong, H.; Song, S. Research on Recovery Mechanism and Process of Waste Thermosetting Phenolic Resins Based on Mechanochemical Method. *Adv. Mater. Sci. Eng.* **2020**, *2020*, 1384194. [CrossRef]
26. Wu, W.; Liu, G.; Cheng, H. Preparation and Performance Analysis of Regenerated Materials for Thermosetting Polyurethane Based on Coupled Thermo-mechanical Model. *Chin. J. Mech. Eng.-Engl. Ed.* **2016**, *27*, 2540–2546, 2555.
27. Li, H.; Xu, B.; Lu, G.; Du, C.; Huang, N. Multi-objective optimization of PEM fuel cell by coupled significant variables recognition, surrogate models and a multi-objective genetic algorithm. *Energy Convers. Manag.* **2021**, *236*, 114063. [CrossRef]
28. Quadrini, F.; Bellisario, D.; Santo, L. Recycling of thermoset polyurethane foams. *Polym. Eng. Sci.* **2013**, *53*, 1357–1363. [CrossRef]

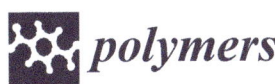

Article

Effect of Engineered Biomaterials and Magnetite on Wastewater Treatment: Biogas and Kinetic Evaluation

Gloria Amo-Duodu, Emmanuel Kweinor Tetteh *, Sudesh Rathilal, Edward Kwaku Armah, Jeremiah Adedeji, Martha Noro Chollom and Maggie Chetty

Green Engineering and Sustainability Research Group, Department of Chemical Engineering, Faculty of Engineering and The Built Environment, Durban University of Technology, Durban 4001, South Africa; gamoduodu04@gmail.com (G.A.-D.); rathilals@dut.ac.za (S.R.); edwardkarmah@gmail.com (E.K.A.); jerry_4real@live.com (J.A.); mnchollom@gmail.com (M.N.C.); chettym@dut.ac.za (M.C.)
* Correspondence: ektetteh34@gmail.com or emmanuelk@dut.ac.za

Abstract: In this study, the principle of sustaining circular economy is presented as a way of recovering valuable resources from wastewater and utilizing its energy potential via anaerobic digestion (AD) of municipality wastewater. Biostimulation of the AD process was investigated to improve its treatability efficiency, biogas production, and kinetic stability. Addressing this together with agricultural waste such as eggshells (CE), banana peel (PB), and calcined banana peels (BI) were employed and compared to magnetite (Fe_3O_4) as biostimulation additives via 1 L biochemical methane potential tests. With a working volume of 0.8 L (charge with inoculum to substrate ratio of 3:5 v/v) and 1.5 g of the additives, each bioreactor was operated at a mesophilic temperature of 40 °C for 30 days while being compared to a control bioreactor. Scanning electron microscopy and energy dispersive X-ray (SEM/EDX) analysis was used to reveal the absorbent's morphology at high magnification of 10 kx and surface pore size of 20.8 µm. The results showed over 70% biodegradation efficiency in removing the organic contaminants (chemical oxygen demand, color, and turbidity) as well as enhancing the biogas production. Among the setups, the bioreactor with Fe_3O_4 additives was found to be the most efficient, with an average daily biogas production of 40 mL/day and a cumulative yield of 1117 mL/day. The kinetic dynamics were evaluated with the cumulative biogas produced by each bioreactor via the first order modified Gompertz and Chen and Hashimoto kinetic models. The modified Gompertz model was found to be the most reliable, with good predictability.

Keywords: anaerobic digestion; biosorbent; biostimulant; magnetite; nanoparticles; kinetic model

1. Introduction

Today's energy-intensive development has led to a surging demand for fossil fuels, which generate environmental pollution and impacts the ecosystem through global warming [1,2]. This has stimulated the search for alternative energy sources that are both sustainable and eco-friendly, to mitigate the environmental crisis [3–5]. Therefore, the exploration of cost-effective technology and sustainable energy resources in wastewater settings, to generate biogas to boost the water economy and its reclamation for reuse, has become important. In addition, the environmental challenge and cost involved in discharging biowaste (banana peels, eggshell, orange peels, sludge, etc.) [6,7], underpin its importance in being engineered as a biostimulant for wastewater treatment and biogas enhancement.

Generally, anaerobic digestion (AD) is considered as one of the most valuable techniques that converts the organic matter present in the biowaste to renewable energy in the form of methane (CH_4)-enriched biogas [8–10]. When the bioreactor is run at optimal conditions, production of the bioenergy such as methane (60–70%) and stabilised digestate by AD creates economic opportunities and eases pollution [8,11,12]. The AD process utilizes microorganism degradation potential in an ecologically sustainable [13–15], odour-reducing, and pathogenic organism-degrading process, especially in reactors running at mesophilic

Citation: Amo-Duodu, G.; Tetteh, E.K.; Rathilal, S.; Armah, E.K.; Adedeji, J.; Chollom, M.N.; Chetty, M. Effect of Engineered Biomaterials and Magnetite on Wastewater Treatment: Biogas and Kinetic Evaluation. *Polymers* **2021**, *13*, 4323. https://doi.org/10.3390/polym13244323

Academic Editors: Alexey Iordanskii and Vetcher Alexandre

Received: 30 September 2021
Accepted: 4 November 2021
Published: 10 December 2021

Publisher's Note: MDPI stays neutral with regard to jurisdictional claims in published maps and institutional affiliations.

Copyright: © 2021 by the authors. Licensee MDPI, Basel, Switzerland. This article is an open access article distributed under the terms and conditions of the Creative Commons Attribution (CC BY) license (https://creativecommons.org/licenses/by/4.0/).

(25–45 °C) and thermophilic (>45 °C) temperatures [14,15]. Furthermore, produced biogas often contains impurities such as H_2S and CO_2, which lower the calorific value of biogas and are detrimental to equipment like pipes and combustion engines [3,16]. Some of the intriguing techniques used in addressing this include co-digestion, pre-treatment, different designs of reactor configurations, and the use of additives to stimulate bacteria growth and prevent inhibitory effects [2,8,16–19].

In recent studies, micro and macro nutrients were found to stimulate methane production and sustain the AD process up to a critical concentration range, after which inhibition occurs [7,9,20]. Other researchers examined the impact of integrating one or two metals into anaerobic biogas production, whereas some elements may have antagonistic or synergistic effects [7,21,22]. Goli et al. [23] increased the production of biodiesel by comparing other homogenous and heterogenous calcium oxide (CaO) catalysts produced from chicken eggshell. Sridhar [24] also studied the use of both calcined and natural eggshell to remove heavy metals (Pb and Cu) from real automotive wastewater, where high efficiency was attained. Amo-Duodu et al. [25] reported 80% increase in biogas yield from sugar refinery wastewater by adding metallic nanoparticles (Fe, Ni, and Cu) at a hydraulic retention time of 10 days and mesophilic temperature of 40 °C. Despite the potential benefits of trace metals on biogas production [2,15,26] and wastewater treatment using the most prevalent materials such as activated carbon, alumina, and silica, [27,28], their extensive use is hampered by cost [29,30]. Therefore, exploring less expensive biomaterials as a source of nutrients and biostimulant for the AD process can make it more economically feasible.

Consequently, advancement of nanomaterials for wastewater treatment is associated with many roadblocks including regulatory challenges, technical hurdles, and public perception [4,31–33]. In addition, there are uncertainties about the impact of nanomaterials on the environment [34–36] and scarcity of comprehensive cost–benefit analyses, as compared to existing technologies [37,38]. In addressing these challenges, agro-wastes such as coconut shells, banana and orange peels, and eggshells are gaining attention for wastewater purification and adsorption as biochar or activated carbons [6,39]. Although agro-wastes are generated in large quantities annually, posing a threat to the environment, landfills can act as biomaterials [6,7,37]. Table 1 presents some reported materials used in wastewater treatment. For instance, banana peels have been used as absorbents in adsorption of heavy metals from wastewater [24,40]. Egg shell has been used in coagulation processes for the removal of heavy metals [23]. On the other hand, magnetite (Fe_3O_4) has been used as a biostimulant for biogas production [41].

Table 1. Biomaterials and magnetite used for wastewater remediation and energy production as compared to current study.

Biosorbent	Waste or Raw Material Used	Treatment Process	Results	Reference
Calcined banana Peels	Synthetic water prepared by diluting concentrated Mn(VII) and Fe(II) with deionised water	Adsorption	The biochar from banana peels was treated with pristine and phosphoric acid; the phosphoric acid pre-treatment had a better absorption efficiency than the pristine pre-treatment.	[40]
Raw Banana Peels	Automotive industrial wastewater Dirty water (river and rainwater)	Primary water treatment Water purification	The process had the highest removal of copper (93.52%) and lead (87.44%). The physical test met the quality conditions except for temperatures that exceeded the quality conditions of the maximum standard value.- Bacteriologically there were a lot of total coliforms exceeding the maximum standard conditions.	[24]

Table 1. Cont.

Biosorbent	Waste or Raw Material Used	Treatment Process	Results	Reference
Egg shell	Real electroplating wastewaters containing Cr, Pb and Cd and synthetic wastewater containing heavy metals (Cr, Pb and Cd)	Jar-test coagulation process	The reuse of waste eggshell in the removal of toxic heavy metals, i.e., Cd and Cr in synthetic wastewater was much enhanced when calcined eggshell was added; however, removal of Pb was rather favourable with natural eggshell.	[23]
Fe_3O_4	Anaerobic sludge acquired from an Anaerobic-Anoxic-Oxic (AAO) reactor	Batch anaerobic digestion process	There was a 28% increase in biogas yield and COD removal of 14,760 mg/L in the reactor with Fe_3O_4	[41]
* Calcined banana Peels	Domestic and municipal wastewater	Biochemical methane potential (BMP) test	32.258 mL/day biogas yield, 73.53%, 71.05% and 88.93% COD, color and turbidity removal, respectively.	This study
* Raw Banana Peels	Domestic and municipal wastewater	BMP	33.226 mL/day biogas yield, 72.69%, 70.35% and 94.13% COD, color and turbidity removal, respectively	This study
* Egg shell	Domestic and municipal wastewater	BMP	32.581 mL/day biogas yield, 73.11%, 69.65% and 94.26% COD, color and turbidity removal, respectively.	This study
* Fe_3O_4	Domestic and municipal wastewater	BMP	37.807 mL/day biogas yield, 92.59%, 74.86% and 94.13% COD, color and turbidity removal, respectively.	This study

South Africa is estimated to produce 54.2 million tons of waste (municipal, commercial, and industrial) annually [41,42]. About 10% of this is recovered and recycled for other purposes, while the remaining 90% is landfilled or discarded [41]. There is a pressing concern for exploring the possibility of improving AD biogas production via the addition of absorbents (banana peels and eggshells) and magnetic biochar (made up of banana peel and powered magnetite). Therefore, this study seeks to explore the feasibility of calcined banana peels (BI), uncalcined banana peels (PB), magnetite (Fe_3O_4), and eggshell (CE) to improve the biogas yield via biochemical methane potential (BMP) tests. In addition, the degree of degrading the organics in the wastewater was kinetically studied using the first order, modified Gompertz, and Chen and Hashimoto kinetic models to establish their performance.

2. Materials and Methods

2.1. Engineered Biomaterials and Characterisation Techniques

The raw eggs and bananas, purchased from a local South African market in Durban, KwaZulu Natal Province, were washed with distilled water. The assay of preparing biochar described by Li et al. [6] was followed. The crushed eggshells (CE) and banana peels (PB) were then oven dried at 80 °C for 24 h. The banana peels-based biochar (BI) was prepared from the dehydrated CE, by soaking 5 g CE in 50 mL 20% vol H_3PO_4 solution for 1 h [6]. This was further calcined at a furnace temperature of 550 °C for 1 h. The physical morphologies and elemental compositions of the biomaterials were analysed using scanning electron microscopy and energy dispersive X-ray (SEM/EDX) analysis. The samples were firstly sputter coated with carbon to do the analysis. This was outsourced by using the University of Cape Town, South Africa SEM/EDX equipment (Nova NanoSEM coupled with EDT and TLD detector) operated at an acceleration voltage of 5 kV with a magnification range of 10–50 k.

2.2. Synthesis and Characteristics of the Magnetite (Fe_3O_4)

The magnetite (Fe_3O_4) used in this study was prepared by following the co-precipitation assay by Tetteh and Rathilal [5]. The chemicals used included sodium hydroxide pellets (NaOH), ferrous sulphate heptahydrate ($FeSO_4 \cdot 7H_2O$), oleic acid (surfactant), and ferrous chloride hexahydrates ($FeCl_3 \cdot 6H_2O$), which were all analytical grade supplied by Sigma Aldrich, South Africa. Stock solutions of 0.4 M Fe^{3+} and 0.2 M Fe^{2+} were first prepared by weighing 108.12 g and 55.61 g respectively and dissolving them with 1 L deionized water. In addition, 3 M NaOH stock solution was prepared by dissolving 199.99 g of NaOH with 1 L deionized water. The magnetite (Fe_3O_4) was then prepared in the volume ratio of 1:1, by using 500 mL of Fe^{3+} and Fe^{2+} stock solutions each. To ensure homogeneity, the solution was then stirred on a magnetic hotplate continuously, while adding 4 mL of oleic acid dropwise. The pH of the solution was then adjusted to a pH of 12 with 250 mL of the 3M NaOH until black precipitate was formed. Afterwards, thick black precipitate was then heated (ageing) at 70 °C. The supernatant was decanted, and the precipitate washed thrice with distilled water and ethanol to get rid of any form of unwanted particles. An oven drying was carried out at 80 °C for 12 h and then furnace calcination of 550 °C for 1 h. Samples obtained were characterized via the Brunauer–Emmett–Teller (BET) analysis by using Micromeritics TriStar II Plus equipment (Durban, South Africa) coupled with Tristar Plus software version 3.01. The carrier gases used were helium and nitrogen. Prior to the analysis, the sample was degassed at a temperature of 400 °C for 24 h. It was allowed to cool and then kept under nitrogen gas at a pressure of 5 mmHg for 24 h. The magnetite surface area of 27.59 m^2/g, pore volume of 0.008 cm^3/g and pore size of 1.48 nm was achieved.

2.3. Wastewater and Inoculum Samples

The wastewater sample (substrate) was collected from a biofiltration sample point of a local South African municipality wastewater treatment plant in the KwaZulu-Natal province. The activated sludge sampled from the anaerobic digester point source was used as the inoculum. An inoculum to substrate ratio of 3:5 (volume basis) was used for each 1 L bio-digester. Standard methods for the examination of water and wastewater [43] were employed to characterise the substrate and inoculum in triplicates with the results shown in Table 2. Color and turbidity were analysed with the spectrophotometer (HACH DR3900, Hach Company, Loveland, CO, USA) and turbidity meter (HACH 210, Hach Company, Loveland, CO, USA), respectively. Using the COD high range vials (HACH), 0.2 mL of the samples were measured and poured into the COD vials. It was then digested at 150 °C for 2 h. After the digestion, the vials were cooled at room temperature and the COD was measured using the spectrophotometer (HACH DR3900, Hach Company, Loveland, CO, USA).

Table 2. Characteristics of wastewater and sludge samples.

Parameters	Results
Chemical oxygen demand (COD) (mg/L)	2380 ± 57.6
Color (Pt.Co)	57 ± 12.5
Turbidity (NTU)	17.32 ± 2.2
Total solids (TS) (mg TS/L)	204.5 ± 24.6
Volatile solids (VS) (mg VS/L)	106 ± 32.6
pH	6.59 ± 1.3

2.4. Biochemical Methane Potential (BMP) Test

A lab-scale batchwise anaerobic system was setup for the BMP test and operated according to Hulsemann et al. [13]. In Figure 1, the BMP test setup consists of a biodigester, a biogas collecting system, and temperature-controlling (water bath) units. Five 1 L Schott bottles with a working volume of 800 mL were used as biodigesters, closed with Teflon caps (three ports). The outlet gas nozzle was connected to the biogas collecting unit via the downward displacement technique using a 1 L graduated cylinder placed upside down in another 5 L cylinder filled with water. Each bioreactor was charged with 1.5 g biomaterials, wastewater (500 mL), and sludge (300 mL) as presented in Table 3. Each bioreactor was purged with nitrogen gas for 2 min, to create the anaerobic conditions. To maintain the temperature at mesophilic conditions (40 ± 2.5 °C) for 30 days, each bioreactor was immersed in the 20 L water bath (WBST0001, United Scientific, Cape town, South Africa) system. After recording the daily biogas produced, each bio-digester was manually shaken for 2 min. All the experiments were done in triplicates, and the results are averagely presented.

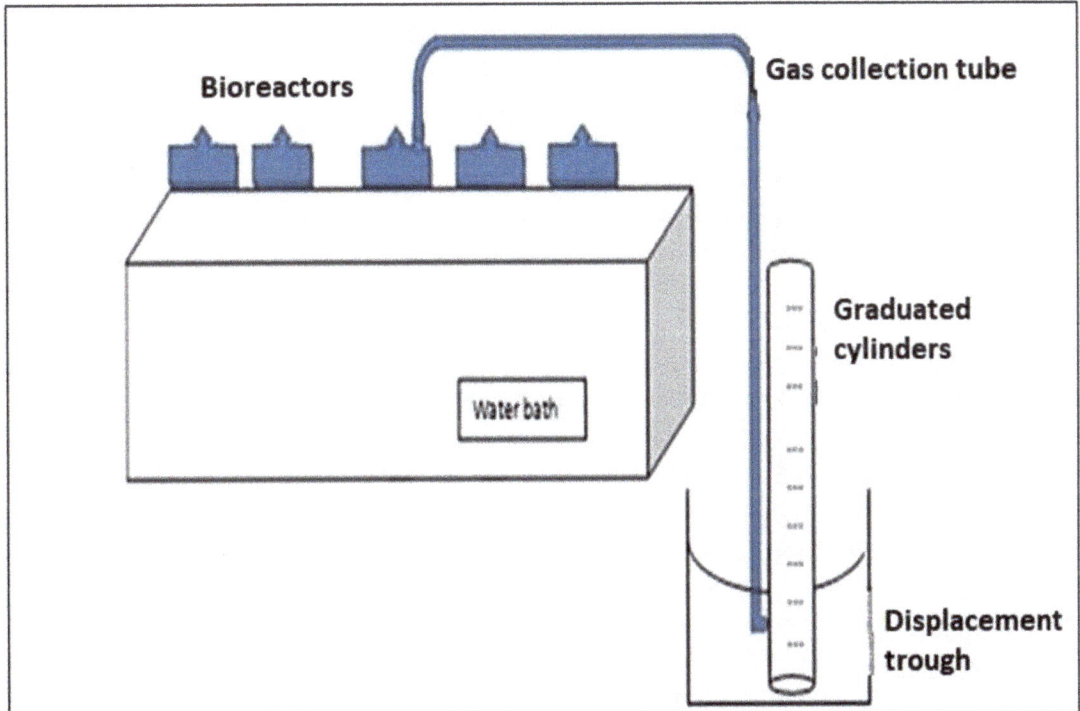

Figure 1. A biochemical methane potential (BMP) test set-up.

Table 3. The biosorbent loading for the BMP test.

Set-Up	Biosorbent Type	Biosorbent Loading (g)	Wastewater (mL)	Inoculum (mL)
A	Calcined banana peels (BI)	1.5	500	300
B	Crushed eggshell (CE)	1.5	500	300
C	Banana peels (PB)	1.5	500	300
D	Magnetite (Fe_3O_4)	1.5	500	300
E	Control (no loading)	n/a	500	300

After 30 days of the experiment, the pH, COD, color, turbidity, TS and VS for each bioreactor was determined. The standard Equation (1) of calculating the efficiency of the contaminant removal was employed.

$$Reactor\ efficiency = \left(\frac{C_i - C_f}{C_i}\right) \times 100 \quad (1)$$

where C_i = contaminant in the influent and C_f = contaminant in the effluent. The cumulative biogas production data obtained was evaluated by first-order and modified Gompertz Equations (2) and (3), respectively.

2.5. Kinetic Study of BMP System

The quantitative comparison of biogas production during the batch mesophilic AD process with different biomaterials was modelled with first order (2), modified Gompertz (3) and Chen and Hashimoto (4) models adapted from Budiyono et al. [44] and Mu et al. [45].

$$Y(t) = Y_m \left[1 - \exp(-kt)\right] \quad (2)$$

$$Y(t) = Y_m \cdot \exp\left(-\exp\left[\frac{2.7183 R_{max.}}{Y_m}[\lambda - t]\right] + 1\right) \quad (3)$$

$$Y(t) = Y_m \left(1 - \frac{K_{CH}}{HRT \times R_{max} + K_{CH} - 1}\right) \quad (4)$$

where
Y(t) is cumulative specific biogas yield (mL/g COD),
Ym is the maximum biogas production (mL/g COD),
λ is lag phase period or minimum time to produce biogas (days),
t is the cumulative time for biogas production (days),
k = R_{max}/ym (1/day),
R_{max} is the maximum specific substrate uptake rate (mL/g COD.day),
k is a first-order rate constant (1/d) and
K_{CH} is the Chen and Hashimoto kinetic constant.

3. Results and Discussion

3.1. Surface Morphology

Figure 2 presents the scanned images of the biomaterials (A) BI- calcined banana peels, (B) CE-crushed egg shell, (C) PB-banana peels, and (D) magnetite (Fe_3O_4). The micrographs (Figure 2A–C) at 5 µm showed the same high magnification of 10 kx and view field of 20.8 µm. This revealed patchy fragmented surfaces with proportioned ridge-like strands of amorphous structures [40]. It was observed that the micrograph (Figure 2D) at the microscale of 10 µm and similar high magnification of 10 kx and view field of 20.8 µm had regular cellular structure. This makes the appearance of the magnetite (Figure 2D) different to that of the biomaterials (Figure 2A–C). The calcination temperature at 550 °C for 1 h changed the banana's organics (BI) to produce brittle and flaky hydrochar, leaving residual concave plates (Figure 2A) to differ from the raw banana (PB) (Figure 2C). The surface area and porosity (6 mm) of the magnetite (Figure 2D) were found to be the highest, followed by BI (Figure 2A) of 5.48 mm as compared to PB (Figure 2C) of 5.21 mm and CE (Figure 2B) of 5.33 mm. The creation of a broad pore size distribution (Fe_3O_4 > BI > CE > PB), ranging from narrow microspores to wide mesopores, may be linked to the high calcination temperature (550 °C), which improved the liquid–solid adsorption capacity [46]. This affirms treating PB with oxidizing solutions before calcination reduces amorphous cellulose transition [24] and therefore improves the surface adhesive properties by removing superfluous impurities from the rough surface [40,46].

Figure 3 presents the EDX spectrum with tabulated elemental distribution of (A) BI, (B) CE, (C) PB and (D) Fe_3O_4. This revealed that carbon (30–55% C) and oxygen (25–50% O) were the most dominant elements in all the biomaterials. Apparently, the presence of the carbon (C) in the magnetite (Figure 3D) EDX spectra was a result of the samples being coated under carbon gases. As revealed by the SEM image (Figure 2D) with bulbous morphology of the aggregates, suggests that the iron minerals were partly covered by carbon particles, which might be due to the calcination treatment and the carbon coating of the sample prior to analysis. Aside from that, CE (Figure 3B) was found to constitute calcium carbonate (25% $CaCO_3$). Likewise, Fe_3O_4 (Figure 3D) had 28% Fe and the PB (Figure 3C) had 2% K. Conversely, the EDX and elemental composition were denatured in the calcined banana peel (BI), which revealed 17% K and 2% Cl as well as less than 1% P and 1% Si. The additional components in the BI could have increased its surface area, facilitating its biosorption and reusability [40].

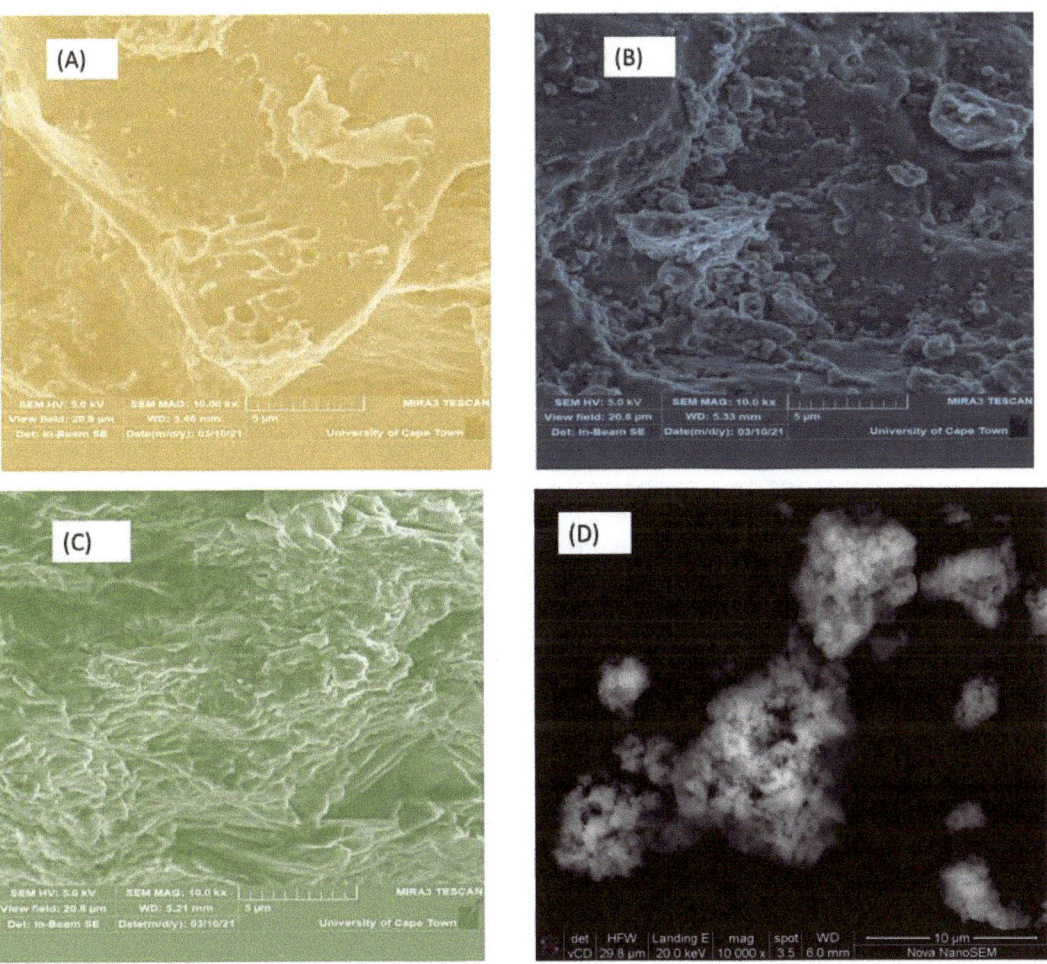

Figure 2. SEM images of biomaterials with view field of 20.8 μm at high magnification of 10 kx; (**A**) BI, (**B**) CE, (**C**) PB and (**D**) Fe_3O_4.

3.2. Biodegradation Efficiency

After 30 days of digestion, the degree of biodegradation was assessed based on the contaminant removal from the wastewater with respect to color, turbidity, and COD for each bioreactor. Figure 4 shows the effect of the biomaterials on the bioreactors (A–D) in removing above 70% contaminants as compared to that of the control bioreactor (E) with <70% efficiency. Bioreactor D showed a tremendous performance with contaminant removal of 92.59%, 74.86%, and 94.13% of COD, color, and turbidity removal, respectively. This was followed by bioreactor B with 72.69% COD, 70.35% color, and 94.13% turbidity > bioreactor

C (73.11% COD, 69.65% color, and 94.26% turbidity) > bioreactor A (73.53% COD, 71.05% color, and 88.93% turbidity) > bioreactor E (59.83% COD, 45.61% color and 78.55% turbidity). This result supports that bioreactor D being charged with the Fe_3O_4 had high surface area and catalytic properties as revealed by the SEM/EDX result (Figure 2D). Evidently, the surface area and adsorptive capacity of the biomaterials (Figure 2A–C) also influenced the biodegradation performance of bioreactors A–C as compared to the control bioreactor E. This agrees with other researchers reporting that biomaterials have high catalytic properties, large surface areas, pore size, and good adsorption properties, all of which can influence treatability performance of wastewater [6,40]. In addition, eggshell and banana peels have been employed by other researchers for remediation of wastewater and removal of heavy metals, and were found to be successful [6,24,46], consistent with the current study.

3.3. Effect of Biosorbent on Biogas Yield

Results of the average and cumulative biogas yield obtained after the 30 days of digestion are summarised in Table 4. The increasing order of the bioreactors' average biogas yield was found as follows: D (40 mL/day) > B (34 mL/day) > C (33 mL/day) > A (32 mL/day) > E (25 mL/day). Bioreactor D was found to be the most efficient, with an average daily biogas production of 40 mL/day and cumulative yield of 1117 mL/day. Thus, the presence of the Fe_3O_4 additives enriched the substrate nutrients, which stimulated the microbial activities of the methanogens to increase the biogas production [47,48]. Again, studies by other researchers have shown similar observations, where the addition of the Fe_3O_4 [49,50] increased biogas and methane yield, as compared to the control.

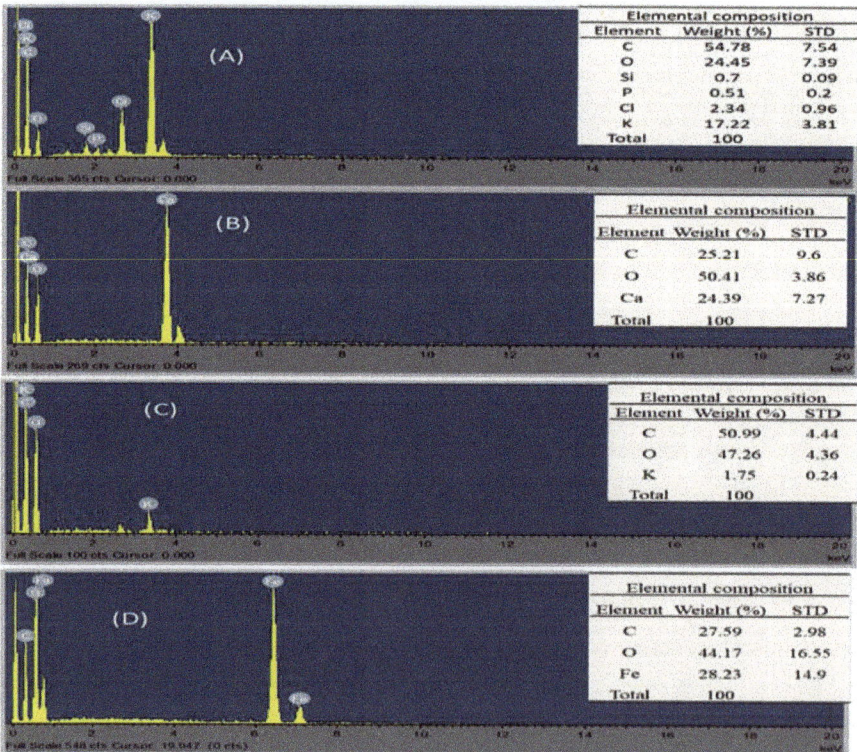

Figure 3. EDX spectrum images and tabulated elemental distribution of biomaterials with view field of 20.8 μm at high magnification of 20 kx. (**A**) BI, (**B**) CE, (**C**) PB and (**D**) Fe_3O_4.

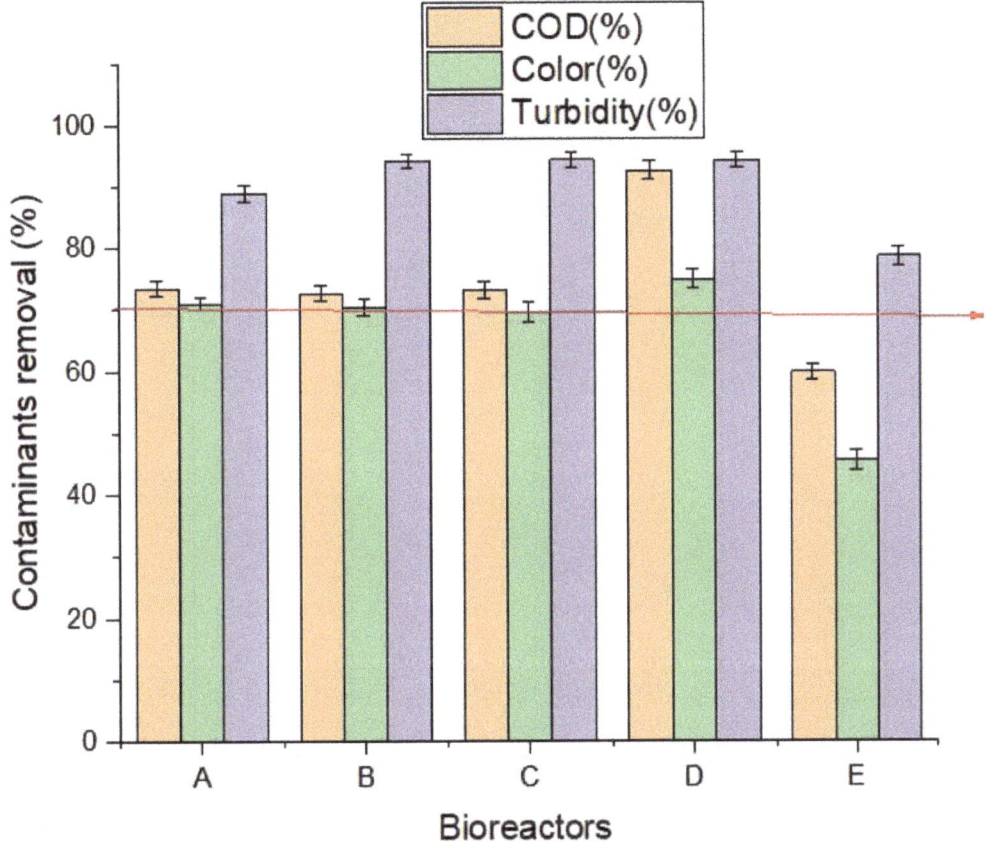

Figure 4. Biodegradation efficiency of bioreactors (A–E) for HRT of 30 days.

Table 4. Average and cumulative biogas yield for HRT 30 days.

Bioreactor	Biosorbent Added (g)	Average Biogas Yield (mL/day)	Cumulative Biogas Yield (mL/day)
A	1.5	32	1000
B	1.5	34	1030
C	1.5	33	1010
D	1.5	40	1117
E	No additives	25	775

Figure 5 presents the specific daily biogas production curve for the bioreactors (A–E). It was observed that during the first five days of the digestion process, the yield was below 20 mL/day which can be attributed to the fact that the microorganisms were still getting acclimatized to the environment (lag phase). From day 6 to 20, it was observed that there was a gradual increase in the biogas yield from 20 mL/day to about 80 mL/day for bioreactors A–D (exponential phase). All the bioreactors (A–E) attained maximum yield (high peak) between days 20–25, before yields started to decline from day 25 to 30 (death phase). From Table 1, comparing this study results to that of previous studies affirms that the use of biomaterials and magnetite can influence the AD process for biogas production and wastewater remediation [50].

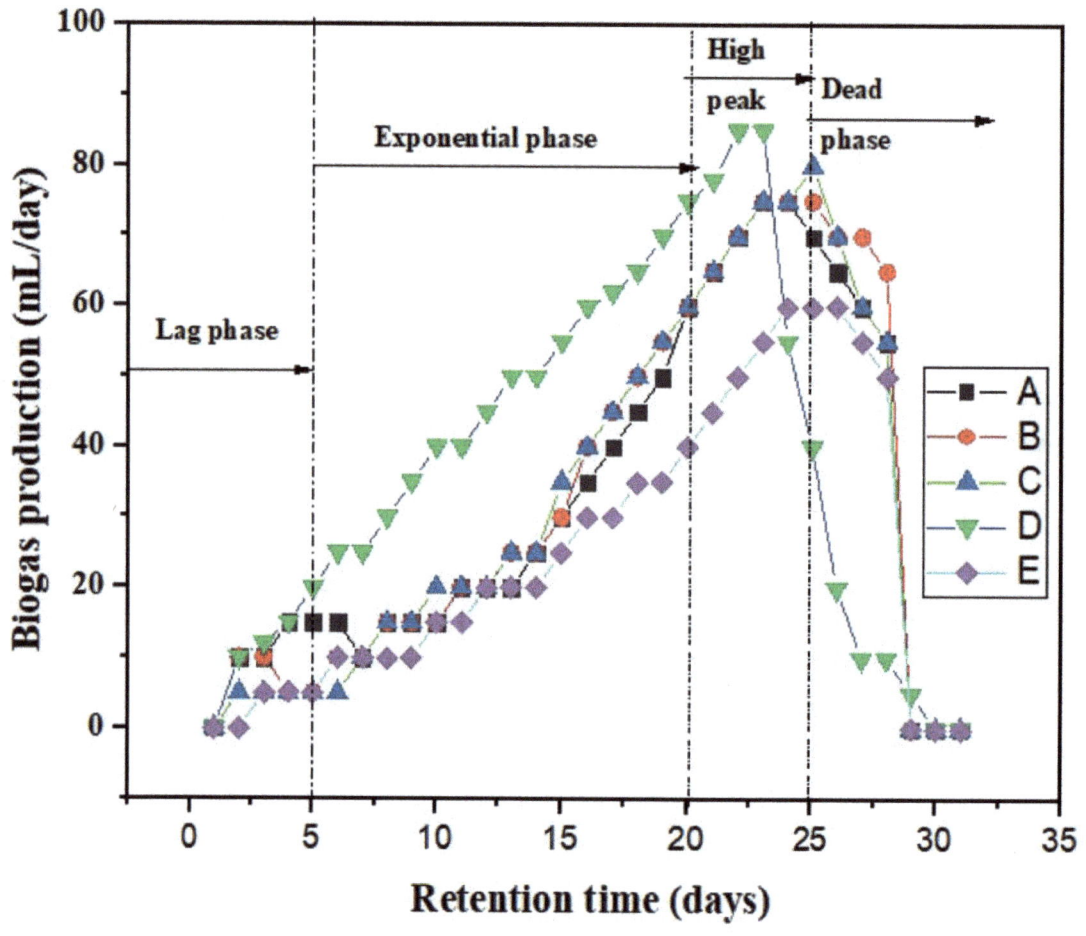

Figure 5. Daily biogas production of bioreactors (A–E) for HRT of 30 days.

3.4. Kinetic Model of the Biogas Production

The kinetic study was carried out to ascertain the impact of the kinetic dynamics of the biomaterials in the AD process. To evaluate the fitness of the first order (2), modified Gompertz (3) and Chen and Hashimoto kinetic (4) models, the cumulative biogas data (sum of daily production) was plotted against time as presented in Figure 6. The results obtained from the models are presented respectively in Tables 5–7. All the models' predictiveness were found to be significant with less than 5% deviation at 95% confidence level. Interestingly, all the kinetic models show that the addition of the Fe_3O_4 additives in bioreactor D increased the biogas production. This is because Fe_3O_4 addition reduced the detrimental effects of sulphides on methanogenesis by forming FeS precipitates. In addition, differences in lag phase times were observed for bioreactors A–E, while that of bioreactor D recorded the lowest at 15 days (Table 6) by the modified Gompertz model. This is explained by the fact that the microbial communities of bioreactor D were well adapted to their environment. Against the background of the AD process mechanism, the initial hydrolysed monomers and subsequent produced volatile fatty acids were rapidly consumed by the acidogenic and the methanogenic bacteria, respectively, with increased biogas production [50]. Among the applied kinetic models (Tables 5–7), the

Gompertz model that predicted the cumulative biogas production with the smallest value (66–92 mL/day) had a perfect fit of regression coefficient (R^2) within 0.9906–0.9939, affirming other reported studies [16,44].

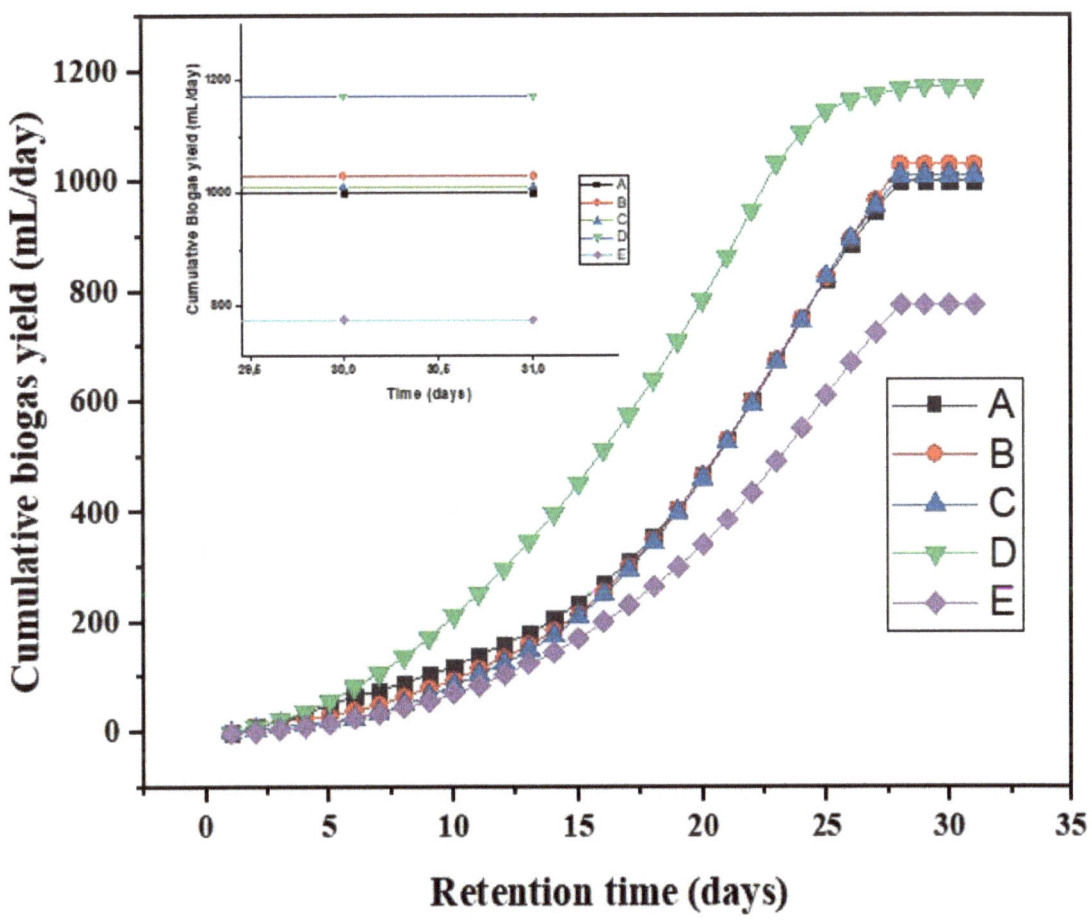

Figure 6. Cumulative biogas yield for bioreactors A–E for HRT of 30 days.

Table 5. Summary of the First order kinetic models for bioreactors A–E.

Set-Up	A	B	C	D	E
Yt (mL/g COD)	1×10^3	1.03×10^3	1.01×10^3	1.17×10^3	775
Ym (mL/g COD)	3.68×10^5	2.06×10^5	3.71×10^5	2.96×10^5	3.08×10^5
k (1/day)	8×10^{-4}	1.4×10^{-4}	8×10^{-4}	1.3×10^{-4}	7×10^{-4}
SSR	5.92×10^5	7.44×10^5	7.64×10^5	4.57×10^5	4.19×10^5
R^2	0.927	0.919	0.919	0.968	0.923
Predicted value (mL/g COD)	874	878	864	1.20×10^3	655
Difference between measured (Yt) and predicted values	126	152	146	31	120

Table 6. Summary of the modified Gompertz kinetic models for bioreactors A–E.

Set-Up	A	B	C	D	E
Y(t) (mL/g COD)	1×10^3	1.03×10^3	1.01×10^3	1.17×10^3	775
Ym (mL/g COD)	1.89×10^3	1.75×10^3	1.57×10^3	1.46×10^3	1.40×10^3
λ(days)	23.58	22.49	21.45	15.79	23.33
k (1/day)	0.081	0.094	0.106	0.123	0.884
SSR	3.79×10^5	3.03×10^5	2.71×10^5	4.16×10^5	1.64×10^5
R^2	0.991	0.993	0.994	0.993	0.993
Predicted value (mL/g COD)	1.09×10^3	1.12×10^3	1.09×10^3	1.25×10^3	841
Difference between measured (Yt) and predicted values (mL)	92	86	81	78	66

Table 7. Summary of the Chen and Hashimoto kinetic model for bioreactors A–E.

Set-Up	A	B	C	D	E
Yt (mL)	1×10^3	1.03×10^3	1.01×10^3	1.17×10^3	7.75×10^2
Ym (mL)	5.33×10^5	2.08×10^5	4.12×10^5	1.81×10^5	1.82×10^5
Rmax (mL/day)	3.6×10^{-5}	2.2×10^{-5}	1.9×10^{-5}	3.7×10^{-5}	0.16
K_{CH} (1/day)	6.73	1.59	0.28	1.71	2×10^{-5}
SSR	5.92×10^5	7.42×10^5	7.65×10^5	4.56×10^5	4.21×10^5
R^2	9.28×10^{-1}	9.19×10^{-1}	9.2×10^{-1}	9.69×10^{-1}	9.24×10^{-1}
Predicted value (mL/g COD)	876	880	864	1.20×10^3	654
Difference between measured (Yt) and predicted values (mL)	124	150	164	31	121

4. Conclusions

The potential of three biomaterials (designated as CE, PB, and BI) as biostimulant additions for anaerobic digestion (AD) of municipal wastewater into biogas was investigated in this study in comparison to magnetite (Fe_3O_4). The additives demonstrated great potential for the abatement of high strength organic wastewater with more than 70% degree of reduction efficiency. Each additive had distinctive adsorptive pores as reflected by the SEM/EDX surface area obtained. This made the Fe_3O_4 with the highest surface area of 6 mm more advantageous for creating a broad pore size distribution than the rest (Fe_3O_4 > BI > CE > PB). In addition, the outcome of assessing the impact of the biosorbent on biogas production and AD process enhancement, conducted at 40 °C and HRT of 30 days was successful. Above all, the addition of 1.5 g of Fe_3O_4 nanoparticles assigned to bioreactor D maximised the average daily biogas (40 mL/day), as compared to CE (34 mL/day) > PB (3381 mL/day) > BI (32 mL/day) > control (25 mL/day). In addition, the modified Gompertz model was found to be appropriate for the prediction of biogas production as compared to the first order and Chen and Hashimoto kinetic models. The prospect of reinforcing this finding is to be encouraged, with optimised lab-scale and pilot scale plants on specified wastewater settings.

Author Contributions: Conceptualization, E.K.T. and G.A.-D.; methodology, E.K.T., G.A.-D., J.A., E.K.A.; validation, E.K.T., E.K.A., M.N.C.; formal analysis, G.A.-D. and J.A.; resources, E.K.T., S.R., G.A.-D.; investigation, E.K.T. and G.A.-D.; data curation, E.K.T., G.A.-D., J.A.; writing of the original draft preparation, E.K.T. and G.A.-D.; writing review and editing, E.K.A., M.N.C., M.C., S.R.; supervision, S.R., M.N.C., M.C.; project administration, E.K.T. and S.R., funding acquisition, E.K.T. and S.R. All authors have read and agreed to the published version of the manuscript.

Funding: This research was funded by the Water Research Commission of South Africa under project identification WRC Project: C2019/2020-00212.

Institutional Review Board Statement: Not applicable.

Informed Consent Statement: Not applicable.

Data Availability Statement: Not applicable.

Acknowledgments: The authors wish to thank the Durban University of Technology, Green Engineering and Sustainability Research Group, and the Water Research Commission of South Africa for their support on the project identification WRC Project: C2019/2020-00212. The authors also wish to thank the National Research Foundation for their scholarship grant number 130143.

Conflicts of Interest: The authors declare no conflict of interest and the funders had no role in the design of the study; in the collection, analyses, or interpretation of data; in the writing of the manuscript, or in the decision to publish the results.

References

1. Abdelwahab, T.A.M.; Mohanty, M.K.; Sahoo, P.K.; Behera, D. Application of nanoparticles for biogas production: Current status and perspectives. *Energy Sources Part A Recover. Util. Environ. Eff.* **2020**. [CrossRef]
2. Zaidi, S.A.A.; RuiZhe, F.; Shi, Y.; Khan, S.Z.; Mushtaq, K. Nanoparticles augmentation on biogas yield from microalgal biomass anaerobic digestion. *Int. J. Hydrogen Energy* **2018**, *43*, 14202–14213. [CrossRef]
3. Msibi, S.S.; Kornelius, G. Potential for domestic biogas as household energy supply in South Africa. *J. Energy S. Afr.* **2017**, *28*, 1–13. [CrossRef]
4. Omara, A.E.-D.; Elsakhawy, T.; Alshaal, T.; El-Ramady, H.; Kovács, Z.; Fári, M. Nanoparticles: A Novel Approach for Sustainable Agro-productivity. *Environ. Biodivers. Soil Secur.* **2019**, *3*, 30–40. [CrossRef]
5. Chollom, M.; Rathilal, S.; Swalaha, F.; Bakare, B.; Tetteh, E. Removal of Antibiotics During the Anaerobic Digestion of Slaughterhouse Wastewater. *Int. J. Sustain. Dev. Plan.* **2020**, *15*, 335–342. [CrossRef]
6. Li, X.; Wang, C.; Zhang, J.; Liu, J.; Liu, B.; Chen, G. Preparation and application of magnetic biochar in water treatment: A critical review. *Sci. Total. Environ.* **2020**, *711*, 134847. [CrossRef] [PubMed]
7. Fermoso, F.G.; van Hullebusch, E.; Collins, G.; Roussel, J.; Mucha, A.P.; Esposito, G. *Trace Elements in Anaerobic Biotechnologies*; IWA Publishing: London, UK, 2019. [CrossRef]
8. Tetteh, E.K.; Rathilal, S. Kinetics and Nanoparticle Catalytic Enhancement of Biogas Production from Wastewater Using a Magnetized Biochemical Methane Potential (MBMP) System. *Catalysts* **2020**, *10*, 1200. [CrossRef]
9. Ángeles, R.; Vega-Quiel, M.J.; Batista, A.; Fernández-Ramos, O.; Lebrero, R.; Muñoz, R. Influence of biogas supply regime on photosynthetic biogas upgrading performance in an enclosed algal-bacterial photobioreactor. *Algal Res.* **2021**, *57*, 102350. [CrossRef]
10. Nethengwe, N.S.; Uhunamure, S.E.; Tinarwo, D. Potentials of biogas as a source of renewable energy: A case study of South Africa. *Int. J. Renew. Energy Res.* **2018**, *8*, 1112–1123.
11. Budzianowski, W.M.; Postawa, K. Renewable energy from biogas with reduced carbon dioxide footprint: Implications of applying different plant configurations and operating pressures. *Renew. Sustain. Energy Rev.* **2017**, *68*, 852–868. [CrossRef]
12. Michailos, S.; Walker, M.; Moody, A.; Poggio, D.; Pourkashanian, M. Biomethane production using an integrated anaerobic digestion, gasification and CO_2 biomethanation process in a real waste water treatment plant: A techno-economic assessment. *Energy Convers. Manag.* **2020**, *209*, 112663. [CrossRef]
13. Hülsemann, B.; Zhou, L.; Merkle, W.; Hassa, J.; Müller, J.; Oechsner, H. Biomethane Potential Test: Influence of Inoculum and the Digestion System. *Appl. Sci.* **2020**, *10*, 2589. [CrossRef]
14. Lohani, S.P.; Havukainen, J. *Anaerobic Digestion: Factors Affecting Anaerobic Digestion Process*; Springer: Singapore, 2018; pp. 343–359.
15. Rehman, M.L.U.; Iqbal, A.; Chang, C.; Li, W.; Ju, M. Anaerobic digestion. *Water Environ. Res.* **2019**, *91*, 1253–1271. [CrossRef]
16. Sarto, S.; Hildayati, R.; Syaichurrozi, I. Effect of chemical pretreatment using sulfuric acid on biogas production from water hyacinth and kinetics. *Renew. Energy* **2019**, *132*, 335–350. [CrossRef]
17. Chen, C.; Guo, W.; Ngo, H.H.; Lee, D.-J.; Tung, K.-L.; Jin, P.; Wang, J.; Wu, Y. Challenges in biogas production from anaerobic membrane bioreactors. *Renew. Energy* **2016**, *98*, 120–134. [CrossRef]
18. Hagos, K.; Zong, J.; Li, D.; Liu, C.; Lu, X. Anaerobic co-digestion process for biogas production: Progress, challenges and perspectives. *Renew. Sustain. Energy Rev.* **2017**, *76*, 1485–1496. [CrossRef]
19. Yoshida, K.; Shimizu, N. Biogas production management systems with model predictive control of anaerobic digestion processes. *Bioprocess Biosyst. Eng.* **2020**, *43*, 2189–2200. [CrossRef]
20. Xu, D.; Ji, H.; Ren, H.; Geng, J.; Li, K.; Xu, K. Inhibition effect of magnetic field on nitrous oxide emission from sequencing batch reactor treating domestic wastewater at low temperature. *J. Environ. Sci.* **2020**, *87*, 205–212. [CrossRef]
21. Thanh, P.M.; Ketheesan, B.; Yan, Z.; Stuckey, D. Trace metal speciation and bioavailability in anaerobic digestion: A review. *Biotechnol. Adv.* **2016**, *34*, 122–136. [CrossRef]
22. Luna-Delrisco, M.; Orupõld, K.; Dubourguier, H.-C. Particle-size effect of CuO and ZnO on biogas and methane production during anaerobic digestion. *J. Hazard. Mater.* **2011**, *189*, 603–608. [CrossRef]

23. Goli, J.; Sahu, O. Development of heterogeneous alkali catalyst from waste chicken eggshell for biodiesel production. *Renew. Energy* **2018**, *128*, 142–154. [CrossRef]
24. Sridhar, N.; Senthilkumar, J.S.; Subburayan, M.R. Removal of Toxic Metals (Lead & Copper) From Automotive Industry Waste Water By Using Fruit Peels. *Int. J. Adv. Inf. Commun. Technol.* **2014**, *1*, 188–191.
25. Amo-Duodu, G.; Rathilal, S.; Chollom, M.N.; Tetteh, E.K. Application of metallic nanoparticles for biogas enhancement using the biomethane potential test. *Sci. Afr.* **2021**, *12*, e00728. [CrossRef]
26. Hassanein, A.; Keller, E.; Lansing, S. Effect of metal nanoparticles in anaerobic digestion production and plant uptake from effluent fertilizer. *Bioresour. Technol.* **2021**, *321*, 124455. [CrossRef] [PubMed]
27. Chen, L.; Feng, W.; Fan, J.; Zhang, K.; Gu, Z. Removal of silver nanoparticles in aqueous solution by activated sludge: Mechanism and characteristics. *Sci. Total Environ.* **2020**, *711*, 135155. [CrossRef] [PubMed]
28. Toor, R.; Mohseni, M. UV-H2O2 based AOP and its integration with biological activated carbon treatment for DBP reduction in drinking water. *Chemosphere* **2007**, *66*, 2087–2095. [CrossRef]
29. Pirilä, M. *Adsorption and Photocatalysis in Water Treatment: Active, Abundant and Inexpensive Materials and Methods*; Acta Universitatis Ouluensis, University of Oulu: Oulu, Finland, 2015. Available online: http://jultika.oulu.fi/files/isbn9789526207629.pdf (accessed on 20 May 2021).
30. Lakshmanan, R. Application of Magnetic Nanoparticles and Reactive Filter Materials for Wastewater Treatment Ramnath Lakshmanan. Ph.D. Thesis, KTH Royal Institute of Technology, Stockholm, Sweden, 2013.
31. Zhang, Y.; Wu, B.; Xu, H.; Liu, H.; Wang, M.; He, Y.; Pan, B. Nanomaterials-enabled water and wastewater treatment. *NanoImpact* **2016**, *3–4*, 22–39. [CrossRef]
32. Qiu, X.; Zhang, Y.; Zhu, Y.; Long, C.; Su, L.; Liu, S.; Tang, Z. Applications of Nanomaterials in Asymmetric Photocatalysis: Recent Progress, Challenges, and Opportunities. *Adv. Mater.* **2021**, *33*, e2001731. [CrossRef]
33. Maksoud, M.I.A.A.; Elgarahy, A.; Farrell, C.; Al-Muhtaseb, A.H.; Rooney, D.W.; Osman, A. Insight on water remediation application using magnetic nanomaterials and biosorbents. *Coord. Chem. Rev.* **2020**, *403*, 213096. [CrossRef]
34. Peeters, K.; Lespes, G.; Zuliani, T.; Ščančar, J.; Milačič, R. The fate of iron nanoparticles in environmental waters treated with nanoscale zero-valent iron, FeONPs and Fe$_3$O$_4$NPs. *Water Res.* **2016**, *94*, 315–327. [CrossRef]
35. Tetteh, E.K.; Amankwa, M.O.; Armah, E.K.; Rathilal, S. Fate of COVID-19 Occurrences in Wastewater Systems: Emerging Detection and Treatment Technologies—A Review. *Water* **2020**, *12*, 2680. [CrossRef]
36. Lombi, E.; Donner, E.; Tavakkoli, E.; Turney, T.; Naidu, R.; Miller, B.W.; Scheckel, K. Fate of Zinc Oxide Nanoparticles during Anaerobic Digestion of Wastewater and Post-Treatment Processing of Sewage Sludge. *Environ. Sci. Technol.* **2012**, *46*, 9089–9096. [CrossRef] [PubMed]
37. Adetunji, A.I.; Olaniran, A.O. Treatment of industrial oily wastewater by advanced technologies: A review. *Appl. Water Sci.* **2021**, *11*, 1–19. [CrossRef]
38. Tetteh, E.K.; Rathilal, S.; Chetty, M.; Armah, E.K.; Asante-Sackey, D. *Treatment of Water and Wastewater for Reuse and Energy Generation-Emerging Technologies*; IntechOpen: London, UK, 2019. [CrossRef]
39. Gunnerson, C.; Stuckey, D. *Integrated Resource Recovery-Anaerobic Digestion*; UNDP Proje: Washington, DC, USA, 1986.
40. Hossain, M.A. Removal of Copper from Water by Adsorption onto Banana Peel as Bioadsorbent. *Int. J. Geomate* **2012**, *2*, 227–234. [CrossRef]
41. Xiang, Y.; Yang, Z.; Zhang, Y.; Xu, R.; Zheng, Y.; Hu, J.; Li, X.; Jia, M.; Xiong, W.; Cao, J. Influence of nanoscale zero-valent iron and magnetite nanoparticles on anaerobic digestion performance and macrolide, aminoglycoside, β-lactam resistance genes reduction. *Bioresour. Technol.* **2019**, *294*, 122139. [CrossRef]
42. Adeleke, O.; Akinlabi, S.; Jen, T.-C.; Dunmade, I. Towards sustainability in municipal solid waste management in South Africa: A survey of challenges and prospects. *Trans. R. Soc. S. Afr.* **2021**, *76*, 53–66. [CrossRef]
43. APHA. *Standard Methods for the Examination of Water and Wastewater*, 22nd ed.; American Public Health Association: Washington, DC, USA, 2012.
44. Syaichurrozi, B.I.; Sumardiono, S. Kinetic Model of Biogas Yield Production from Vinasse at Various Initial pH: Comparison between Modified Gompertz Model and First Order Kinetic Model. *Res. J. Appl. Sci. Eng. Technol.* **2014**, *7*, 2798–2805. [CrossRef]
45. Mu, Y.; Wang, G.; Yu, H.-Q. Kinetic modeling of batch hydrogen production process by mixed anaerobic cultures. *Bioresour. Technol.* **2006**, *97*, 1302–1307. [CrossRef]
46. Sibiya, N.P.; Rathilal, S.; Tetteh, E.K. Coagulation Treatment of Wastewater: Kinetics and Natural Coagulant Evaluation. *Molecules* **2021**, *26*, 698. [CrossRef]
47. Goswami, R.K.; Mehariya, S.; Karthikeyan, O.P.; Verma, P. Advanced microalgae-based renewable biohydrogen production systems: A review. *Bioresour. Technol.* **2020**, *320*, 124301. [CrossRef]
48. Pessoa, M.; Sobrinho, M.M.; Kraume, M. The use of biomagnetism for biogas production from sugar beet pulp. *Biochem. Eng. J.* **2020**, *164*, 107770. [CrossRef]
49. Abdelradi, F. Food waste behaviour at the household level: A conceptual framework. *Waste Manag.* **2018**, *71*, 485–493. [CrossRef] [PubMed]
50. Baek, G.; Kim, J.; Cho, K.; Bae, H.; Lee, C. The biostimulation of anaerobic digestion with (semi)conductive ferric oxides: Their potential for enhanced biomethanation. *Appl. Microbiol. Biotechnol.* **2015**, *99*, 10355–10366. [CrossRef] [PubMed]

Article

Pyrolytic Behavior of Polyvinyl Chloride: Kinetics, Mechanisms, Thermodynamics, and Artificial Neural Network Application

Mohammed Al-Yaari and Ibrahim Dubdub *

Chemical Engineering Department, King Faisal University, P.O. Box 380, Al-Ahsa 31982, Saudi Arabia; malyaari@kfu.edu.sa
* Correspondence: idubdub@kfu.edu.sa; Tel.: +966-13-589-6989

Abstract: Pyrolysis of waste polyvinyl chloride (PVC) is considered a promising and highly efficient treatment method. This work aims to investigate the kinetics, and thermodynamics of the process of PVC pyrolysis. Thermogravimetry of PVC pyrolysis at three heating rates (5, 10, and 20 K/min) showed two reaction stages covering the temperature ranges of 490–675 K, and 675–825 K, respectively. Three integral isoconversional models, namely Flynn-Wall-Qzawa (FWO), Kissinger-Akahira-Sunose (KAS), and Starink, were used to obtain the activation energy (E_a), and pre-exponential factor (A) of the PVC pyrolysis. On the other hand, the Coats-Redfern non-isoconversional model was used to determine the most appropriate solid-state reaction mechanism/s for both stages. Values of E_a, and A, obtained by the isoconversional models, were very close and the average values were, for stage I: E_a = 75 kJ/mol, A = 1.81 × 10^6 min^{-1}; for stage II: E_a = 140 kJ/mol, A = 4.84 × 10^9 min^{-1}. In addition, while the recommended mechanism of the first stage reaction was P2, F3 was the most suitable mechanism for the reaction of stage II. The appropriateness of the mechanisms was confirmed by the compensation effect. Thermodynamic study of the process of PVC pyrolysis confirmed that both reactions are endothermic and nonspontaneous with promising production of bioenergy. Furthermore, a highly efficient artificial neural network (ANN) model has been developed to predict the weight left % during the PVC pyrolysis as a function of the temperature and heating rate. The 2-10-10-1 topology with TANSIG-LOGSIG transfer function and feed-forward back-propagation characteristics was used.

Keywords: polyvinyl chloride (PVC); pyrolysis; thermogravimetric analysis (TGA); kinetics; thermodynamics; artificial neural networks (ANN)

1. Introduction

Plastics are widely used because of their distinguished properties including degradation resistance, flexibility, and low weight and cost [1]. Therefore, the global production rate of plastics is increasing dramatically, and thus massive plastic waste is generated. Unfortunately, most of the plastic waste is either incinerated or disposed of in landfills [2] which causes major environmental concerns. Pyrolysis has been reported as a very promising thermochemical method to treat plastic waste and produce bioenergy and/or valuable chemicals [3,4].

Municipal plastic waste (MPW) mainly comprises low-density polyethylene (LDPE), high-density polyethylene (HDPE), polypropylene (PP), polystyrene (PS), polyethylene terephthalate (PET), polyvinyl chloride (PVC), and other plastics. PVC represents around 11 wt% of the MPW, however, the composition may change from one location to another [5]. In this work, the pyrolysis process of PVC is targeted for investigation.

Kim (2001) [6] studied the pyrolysis of polyvinyl chloride (PVC) using thermogravimetric analysis (TGA) data at three different heating rates (5, 10, and 30 K/min). Two pyrolysis stages were observed and attributed to the production of volatiles and intermediates, respectively. The Freeman-Carroll model was used to obtain the kinetics parameters. The average values of the triple kinetic parameters (activation energy (E_a), pre-exponential factor (A), and reaction order (n)) were reported as E_a = 129.95 kJ/mol, A = 3.04 × 10^{11} min^{-1}, and n = 1.49 for the first stage, and E_a = 282.05 kJ/mol, A = 2.63 × 10^{20} min, and n = 2.07 for the second stage.

Karayildirim et al. (2006) [7] used thermogravimetry (TG)/mass spectrometry (MS) to investigate the pyrolysis of PVC at the heating rates of 2, 5, 10, and 15 K/min. Runge–Kutta, and Flynn-Wall-Ozawa (FWO) methods were used to obtain the kinetics parameters. The reported values were E_a = 190 kJ/mol, A = 4.40 × 10^7 min^{-1}, and n = 1.5 for the first stage, respectively, and E_a = 250 kJ/mol, A = 6.57 × 10^7 min^{-1}, and n = 1.6 for the second stage.

Wu et al. (2014) [8] studied the pyrolysis and co-pyrolysis of polyethylene (PE), PS, and PVC using TG/Fourier transform infrared (FTIR) at the heating rate of 40 K/min. Pyrolysis of PVC was reported to occur in two stages starting at 594 K and 770 K, respectively.

Yu et al. (2016) [9] reviewed different recycling methods used for the chemical treatment of PVC waste. The onset temperature of PVC pyrolysis was reported to start much earlier than PE, PP, PS, and PET. Two pyrolysis stages of PVC were confirmed. While the first stage was observed in the temperature range of 523–623 K and attributed to the dehydrochlorination of PVC to produce volatile and de-HCl PVC, the second one was reported in the range of 623–798 K and attributed to the pyrolysis of the de-HCl PVC.

Xu et al. (2018) [10] examined the pyrolysis of PVC at high heating rates (100, 300, and 500 K/min) using TGA data. The activation energy for both stages was obtained by three model-free methods namely FWO, Kissinger–Akahira–Sunose (KAS), and Friedman models. For the first stage, the mean E_a values were 48.62, 48.11, and 49.79 kJ/mol obtained by the three methods, respectively, and for the second stage, the obtained values were 113.59, 109.78, and 117.33 kJ/mol, respectively. In addition, two model-fitting methods, namely Coats-Redfern and Criado models, were used to predict the mechanism of the pyrolysis.

Ma et al. (2019) [11] performed a PVC pyrolysis by TG at 20 K/min and two pyrolysis stages were reported. The first stage was observed between 473 and 633 K (dehydrochlorination and polyene chains formation), and the second stage was between 633 and 773 K (degradation of the polyene chains). They used the Coats-Redfern model with the assumption of the first-order reaction mechanism. The values of 114.57 and 7.73 kJ/mol were reported as the activation energy values of both stages, respectively.

Özsin and Pütün (2019) [12] investigated the PVC pyrolysis using between 298 and 1273 K at heating rates of 5, 10, 20, and 40 K/min. To obtain the kinetic parameters, Friedman, FWO, Vyazovkin, and distributed activation energy (DAEM) models were used. Three stages of pyrolysis were reported and attributed to the dehydrochlorination, the formation of alkyl aromatics, and main pyrolysis of PVC, respectively. Activation energy values ranging between 93.2 kJ/mol and 263.7 kJ/mol were reported.

Zhou et al. (2019) [13] studied the pyrolysis of chlorinated polyvinyl chloride (CPVC) using TGA at 10, 20, 30, and 60 K/min heating rates. The FWO model was used to obtain the values of E_a and A. However, the Coats-Redfern model was used to predict the reaction mechanism. The reported average values of E_a of Stage I and Stage II were 140.27, and 246.07 kJ/mol, respectively. In addition, F1, and F4 were reported as the most suitable reaction mechanisms of stages I, and II, respectively.

Currently, many researchers are aiming to develop efficient artificial neural network (ANN) models, as an alternative option, to forecast experimental data for different engineering applications. Specifically, ANN modeling has been used to predict the TGA data of the pyrolysis of biomass.

Kinetic data of the thermal decomposition of different materials, including blends, refuse-derived fuel, polycarbonate/$CaCO_3$ composites, and high-ash sewage sludge, were predicted by high-efficient ANN models [14–19].

Recently, Dubdub and Al-Yaari (2020, and 2021) developed highly efficient ANN models to predict the TGA data of the pyrolysis of LDPE at 5, 10, 20, and 40 K/min [20], pyrolysis of HDPE at 5, 10, 20, and 40 K/min [21], and the co-pyrolysis of PS, PP, LDPE, and HDPE at 60 K/min [22]. In addition, Al-Yaari and Dubdub (2020) [23] developed an ANN model to predict the thermal behavior of the catalytic pyrolysis of HDPE at 5,10, and 15 K/min.

This study aims to build knowledge on PVC pyrolysis using TGA experimental data. The kinetic triplet (activation energy, pre-exponential factor, and reaction mechanism) of the pyrolysis process were obtained by FWO, KAS, Starink, and Coats-Redfern models. In addition, thermodynamic properties of the process of PVC pyrolysis have been investigated. Furthermore, a highly efficient ANN model has been developed to predict the pyrolytic behavior of PVC.

2. Materials and Methods

2.1. Proximate and Ultimate Analyses

Polymeric materials (PVC) were produced by Ipoh SY Recycle Plastic Sdn. Bhd., Perak, Malaysia. Proximate and ultimate analyses were performed to identify the physicochemical properties of the PVC samples. While the proximate analysis aims to determine the moisture, volatile matter, fixed carbon, and ash contents using Simultaneous Thermal Analyzer STA-6000, manufactured by PerkinElmer, Waltham, MA, USA, the ultimate analysis was performed to determine the % of carbon (C), hydrogen (H), nitrogen (N), and sulfur (S) using 2400 Series II CHNS Elemental Analyzer, manufactured by PerkinElmer, Waltham, MA, USA. Details of both analyses are fully described elsewhere [24].

2.2. Thermogravimetry of PVC

PVC pellets were ground into powder by a grinding mill before feeding to the thermogravimetric analyzer. Ten mg of PVC powder samples were used throughout the study. A Thermogravimetric Analyzer TGA-7, manufactured by PerkinElmer Co., Waltham, MA, USA, was used. Thermogravimetric experiments were conducted in an inert atmosphere of pure N_2 at three different heating rates (5, 10, and 20 K/min).

2.3. Determination of the Kinetic Triplet of the PVC Pyrolysis

The reaction kinetics of the PVC pyrolysis can be expressed as follows:

$$\frac{d\alpha}{dt} = A \exp\left(-\frac{E_a}{RT}\right) f(\alpha) \tag{1}$$

where α is the reaction conversion, t is time, A is the pre-exponential (frequency) factor, E_a is the reaction activation energy, R is the universal gas constant, T is the absolute temperature, and $f(\alpha)$ is the conversion-dependent reaction model in its differential form.

For non-isothermal pyrolysis, Equation (1) can be re-written as:

$$\beta \frac{d\alpha}{dT} = A \exp\left(-\frac{E_a}{RT}\right) f(\alpha) \tag{2}$$

where β is the heating rate expressed as the change in temperature with time (dT/dt). If A, and E_a are assumed to be independent of α and $f(\alpha)$, A, and E_a are independent of T. Equation (2) can be integrated and rearranged as follows:

$$\int_0^\alpha \frac{d\alpha}{f(\alpha)} = \frac{A}{\beta} \int_{T_o}^T \exp\left(-\frac{E_a}{RT}\right) dT \tag{3}$$

or

$$g(\alpha) = \frac{A}{\beta} \int_{T_o}^T \exp\left(-\frac{E_a}{RT}\right) dT \tag{4}$$

where $g(\alpha)$ is the integral form of the conversion-dependent reaction model, and T_o is the initial absolute temperature of the PVC pyrolysis.

The temperature integral does not have an analytical solution, and Equation (4) can be approximated and expressed as:

$$g(\alpha) \cong \frac{A\,E_a}{\beta\,R} P\left(-\frac{E_a}{RT}\right) \quad (5)$$

Numerical methods need to be used to obtain the polynomial of $\left(-\frac{E_a}{RT}\right)$ and a series expansion can give different approximations of the polynomial term.

The FWO model used the following Doyle's approximation [25] of the temperature integral:

$$P\left(-\frac{E_a}{RT}\right) = Exp\left(-5.331 - 1.052\,\frac{E_a}{RT}\right) \quad (6)$$

and thus Equation (5) became:

$$g(\alpha) = \frac{A\,E_a}{\beta\,R} Exp\left(-5.331 - 1.052\,\frac{E_a}{RT}\right) \quad (7)$$

or

$$\ln \beta = \ln \frac{A\,E_a}{R\,g(\alpha)} - 5.331 - 1.052\,\frac{E_a}{RT} \quad (8)$$

However, the KAS model used the Murry-White approximation [26] of the temperature integral where:

$$P\left(-\frac{E_a}{RT}\right) = \frac{Exp\left(-\frac{E_a}{RT}\right)}{\left(\frac{E_a}{RT}\right)^2} \quad (9)$$

and thus Equation (5) became:

$$\ln \frac{\beta}{T^2} = \ln \frac{A\,R}{E_a\,g(\alpha)} - \frac{E_a}{RT} \quad (10)$$

In 2003, Starink [27] used the following approximation of the temperature integral:

$$P\left(-\frac{E_a}{RT}\right) = \left(\frac{T\,R}{E_a}\right)^{1.92} \exp\left(-1.0008\,\frac{E_a}{RT} - 0.312\right) \quad (11)$$

and thus Equation (5) became:

$$\ln \frac{\beta}{T^{1.92}} = Constant - 1.0008\,\frac{E_a}{RT} \quad (12)$$

where:

$$Constant = \ln \frac{A\,R^{0.92}}{E_a^{0.92}\,g(\alpha)} - 0.312 \quad (13)$$

The FWO (Equation (8)), KAS (Equation (10)), and Starink (Equation (12)) models can be expressed in the following general form:

$$\ln \frac{\beta}{T^a} = b + c\,\frac{E_a}{RT} \quad (14)$$

where a, b, and c constants are presented in Table 1.

Table 1. Parameters of the generalized form of the isoconversional models.

Model	a	b	c
FWO	0	$\ln \frac{A\,E_a}{R\,g(\alpha)} - 5.331$	−1.052
KAS	2	$\ln \frac{A\,R}{E_a\,g(\alpha)}$	−1
Starink	1.92	$\ln \frac{A\,R^{0.92}}{E_a^{0.92}\,g(\alpha)} - 0.312$	−1.0008

In this work, the values of the apparent activation energy of the PVC pyrolysis have been obtained by plotting $\ln \frac{\beta}{T^a}$ vs. $\frac{1}{T}$ for FWO, KAS, and Starink models using the TGA experimental data. The obtained values of E_a by isoconversional methods are independent of the reaction mechanism.

The Coats-Redfern (CR) model, expressed by Equation (15), has been used to determine the most suitable reaction mechanism among 15 solid-state reaction models presented in Table 2. Solid-state kinetic models are categorized based on their mechanistic basis as reaction-order, diffusion, nucleation, and geometrical contraction models [28].

The E_a values obtained by the CR model for different reaction mechanisms were compared with the average E_a values obtained by the isoconversional models. The most appropriate mechanism provides the closest values of activation energy.

$$\ln \frac{g(\alpha)}{T^2} = \ln \frac{A\,R}{E_a\,\beta} - \frac{E_a}{RT} \quad (15)$$

Then, the values of the pre-exponential factor can be obtained from the slope of the linear relationships of Equations (8), (10), and (12) when the reaction mechanism has been determined. If the reaction mechanism has been determined well, the following linear relationship must be retained (compensation effect).

$$\ln A = dE_a + e \quad (16)$$

where d, and e are the compensation parameters that can be obtained from the plot of $\ln A$ vs. E_a.

Table 2. List of the most used solid-state reaction models.

Model	Mechanism	$g(\alpha)$
Reaction-Order Models		
F1	First-order reaction	$-\ln(1-\alpha)$
F2	Second-order reaction	$(1-\alpha)^{-1} - 1$
F3	Third-order reaction	$[(1-\alpha)^{-1} - 1]/2$
Diffusion Models		
D1	One-dimensional diffusion	α^2
D2	Two-dimensional diffusion	$(1-\alpha)\ln(1-\alpha) + \alpha$
D3	Three-dimensional diffusion	$\left[1-(1-\alpha)^{1/3}\right]^2$
Nucleation Models		
P2	Power law ($n = \frac{1}{2}$)	$\alpha^{1/2}$
P3	Power law ($n = \frac{1}{3}$)	$\alpha^{1/3}$
P4	Power law ($n = \frac{1}{4}$)	$\alpha^{1/4}$
A2	Avrami-Erofeev ($n = \frac{1}{2}$)	$[-\ln(1-\alpha)]^{1/2}$
A3	Avrami-Erofeev ($n = \frac{1}{3}$)	$[-\ln(1-\alpha)]^{1/3}$
A4	Avrami-Erofeev ($n = \frac{1}{4}$)	$[-\ln(1-\alpha)]^{1/4}$
Geometrical Contraction Models		
R1	Prout-Tompkins	α
R2	Contracting cylinder	$1-(1-\alpha)^{1/2}$
R3	Contracting sphere	$1-(1-\alpha)^{1/3}$

2.4. Estimation of the Thermodynamic Parameters of the PVC Pyrolysis

Based on the obtained values of activation energy, pre-exponential factor, and the maximum peak temperature (T_p), some of the thermodynamic characteristics of the PVC pyrolysis can be determined using the following Equations:

$$\Delta H = E_a - R\, T_p \tag{17}$$

$$\Delta G = E_a + R\, T_p \ln\left(\frac{k_B\, T_p}{h\, A}\right) \tag{18}$$

$$\Delta S = \frac{\Delta H - \Delta G}{T_p} \tag{19}$$

where:

ΔH: is the change in enthalpy,
ΔG: is the change in Gibbs free energy,
ΔS: is the change in entropy,
T_p: is the maximum peak temperature obtained from the derivative thermogravimetric curves,
k_B: is the Boltzmann constant (1.381×10^{-23} J/K),
h: is the Planck constant (6.626×10^{-34} J/s).

The thermodynamic parameters (ΔH, ΔG and ΔS) are of great importance to the optimization of the large-scale reactor used for pyrolysis.

2.5. Performance of Artificial Neural Networks

As mentioned previously, process modeling using artificial neural networks (ANNs) has attracted the attention of researchers due to its robustness, easiness, and cost-effectiveness, especially when the system becomes more complex and non-linear relationships between parameters are adopted.

Typically, the datasets are divided randomly into three subsets: training, validation, and test sets. During the training stage, network learning is established and parameters wight is corrected. However, the network performance is checked during the validation stage and the network is generalized in the test stage [29].

For the best performance of ANNs, a genetic algorithm should be implemented to optimize some topological features such as the number of hidden layers, the number of neurons in the hidden layers, and the transfer functions. The following statistical parameters, expressed by Equations (20)–(23), are used to evaluate the performance of the developed ANN models [30].

$$\text{Correlation coefficient (R)} = \frac{\sum_{i=1}^{N}(x_i - \overline{x_i})(y_i - \overline{y_i})}{\sqrt{\sum_{i=1}^{N}(x_i - \overline{x_i})^2 \sum_{i=1}^{N}(y_i - \overline{y_i})^2}} \tag{20}$$

$$\text{Root mean square error (RMSE)} = \sqrt{\frac{1}{N}\sum (y - x)^2} \tag{21}$$

$$\text{Mean absolute error (MAE)} = \frac{1}{N}\sum |y - x| \tag{22}$$

$$\text{Mean bias error (MBE)} = \frac{1}{N}\sum (y - x) \tag{23}$$

where:

x: is the experimental value of the weight left %,
y: is the predicted value of the weight left %,
$\overline{x}$: is the mean values of the experimental weight left %, and
$\overline{y}$: is the mean values of the experimental weight left %.

In this investigation, the TGA data of the mass left % during the PVC pyrolysis has been targeted to be predicted by developing an efficient ANN model.

3. Results and Discussion

3.1. Proximate and Ultimate Analysis

The characterization results of proximate and ultimate analyses of the PVC samples are presented in Table 3. Proximate analysis showed 0.146 wt% of moisture, 88.765 wt% of volatile matter (VM), 10.566 wt% of fixed carbon (FC), and 2.12 wt% of ash. The high valuable contents (VM, and FC) and low ash content indicate the suitability of the production of bioenergy from the pyrolysis of PVC.

In addition, as obtained from the ultimate analysis, low nitrogen and sulfur contents are preferable to avoid the production of toxic gasses such as NO_x and SO_x and thus benefit the environment.

Table 3. Some characteristics of the PVC samples.

Proximate Analysis, wt%				Ultimate Analysis, wt%				
Moisture	Volatile Matter	Fixed Carbon	Ash	C	H	N	S	O [a]
0.146	88.765	10.566	0.523	83.75	13.70	0.14	0.78	1.63

[a] by difference.

3.2. Thermogravimetry of PVC

The corresponding thermogravimetric (TG) and derivative thermogravimetric (DTG) curves of the PVC pyrolysis at 5, 10, and 20 K/min heating rates are illustrated in Figures 1 and 2, respectively. Although all curves show similarities in their appearance, they were shifted to higher temperatures as the heating rate increased. As the heating rate increased, the mass loss at a specific temperature decreased (see Figure 1). On the other hand, as the heating rate increases, the mass-loss rate increases, and thus the size of the DTG-peak increases (see Figure 2). This finding can be attributed to the thermal lag and/or heat transfer limitations [31].

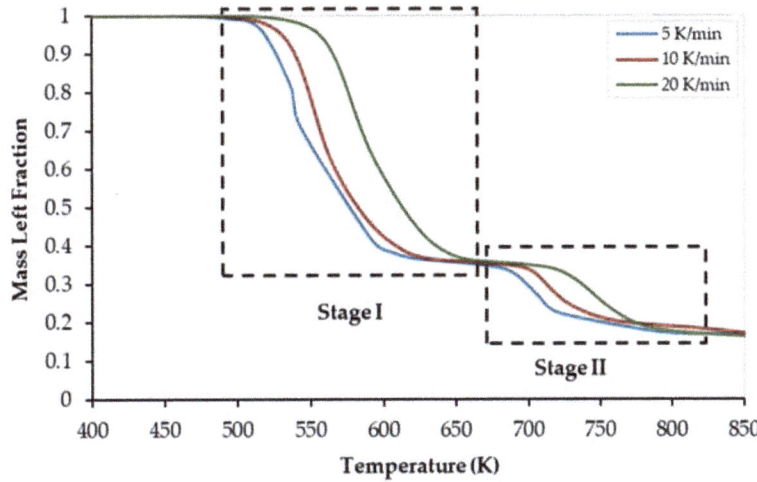

Figure 1. Thermogravimetric (TG) curves of the PVC pyrolysis at different heating rates.

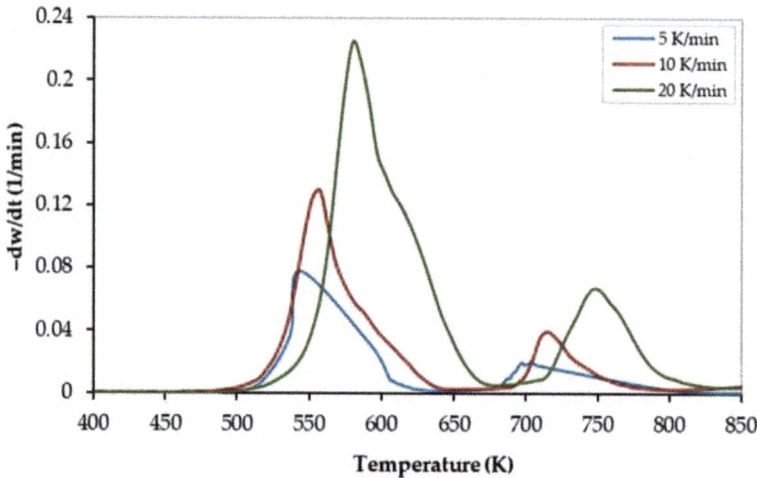

Figure 2. Derivative-thermogravimetric (DTG) curves of the PVC pyrolysis at different heating rates.

As shown in both figures, the thermal degradation zone of PVC is in the temperature range of 490 K to 825 K with almost 20 wt% of pyrolysis residues. In addition, both curves revealed two reaction stages covering the temperature ranges of 490–675 K, and 675–825 K, for stages I, and II, respectively. Thus, the PVC pyrolysis occurs a multi-stage mechanism as reported elsewhere [8,9,11,32], The pyrolysis characteristic temperatures of both stages are presented in Table 4. In the first and second degradation stages, characteristic peaks were observed at temperatures of 599 ± 16.4 K. and 724.3 ± 21.9 K, respectively.

Table 4. Characteristic temperatures of the PVC pyrolysis at different heating rates.

Heating Rate (K/min)	Stage I				Stage II			
	On-Set (K)	Peak (K)	Final (K)	Mass Loss (%)	On-Set (K)	Peak (K)	Final (K)	Mass Loss (%)
5	490	540	630	64	675	700	810	82
10	495	557	640	64	690	720	815	82
20	500	580	675	64	720	753	825	82
Average *	495 ± 4.08	559 ± 16.4	648.3 ± 19.3	64	695 ± 18.7	724.3 ± 21.9	816.7 ± 6.2	82

* mean ± standard deviation Determination of the kinetic triplet of the PVC pyrolysis.

While the first reaction stage involves the dehydrochlorination reaction to produce de-HCl PVC and volatiles, the pyrolysis of the de-HCl PVC occurs during the second stage [29]. The first reaction (dehydrochlorination) requires less energy to break the C–Cl bond when compared to the energy required for the second reaction (de-HCl PVC pyrolysis) where the C–C stable bond is broken. Thus, as illustrated in Figure 2, the peak of the first main reaction is much bigger than the peak of the second one which is in full agreement with the available literature [10].

The apparent activation energy has been obtained using TGA data along with FWO, KAS, and Starink models, expressed by Equations (8), (10), and (12), respectively. From the slope of the plots of $(\ln \frac{\beta}{T^a})$ versus $(\frac{1}{T})$, where the exponent a is defined in Table 1, the values of E_a were determined at a conversion range of 0.1–0.8. While stage I covers the conversion range of 0.1–0.6, stage II covers the range of 0.7–0.8 and this is in full agreement with published data [10]. Regression lines of all plots for both pyrolytic stages are presented in Figures 3 and 4, respectively, and the values of E_a obtained by all models are presented in Table 5.

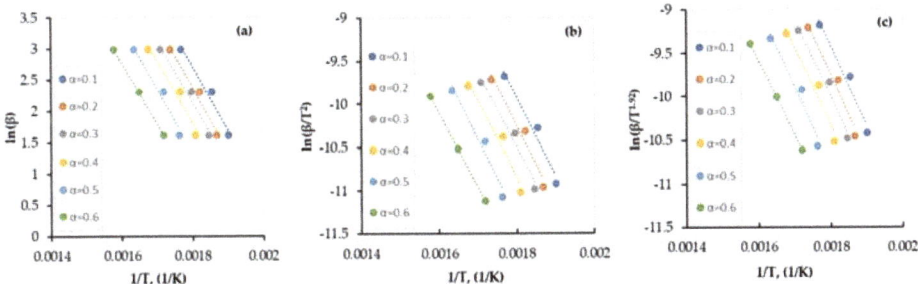

Figure 3. Regression lines of the experimental data of the PVC pyrolysis (stage I) by: (**a**) FWO, (**b**) KAS, and (**c**) Starink models.

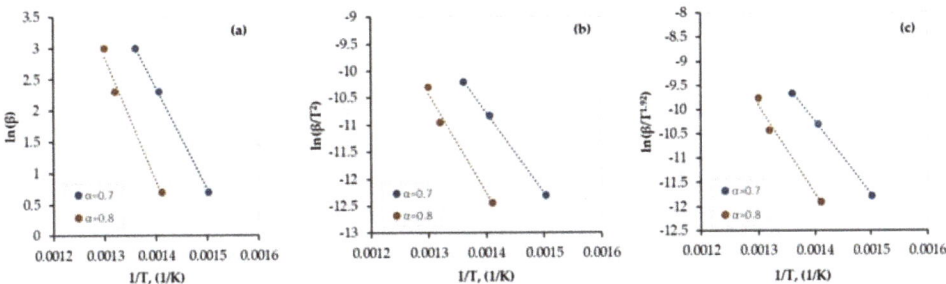

Figure 4. Regression lines of the experimental data of the PVC pyrolysis (stage II) by: (**a**) FWO, (**b**) KAS, and (**c**) Starink models.

Table 5. Activation energy values obtained by isoconversional models.

Conversion	FWO		KAS		Starink		Average	
	E_a (kJ/mol)	R^2	E_a (kJ/mol)	R^2	E_a (kJ/mol)	R^2	E_a (kJ/mol)	R^2
Stage I								
0.1	80	0.9704	75	0.9628	75	0.9631	77	0.9654
0.2	81	0.9708	75	0.9631	76	0.9634	76	0.9658
0.3	78	0.9752	73	0.9681	73	0.9685	73	0.9706
0.4	79	0.9635	74	0.9533	74	0.9538	74	0.9569
0.5	82	0.9682	76	0.9595	77	0.9599	77	0.9625
0.6	77	1	71	1	71	1	71	1.0000
Average	79	0.9747	74	0.9678	74	0.9681	75	0.9702
Stage II								
0.7	129	0.9995	124	0.9993	124	0.9993	125	0.9994
0.8	157	0.9827	153	0.98	153	0.9802	154	0.9810
Average	143	0.9911	138	0.9897	139	0.9898	140	0.9902

As shown in Figures 3 and 4, although all regression lines for the FWO, KAS, and Starink models are parallel, the gap between those of stage I is smaller than those of stage II which indicates that the reaction that occurs in stage I (dehydrochlorination) is faster than that of stage II (de-HCl PVC pyrolysis). In addition, the gap between the regression lines of stage I increased as the conversion increased, which indicates a decrease in the conversion rate of the dehydrochlorination process with time. This is also confirmed by the TG and DTG curves (Figures 1 and 2, respectively) when the plateau (shoulder) zone between the first and the second stages is approached. Furthermore, as the gap between the regression lines increased, a higher variation in the value of activation energy was observed (see Table 5).

As shown in Table 5, TGA data were fitted well ($R^2 > 0.95$) by the FWO, KAS, and Starink models and the obtained values were very close and thus indicate that all three models are suitable to be used. The average value of activation energy of the first stage of the PVC pyrolysis is 75 kJ/mole with a regression coefficient (R^2) of 0.9702 and that of the second stage is 140 kJ/mole with an R^2 value of 0.9902.

For stage I, at low conversion ($\alpha < 0.3$), the obtained values of E_a are larger than the average value of E_a which can be attributed to the low energy provided initially to the dehydrochlorination reaction. However, at high conversion ($\alpha > 0.5$), the provided energy is higher than the required one and thus there is a reduction in the E_a values. Similarly, for stage II, the E_a value at $\alpha = 0.7$ is less than that at $\alpha = 0.8$ which could be explained using the same concept. However, as mentioned earlier, the E_a values of stage II are larger than those of stage I which are due to the difference in the amount of the required energy for reactions that occur in both stages. A similar trend was reported earlier for the PVC pyrolysis by Mumbach et al. [33], but with different values which can be attributed to the difference in the PVC compositions of both works.

Then, the Coat-Redfern model, expressed by Equation (15), was used to obtain the most suitable reaction mechanism/s for both stages of the PVC pyrolysis. Values of activation energy and pre-exponential factor at different heating rates for 15 solid-state reaction mechanisms, defined in Table 2, were obtained from the slope and the intercept of the plots of $\ln \frac{g(\alpha)}{T^2}$ versus $\frac{1}{T}$. The obtained kinetic parameters are presented in Tables 6 and 7 for stage I, and stage II, respectively.

As shown in Tables 6 and 7, the Coats-Redfern model fitted well the TGA data of the PVC pyrolysis with a regression coefficient of almost $R^2 > 0.99$. The average E_a values obtained by Coats-Redfern were then compared with the average values of E_a obtained by the isoconversional models for both stages. The most suitable reaction mechanism should give a closer value of the activation energy. Based on this criterion, stage I was suitably represented by the power-law nucleation reaction model (P2) ($E_a = 79$ kJ/mol, $R^2 = 0.9962$), and stage II was best represented by the 3rd order reaction mechanism (F3) ($E_a = 125$ kJ/mol, $R^2 = 0.9986$).

Table 6. Kinetic parameters obtained by Coats-Redfern model for Stage I.

Reaction Mechanism	Heating Rates									Average		
	5 K/min			10 K/min			20 K/min					
	E_a (kJ/mol)	$\ln A$ ($\min^{-1}$)	R^2	E_a (kJ/mol)	$\ln A$ ($\min^{-1}$)	R^2	E_a (kJ/mol)	$\ln A$ ($\min^{-1}$)	R^2	E_a (kJ/mol)	$\ln A$ ($\min^{-1}$)	R^2
F1	187	38.66	0.9999	173	36.03	0.9971	182	36.73	0.9981	181	37	0.9984
F2	202	42.25	0.9999	189	39.63	0.9988	198	40.26	0.9991	196	41	0.9993
F3	218	46.06	0.9991	205	43.44	0.9996	215	43.98	0.9997	213	44	0.9995
D1	355	75.25	0.9992	327	68.49	0.9947	343	69.31	0.9967	342	71	0.9969
D2	364	76.69	0.9995	336	69.94	0.9957	353	70.74	0.9973	351	72	0.9975
D3	374	77.4	0.9998	346	70.67	0.9965	363	71.45	0.9977	361	73	0.9980
D4	368	75.93	0.9996	339	69.18	0.996	356	69.98	0.9974	354	72	0.9977
A2	89	16.67	0.9999	82	16.09	0.9968	86	16.76	0.9978	86	17	0.9982
A3	57	9.9	0.9999	52	12.85	0.9968	54	13.66	0.9975	54	12	0.9981
A4	89	16.67	0.9999	37	15.71	0.9958	38	16.51	0.9972	55	16	0.9976
R1	173	35.31	0.9991	159	32.67	0.9944	167	33.4	0.9965	166	34	0.9967
R2	180	36.27	0.9996	166	33.63	0.9959	174	34.35	0.9974	173	35	0.9976
R3	183	36.43	0.9997	168	33.79	0.9963	177	34.5	0.9976	176	35	0.9979
P2	82	14.96	0.999	75	14.37	0.9936	79	15.05	0.9961	79	15	0.9962
P3	52	10.9	0.9989	47	13.84	0.9927	49	14.64	0.9955	49	13	0.9957
P4	37	13.83	0.9987	33	16.41	0.9916	35	17.24	0.9948	35	16	0.9950

Table 7. Kinetic parameters obtained by Coats-Redfern model for Stage II.

Reaction Mechanism	Heating Rates									Average		
	5 K/min			10 K/min			20 K/min					
	E_a	$\ln A$	R^2	E_a	$\ln A$	R^2	E_a	$\ln A$	R^2	E_a	$\ln A$	R^2
	(kJ/mol)	$\ln$ (min^{-1})		(kJ/mol)	$\ln$ (min^{-1})		(kJ/mol)	$\ln$ (min^{-1})		(kJ/mol)	$\ln$ (min^{-1})	
F1	41	14.81	0.9981	31	18.23	0.9981	29	19.34	0.9987	34	17	0.9983
F2	93	14.35	0.9978	63	12.81	0.999	63	13.73	0.9992	73	14	0.9987
F3	162	45.76	0.9977	105	18.02	0.999	108	19.01	0.999	125	28	0.9986
D1	31	17.17	0.9989	31	19.11	0.9959	50	20.63	0.9977	37	19	0.9975
D2	45	15.33	0.9989	42	17.93	0.9972	59	19.52	0.998	49	18	0.9980
D3	67	12.91	0.9987	57	16.8	0.9982	69	18.4	0.9975	64	16	0.9981
D4	52	15.57	0.9988	47	18.58	0.9977	62	20.17	0.997	54	18	0.9978
A2	15	18.55	0.9198	10	20.81	0.9134	8	21.7	0.9959	11	20	0.9430
A3	6	19.21	0.9901	2	20.45	0.9635	1	21.37	0.9427	3	20	0.9654
A4	1	18.21	0.9964	1	20.38	0.9948	2	21.92	0.9788	1	20	0.9900
R1	10	19.46	0.9973	10	21.26	0.9888	7	22.12	0.9976	9	21	0.9946
R2	22	18.38	0.9981	19	20.76	0.9962	16	21.87	0.9965	19	20	0.9969
R3	28	17.97	0.9981	23	20.63	0.9971	20	21.76	0.9825	24	20	0.9926
P2	1	18.86	0.9455	1	20.6	0.8186	3	15.05	0.9656	2	18	0.9099
P3	5	21.04	0.9988	5	22.75	0.9939	6	14.64	0.9964	5	19	0.9964
P4	1	18.86	0.9996	7	23.36	0.998	7	17.24	0.9986	5	20	0.9987

The power-law (P2) reaction model is among the simplest cases of nucleation models where the growth of the reaction nuclei is assumed constant, and the reaction rate follows the power law ($f(\alpha) = 2\alpha^{1/2}$) [28]. However, in reaction-order models, the reaction rate is directly proportional to the reactants remaining fraction raised to the reaction order. For F3, $f(\alpha) = (1-\alpha)^3$.

After the determination of the most suitable reaction mechanisms for both stages, the values of the pre-exponential factor were calculated by the isoconversional models. These values are presented in Table 8. The obtained values by isoconversional models are comparable with those obtained by Coats-Redfern for the selected reaction mechanism. Alternatively, compensation effect parameters ($d = 0.1244$, and $e = 4.9338$) can be obtained from the regression line of Figure 5. Then, the pre-exponential factor can be calculated using Equation (16) for each value of activation energy obtained by the isoconversional models.

Table 8. Pre-exponential factor values obtained by isoconversional models.

Conversion	$\ln [A$ (min^{-1})]			
	FWO	KAS	Starink	Average
Stage I (P2 reaction mechanism)				
0.1	16.0	14.3	14.5	14.9
0.2	16.1	14.4	14.6	15.0
0.3	15.5	13.7	13.9	14.3
0.4	15.5	13.7	13.9	14.4
0.5	15.8	14.0	14.2	14.7
0.6	14.3	12.4	12.6	13.1
Average	15.5	13.8	14.0	14.4
Stage II (F3 reaction mechanism)				
0.7	21.0	19.8	20.0	20.3
0.8	24.8	23.9	24.1	24.3
Average	22.9	21.85	22.05	22.3

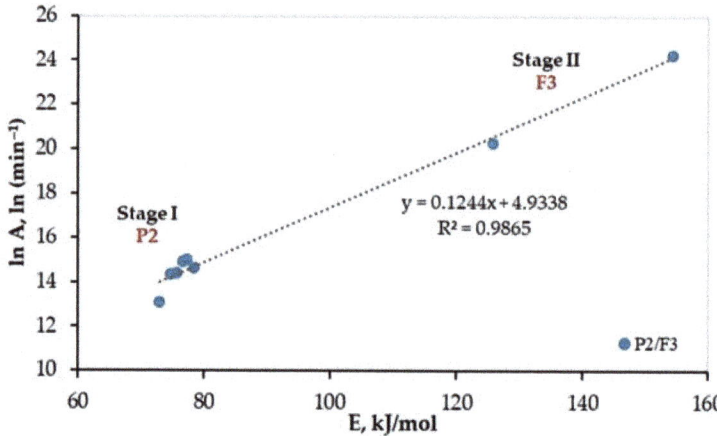

Figure 5. Linear fitted curve for the compensation effect.

To check the proposed mechanism, the Criado model [24] or compensation effect (i.e., Equation (16)) [34,35] can be used. In this work, the linearity between lnA and E_a was checked at both stages. As shown in Figure 5, a linear relationship was confirmed with a regression coefficient of 0.9865. This implies the suitability of the proposed reaction models for both stages, and the average values of A were 1.79×10^6 min^{-1}, and 4.84×10^9 min^{-1} for stages I, and II, respectively.

3.3. Estimation of the Thermodynamic Parameters of the PVC Pyrolysis

Thermodynamic properties along with the kinetic triplet are very important to the optimization of the large-scale pyrolytic reactor. Therefore, the thermodynamic parameters (ΔH, ΔG and ΔS) were obtained at different heating rates (5, 10, and 20 K/min) for both stages of the PVC pyrolysis as presented in Table 9.

Table 9. Thermodynamic parameters estimated for the process of the pyrolysis of PVC.

Stage	I			II		
Heating rates (K/min)	5	10	20	5	10	20
Kinetic Parameters						
E_a (kJ/mol)		75			140	
A (min^{-1})		1.79×10^6			4.84×10^9	
Kinetic Equation		$\frac{d\alpha}{dt} = 1.79 \times 10^6 \, e^{\frac{-75000}{RT}} \left(2\alpha^{\frac{1}{2}}\right)$			$\frac{d\alpha}{dt} = 4.84 \times 10^9 \, e^{\frac{-140000}{RT}} (1-\alpha)^3$	
T_p (K)	540	557	580	700	720	753
Thermodynamic Parameters						
ΔH (kJ/mol)	70.5	70.4	70.2	134.2	134.0	133.7
ΔG (kJ/mol)	145.2	147.6	150.8	186.6	188.1	190.6
ΔS (kJ/mol.K)	−0.14	−0.14	−0.14	−0.07	−0.08	−0.08
Potential Energy Barrier						
E_a–ΔH (kJ/mol) *		4.6			6	

* Based on the mean values of ΔH

As presented in Table 9, the PVC pyrolysis has positive ΔH values (stage I: 70.4 ± 0.17 kJ/mol, and stage II: 134 ± 0.22 kJ/mol). The positive sign of ΔH values indicates that both stages include endothermic reactions. In addition, these results reveal that higher energy is needed for stage II when compared with stage I. Furthermore, a small energy barrier (E_a-ΔH) of (stage I: 4.6 kJ/mole, and stage II: 6 kJ/mole) was observed. This amount of energy must be added for the reactions to take place.

Moreover, the positive and negative signs of ΔG and ΔS, respectively, indicate that the PVC pyrolysis is a nonspontaneous process (i.e., products have a lower disorder degree than the reactants). Additionally, the values of ΔS can indicate the reactivity order (stage I presented a slightly lower reactivity when compared to stage II). Furthermore, the positive values of ΔG reflect the amount of available bioenergy that can be produced from the PVC pyrolysis in each stage. The thermodynamic results reveal the promising potential of the PVC pyrolysis to efficiently produce bioenergy.

3.4. Pyrolysis Prediction by ANN Model

The experimental TGA datasets (403 datasets) were automatically and randomly divided into three sets: 70% (283 data sets) were used for training, 15% (60 data sets) were used for validation, and 15% (60 data sets) were used for testing.

To find the best topology of the ANN model aiming to predict the TGA data of the PVC pyrolysis, the number of hidden layers, number of neurons in each layer, and transfer functions have been optimized. Table 10 shows the performance of different ANN structures. The value of correlation coefficient (R) was considered as the main criterion for the selection of the most efficient network structure to estimate the weight left % as the output variable. As presented in Table 10, the ANN7 model shows the best performance (R = 0.99999) with the minimum no. of hidden layers and neurons, and the topology of the selected network (2-10-10-1) is presented in Figure 6. The ANN7 model has two input parameters (temperature, and heating rate), two hidden layers having 10 neurons in each layer, and an output parameter (PVC weight left %). In addition, the model has a feed-forward back-propagation characteristic and the TANSIG-LOGSIG transfer function was recommended.

Table 10. Prediction performance of different ANN structures.

Model	Network Topology	1st Transfer Function (Hidden Layer 1)	2nd Transfer Function (Hidden Layer 2)	R
ANN1	NN-2-5-1	TANSIG	-	0.99885
ANN2	NN-2-5-1	LOGSIG	-	0.99808
ANN3	NN-2-10-1	TANSIG	-	0.99956
ANN4	NN-2-10-1	LOGSIG	-	0.99940
ANN5	NN-2-15-1	TANSIG	-	0.99957
ANN6	NN-2-15-1	LOGSIG	-	0.99985
ANN7	NN-2-10-10-1	TANSIG	LOGSIG	0.99999
ANN8	NN-2-10-10-1	LOGSIG	LOGSIG	0.99996
ANN9	NN-2-10-15-1	LOGSIG	LOGSIG	0.99999
ANN10	NN-2-15-15-1	LOGSIG	LOGSIG	0.99979
ANN11	NN-2-10-15-1	TANSIG	LOGSIG	0.99996
ANN12	NN-2-15-15-1	LOGSIG	TANSIG	0.99988
ANN13	NN-2-15-15-1	LOGSIG	LOGSIG	0.99999

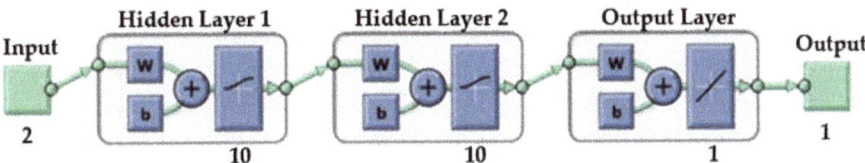

Figure 6. Topology of the best-selected network.

Then, the performance of the developed model was tested. As shown in Figure 7, a full agreement between the ANN-predicted values (Y-axis) and the experimental values (X-axis) has been guaranteed (R = 1.0). In addition, as presented in Table 11, RMSE, MAE, and MBE were significantly low. This implies the robustness of the developed model to predict the TGA data of the PVC pyrolysis.

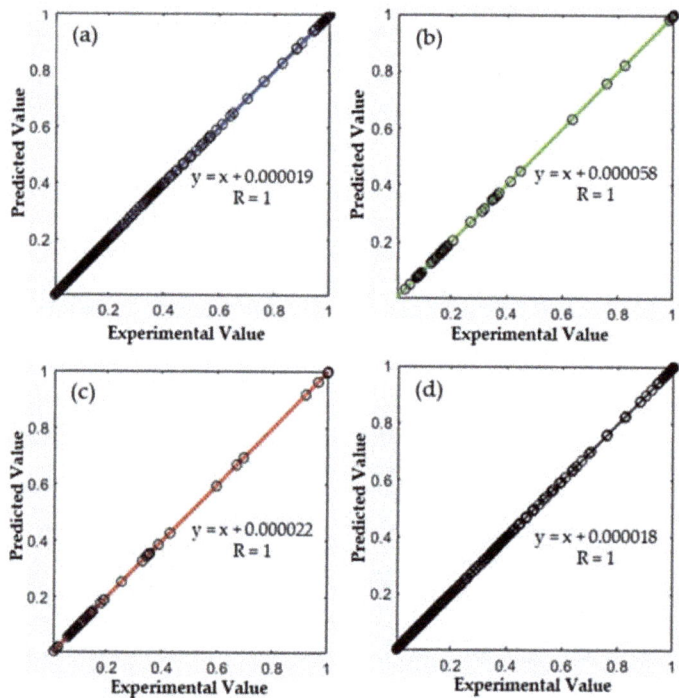

Figure 7. Regression plots of (**a**) training data, (**b**) validation data, (**c**) test data, and (**d**) complete datasets of the (2-10-10-1) ANN model.

Table 11. Statistical parameters of the (2-10-10-1) ANN network.

Set	Statistical Parameters			
	R	RMSE	MAE	MBE
Training	1.00000	0.000507	0.000315	0.000037
Validation	1.00000	0.000547	0.000319	0.000026
Test	1.00000	0.000388	0.000255	0.000005
All	1.00000	0.000481	0.000296	0.000027

Additionally, the performance of the developed ANN model was checked using new datasets. This step is known as 'the simulation step'. For this purpose, nine extra datasets were utilized. Table 12 presents the input and the targeted-output data of this step.

Table 12. Input and targeted-output data used during the simulation step.

No.	Input Data		Targeted-Output Data
	Heating Rate (K/min)	Temperature (K)	Weight Left (Fraction)
1	2	540.5	0.74
2	2	599.2	0.41
3	2	758.6	0.10
4	10	547.1	0.82
5	10	577.8	0.53
6	10	761.3	0.20
7	20	563.0	0.92
8	20	596.5	0.61
9	20	735.5	0.31

As presented in Figure 8, the full agreement between the experimental and the predicted values indicates the high performance of the developed model. In addition, the R-value of one and very low values of RMSE, MAE, and MBE (see Table 13) were obtained and additionally confirmed the robustness of the developed model to predict the TGA data of the PVC pyrolysis.

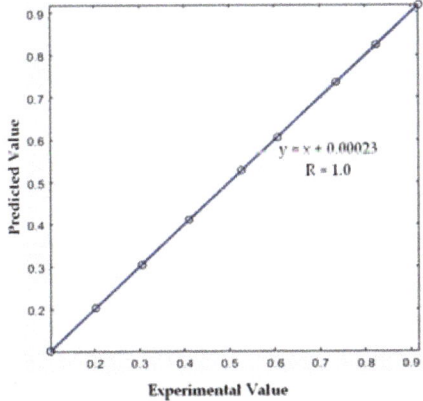

Figure 8. Comparison between predicted and experimental data using the simulation data.

Table 13. Statistical parameters of the (2-10-10-1) model during the simulation step.

Set	Statistical Parameters			
	R	RMSE	MAE	MBE
Simulated	1.00000	0.000576	0.000479	−0.000118

4. Conclusions

In this work, a comprehensive investigation of the pyrolysis of PVC at heating rates of 5, 10, and 20 K/min using thermogravimetric analysis and artificial neural network modeling was performed. Three isoconversional integral models (FWO, KAS, and Starink) and the Coats-Redfern non-isoconversional model were used to obtain the kinetic triplet (activation energy, pre-exponential factor, and reaction mechanism) of the PVC pyrolysis. Based on the reported results, the following conclusions can be drawn:

- The proximate analysis of PVC samples indicates the suitability of the production of bioenergy from the pyrolysis of PVC. For future work, products of the PVC pyrolysis should be identified, and their caloric values are to be obtained and compared with those of conventional fuels.

- The ultimate analysis of PVC samples showed low nitrogen and sulfur contents which are preferable to avoid the production of toxic gasses such as NOx and Sox, and thus benefit the environment.
- Thermogravimetric and derivative-thermogravimetric curves revealed that the PVC pyrolysis occurred in two stages covering the temperature range of 490–825 K.
- The kinetic triplets of both stages of the PVC pyrolysis were obtained and thus step-by-step guidance was outlined. This procedure can be followed to obtain the kinetic triplet of the pyrolysis of different wastes.
- The thermodynamic properties (ΔH, ΔG, and ΔS) of the process of the PVC pyrolysis showed that the reactions of both stages are endothermic and nonspontaneous, and confirmed the suitability of the production of bioenergy by the pyrolysis process.
- A highly efficient ANN model to predict the TGA data of the PVC pyrolysis was developed. It has the following characteristics: a feed-forward back-propagation algorithm, TANSIG-LOGSIG transfer function, and 2-10-10-1 network topology. For future studies, other artificial intelligence (AI) algorithms can be developed to predict the TGA data of the PVC pyrolysis, and their performance can be tested.

Author Contributions: Both authors contributed significantly to the completion of this article, but they had different roles in all aspects. Conceptualization, M.A.-Y. and I.D.; data curation, M.A.-Y. and I.D.; formal analysis, M.A.-Y. and I.D.; funding acquisition, M.A.-Y.; investigation, M.A.-Y. and I.D.; methodology, M.A.-Y. and I.D.; project administration, M.A.-Y.; software, I.D.; validation, M.A.-Y.; visualization, M.A.-Y. and I.D.; writing—original draft, M.A.-Y. and I.D.; writing—review and editing, M.A.-Y. All authors have read and agreed to the published version of the manuscript.

Funding: This research and the APC were funded by the Deanship of Scientific Research at King Faisal University (Saudi Arabia), Annual Research Program (Grant No. 170083).

Institutional Review Board Statement: Not applicable.

Informed Consent Statement: Not applicable.

Data Availability Statement: The authors confirm that the data supporting the findings of this study are available within the article.

Acknowledgments: The authors gratefully thank the Deanship of Scientific Research at King Faisal University (Saudi Arabia) for the financial support under the Annual Research Program (Grant No. 170083).

Conflicts of Interest: The authors declare no conflict of interest.

References

1. Chen, R.; Li, Q.; Xu, X.; Zhang, D. Comparative pyrolysis characteristics of representative commercial thermosetting plastic waste in inert and oxygenous atmosphere. *Fuel* **2019**, *246*, 212–221. [CrossRef]
2. Silvarrey, L.D.; Phan, A.N. Kinetic study of municipal plastic waste. *Int. J. Hydrogen Energy* **2016**, *41*, 16352–16364. [CrossRef]
3. Ding, Y.; Zhang, W.; Yu, L.; Lu, K. The accuracy and efficiency of GA and PSO optimization schemes on estimating reaction kinetic parameters of biomass pyrolysis. *Energy* **2019**, *176*, 582–588. [CrossRef]
4. Bach, Q.-V.; Chen, W.-H.; Eng, C.F.; Wang, C.-W.; Liang, K.-C.; Kuo, J.-Y. Pyrolysis characteristics and non-isothermal torrefaction kinetics of industrial solid wastes. *Fuel* **2019**, *251*, 118–125. [CrossRef]
5. Cardona, S.C.; Corma, A. Tertiary recycling of polypropylene by catalytic cracking in a semibatch stirred reactor: Use of spent equilibrium FCC commercial catalyst. *Appl. Catal. B Environ.* **2000**, *25*, 151–162. [CrossRef]
6. Kim, S. Pyrolysis kinetics of waste PVC pipe. *Waste Manag.* **2001**, *21*, 609–616. [CrossRef]
7. Karayildirim, T.; Yanik, J.; Yuksel, M.; Saglam, M.; Vasile, C.; Bockhorn, H. The effect of some fillers on PVC degradation. *J. Anal. Appl. Pyrolysis* **2006**, *75*, 112–119. [CrossRef]
8. Wu, J.; Chen, T.; Luo, X.; Han, D.; Wang, Z.; Wu, J. TG/FTIR analysis on co-pyrolysis behavior of PE, PVC and PS. *Waste Manag.* **2014**, *34*, 676–682. [CrossRef] [PubMed]
9. Yu, J.; Sun, L.; Ma, C.; Qiao, Y.; Yao, H. Thermal degradation of PVC: A review. *Waste Manag.* **2016**, *48*, 300–314. [CrossRef]
10. Xu, F.; Wang, B.; Yang, D.; Hao, J.; Qiao, Y.; Tian, Y. Thermal degradation of typical plastics under high heating rate conditions by TG-FTIR: Pyrolysis behaviors and kinetic analysis. *Energy Convers. Manag.* **2018**, *171*, 1106–1115. [CrossRef]
11. Ma, W.; Rajput, G.; Pan, M.; Lin, F.; Zhong, L.; Chen, G. Pyrolysis of typical MSW components by Py-GC/MS and TG-FTIR. *Fuel* **2019**, *251*, 693–708. [CrossRef]

12. Özsin, G.; Pütün, A.E. TGA/MS/FT-IR study for kinetic evaluation and evolved gas analysis of a biomass/PVC co-pyrolysis process. *Energy Convers. Manag.* **2019**, *182*, 143–153. [CrossRef]
13. Zhou, R.; Huang, B.; Ding, Y.; Li, W.; Mu, J. Thermal Decomposition Mechanism and Kinetics Study of Plastic Waste Chlorinated Polyvinyl Chloride. *Polymers* **2019**, *11*, 2080. [CrossRef]
14. Conesa, J.A.; Caballero, J.A.; Reyes-Labarta, J.A. Artificial neural network for modelling thermal decompositions. *J. Anal. Appl. Pyrolysis* **2004**, *71*, 343–352. [CrossRef]
15. Yıldız, Z.; Uzun, H.; Ceylan, S.; Topcu, Y. Application of artificial neural networks to co-combustion of hazelnut husk–lignite coal blends. *Bioresour. Technol.* **2016**, *200*, 42–47. [CrossRef]
16. Çepelioğullar, Ö.; Mutlu, İ.; Yaman, S.; Haykiri-Acma, H. A study to predict pyrolytic behaviors of refuse-derived fuel (RDF): Artificial neural network application. *J. Anal. Appl. Pyrolysis* **2016**, *122*, 84–94. [CrossRef]
17. Charde, S.J.; Sonawane, S.S.; Sonawane, S.H.; Shimpi, N.G. Degradation Kinetics of Polycarbonate Composites: Kinetic Parameters and Artificial Neural Network. *Chem. Biochem. Eng. Q.* **2018**, *32*, 151–165. [CrossRef]
18. Chen, J.; Xie, C.; Liu, J.; He, Y.; Xie, W.; Zhang, X.; Chang, K.; Kuo, J.; Sun, J.; Zheng, L.; et al. Co-combustion of sewage sludge and coffee grounds under increased O_2/CO_2 atmospheres: Thermodynamic characteristics, kinetics and artificial neural network modeling. *Bioresour. Technol.* **2018**, *250*, 230–238. [CrossRef] [PubMed]
19. Naqvi, S.R.; Tariq, R.; Hameed, Z.; Ali, I.; Taqvi, S.A.; Naqvi, M.; Niazi, M.B.K.; Noor, T.; Farooq, W. Pyrolysis of high-ash sewage sludge: Thermo-kinetic study using TGA and artificial neural networks. *Fuel* **2018**, *233*, 529–538. [CrossRef]
20. Dubdub, I.; Al-Yaari, M. Pyrolysis of Low Density Polyethylene: Kinetic Study Using TGA Data and ANN Prediction. *Polymers* **2020**, *12*, 891. [CrossRef] [PubMed]
21. Dubdub, I.; Al-Yaari, M. Pyrolysis of high-density polyethylene: II. Artificial Neural Networks Modeling. In Proceedings of the 9th Jordan International Chemical Engineering Conference (JICHEC9), Amman, Jordan, 12–14 October 2021.
22. Dubdub, I.; Al-Yaari, M. Pyrolysis of Mixed Plastic Waste: II. An Artificial Neural Networks Prediction and Sensitivity Analysis. *Appl. Sci.* **2021**, *11*, 8456. [CrossRef]
23. Al-Yaari, M.; Dubdub, I. Application of Artificial Neural Networks to Predict the Catalytic Pyrolysis of HDPE Using Non-Isothermal TGA Data. *Polymers* **2020**, *12*, 1813. [CrossRef] [PubMed]
24. Dubdub, I.; Al-Yaari, M. Pyrolysis of Mixed Plastic Waste: I. Kinetic Study. *Materials* **2020**, *13*, 4912. [CrossRef] [PubMed]
25. Doyle, C.D. Series Approximations to the Equation of Thermogravimetric Data. *Nature* **1965**, *207*, 290–291. [CrossRef]
26. Murray, P.; White, J. Kinetics of thermal dehydration characteristics of the clay minerals. *Trans. Br. Ceram. Soc.* **1955**, *54*, 204–238.
27. Starink, M.J. The determination of activation energy from linear heating rate experiments: A comparison of the accuracy of isoconversion methods. *Thermochim. Acta* **2003**, *404*, 163–176. [CrossRef]
28. Khawam, A.; Flanagan, D.R. Solid-State Kinetic Models: Basics and Mathematical Fundamentals. *J. Phys. Chem. B* **2006**, *110*, 17315–17328. [CrossRef] [PubMed]
29. Jiménez, A.; López, J.; Vilaplana, J.; Dussel, H.-J. Thermal degradation of plastisols. Effect of some additives on the evolution of gaseous products. *J. Anal. Appl. Pyrolysis* **1997**, *40–41*, 201–215. [CrossRef]
30. Govindan, B.; Jakka, S.C.B.; Radhakrishnan, T.K.; Tiwari, A.K.; Sudhakar, T.M.; Shanmugavelu, P.; Kalburgi, A.K.; Sanyal, A.; Sarkar, S. Investigation on Kinetic Parameters of Combustion and Oxy-Combustion of Calcined Pet Coke Employing Thermogravimetric Analysis Coupled to Artificial Neural Network Modeling. *Energy Fuels* **2018**, *32*, 3995–4007. [CrossRef]
31. Al-Salem, S.M.; Antelava, A.; Constantinou, A.; Manos, G.; Dutta, A. A review on thermal and catalytic pyrolysis of plastic solid waste (PSW). *J. Environ. Manag.* **2017**, *197*, 177–198. [CrossRef] [PubMed]
32. López, A.; de Marco, I.; Caballero, B.M.; Laresgoiti, M.F.; Adrados, A. Dechlorination of fuels in pyrolysis of PVC containing plastic wastes. *Fuel Process. Technol.* **2011**, *92*, 253–260. [CrossRef]
33. Mumbach, G.D.; Alves, J.L.F.; Da Silva, J.C.G.; De Sena, R.F.; Marangoni, C.; Machado, R.A.F.; Bolzan, A. Thermal investigation of plastic solid waste pyrolysis via the deconvolution technique using the asymmetric double sigmoidal function: Determination of the kinetic triplet, thermodynamic parameters, thermal lifetime and pyrolytic oil composition for clean energy recovery. *Energy Convers. Manag.* **2019**, *200*, 112031. [CrossRef]
34. Dubdub, I.; Al-Yaari, M. Thermal Behavior of Mixed Plastics at Different Heating Rates: I. Pyrolysis Kinetics. *Polymers* **2021**, *13*, 3413. [CrossRef] [PubMed]
35. Al-Yaari, M.; Dubdub, I. Pyrolysis of high-density polyethylene: I. Kinetic Study. In Proceedings of the 9th Jordan International Chemical Engineering Conference (JICHEC9), Amman, Jordan, 12–14 October 2021.

Review

Parameters Influencing Moisture Diffusion in Epoxy-Based Materials during Hygrothermal Ageing—A Review by Statistical Analysis

Camille Gillet [1,2,*], Ferhat Tamssaouet [3], Bouchra Hassoune-Rhabbour [1], Tatiana Tchalla [2] and Valérie Nassiet [1,*]

1. Laboratoire Génie de Production, INP-ENIT, Université de Toulouse, 47 Av. d'Azereix, 65000 Tarbes, France; bouchra.hassoune-rhabbour@enit.fr
2. Safran Aircraft Engines, site de Villaroche, 77550 Moissy-Cramayel, France; tatiana.tchalla@safrangroup.com
3. Laboratoire PROMES-CNRS (UPR 8521), Université de Perpignan, TECNOSUD, 66100 Perpignan, France; ferhat.tamssaouet@univ-perp.fr
* Correspondence: camille.gillet@enit.fr (C.G.); valerie.nassiet@enit.fr (V.N.)

Abstract: The hygrothermal ageing of epoxy resins and epoxy matrix composite materials has been studied many times in the literature. Models have been developed to represent the diffusion behaviour of the materials. For reversible diffusions, Fick, Dual–Fick and Carter *and* Kibler models are widely used. Many parameters, correlated or not, have been identified. The objectives of this review by statistical analysis are to confirm or infirm these correlations, to highlight other correlations if they exist, and to establish which are the most important to study. This study focuses on the parameters of the Fick, Dual–Fick and Carter *and* Kibler models. For this purpose, statistical analyses are performed on data extracted and calculated from individuals described in the literature. Box plot and PCA analyses were chosen. Differences are then noticeable according to the different qualitative parameters chosen in the study. Moreover, correlations, already observed in the literature for quantitative variables, are confirmed. On the other hand, differences appear which may suggest that the models used are inappropriate for certain materials.

Keywords: epoxy resin; composite material; hygrothermal ageing; water diffusion; Fick model deviation; statistical analysis; box plot; PCA

1. Introduction

For many decades, epoxy-matrix composites have been widely used in many industries, especially in aeronautics. Airframers seek to lighten the onboard aircraft mass as much as possible to reduce their fossil energy consumption. From that perspective, metals are substituted by polymer-matrix composite materials, which offer very high mechanical properties for a significantly lower density. Epoxy is a commonly used thermosetting polymer family, whether as a matrix in composite materials or as a structural adhesive for bonded joints and repairs in the aeronautical industry. Epoxy prepolymers and their hardeners form non-cross-linked systems with low viscosity, which ease their processing. They cover a wide range of cross-linking temperatures (from 5 to over 200 °C), depending on their chemical composition and the intended utilization. The epoxy system shrinkage after cross-linking is low compared to other thermosetting materials such as phenolic resins, making them good candidates as adhesives. In addition, the presence of polar functional groups such as hydroxyl gives them the ability to adhere to substrates or fibres. Besides, epoxies are commonly used as a matrix for composite materials because of their good mechanical properties and chemical resistance to many corrosive substances and acids. Epoxies also have good dielectric properties, allowing them to be electrical insulators [1,2].

However, despite their many beneficial properties, epoxy-based materials are sensitive to wet exposure as water molecules can penetrate their macromolecular networks. Water molecule action can be reversible, as in the case of plasticization, or irreversible, such as hydrolysis. In the first case, the water molecules break the secondary bonds between neighbouring chains and partially destroy the mechanical cohesion of the polymer, which results in: (1) a decrease in the glass transition temperature T_g, which can be reversible; as well as (2) a loss of additives and the material swelling [3–5]. These swellings cause concentrations of mechanical stresses that can eventually lead to irreversible damage, such as decohesion between the fibre and the matrix or microcracks. Due to the difference in elasticity between the fibres and the matrix, and the water absorption, stresses develop along with the fibre/matrix interface. Within the sample, stress equilibrium is more easily maintained. However, at the surface, edge stresses are high enough to produce cracks. After cyclic exposure to a moist environment, micro-cracks and disbonds at the fibre/matrix interface tend to coalesce and expand [6–10]. In the case of hydrolysis, the water molecules penetrate directly to the macromolecule skeleton causing chain breaks, which destroy the cohesion of the material and allow the formation and propagation of cracks [11,12]. This decrease in mechanical properties is greatly impacting and has been widely demonstrated in the literature [13–21].

The literature has proved several times that hygrothermal ageing is governed by many parameters. These latter include, firstly: material parameters such as prepolymer type [22–25], hardener type [22,23,26], reinforcement type [13,27–29] or thickness [30,31]; secondly: conditioning parameters such as temperature [4,6,32] and relative humidity [22,33,34]; and thirdly: the resulting diffusion parameters such as saturation water mass uptake, saturation time and diffusivity [35]. Some of these parameters are correlated, meaning that they evolve together, while others are independent. The objective of this paper is to explore how these parameters influence each other and determine the principles behind this. Comparisons of water mass uptake by epoxy resin and epoxy-based composites have already been made in several publications, but without considering the variation of all the parameters [24]. A classical approach to comparing the different ageing mechanisms does not seem to be sufficient for a large dataset, because of the heterogeneity of the results.

This paper aims to confirm the correlations found by the literature between the different variables influencing the hygrothermal ageing and emphasize less apparent correlations. The objective is to highlight the most important variables of the problem, i.e., the most important to study.

Therefore, statistical studies on a large number of data were carried out. They include information on material type, ageing conditioning and parameters describing the diffusion curve. Data were extracted and calculated from *90 publications* on hygrothermal ageing of epoxy and epoxy-based composites: [4–7,10,13,15,16,18,19,22,24,26–34,36–104]. This study, which includes *448 individuals*, focuses on reversible gravimetric moisture uptake curves exhibiting Fick or Fick-derived two-step diffusion behaviour; water uptake is a function of the root of time. Irreversible behaviours, in particular those with mass losses during wet ageing or high mass uptake without asymptote, have been excluded from this study in order to avoid bringing together too different phenomena [22,32,105].

First, the diffusion parameters and the models to which they are related are presented. Then, the dataset is represented in order to study the relationships between the variables. For this purpose, box plots and scatter plots are generated to observe the variables' dispersion. However, the wide dispersion of the data could make it difficult to draw conclusions by simply observing the evolution of the variables between them. These data could present a great heterogeneity caused by the great diversity of the studied materials due to their different nature and process and the different conditioning parameters. All these parameters vary at the same time from one individual to another, which leads to different absorption behaviours, knowing that irreversible behaviours have already been ruled out. To complete the statistical study, the principal component analysis (PCA) is used. This analysis is effective for deducing correlations when a large number of quantitative variables is involved. By reducing and centering the variables, PCA facilitates observations despite their large dispersion. This includes:

- Highlighting material and ageing parameters influencing the diffusion parameters, by box plots and scatter plots;
- The correlation of the different quantitative parameters of the study by PCA;
- The separation of the data into four classified PCAs in order to propose more efficient correlations;
- Evidence of differences between PCA depending on the diffusion model and the type of material (resin or composite);
- Discussion of the limitations of using the diffusion models on epoxy-based materials.

2. Studied Parameters

The determination of diffusion parameters in the case of composites has been extensively study in the literature, with respect to the material configuration, the diffusion duration, and/or the diffusivity linearity, etc. [35,37,45,65,106–108].

Various models have been developed to describe the reversible behaviour of water absorption. The most well-known is the Fick model, for diffusion behaviour without anomaly, with a linear slope followed by saturation [109]. However, many epoxies show behaviours that deviates from the Fick model with the presence of two diffusion steps. Various models have been proposed to consider these anomalies, including the Dual–Fick model [20,110,111] and the Carter and Kibler model [19,21,45,51,112,113]. The equations of these three models are given below: Fick (Equation (1)), Dual–Fick (Equation (2)) and Carter and Kibler (Equation (3)).

$$M(t) = M_{sat}\left(1 - \frac{8}{\pi^2}\sum_{n=0}^{\infty}\frac{1}{(2n+1)^2} \cdot exp\left[-(2n+1)^2 \cdot \pi^2 \cdot \frac{D \cdot t}{h^2}\right]\right) \quad (1)$$

$$M(t) = \sum_{i=1}^{2}\left[M_{sat_i}\left(1 - \frac{8}{\pi^2}\sum_{n=0}^{\infty}\frac{1}{(2n+1)^2} \cdot exp\left[-(2n+1)^2 \cdot \pi^2 \cdot \frac{D_i \cdot t}{h^2}\right]\right)\right] \quad (2)$$

$$M(t) = M_{sat}\left(1 - \frac{8}{\pi^2}\sum_{l=1}^{\infty}\frac{r_l^+ exp(-r_l^- t) - r_l^- exp(-r_l^+ t)}{l^2(r_l^+ - r_l^-)} + \frac{8}{\pi^2}\frac{\kappa\beta}{l(\gamma+\beta)}\sum_{l=1}^{\infty}\frac{exp(-r_l^- t) - exp(-r_l^+ t)}{r_l^+ - r_l^-}\right). \quad (3)$$

In this study, the different parameters defining the diffusion are taken from those three models. For Fick behaviour individuals, the parameters selected are the saturation water mass uptake M_{sat}, the saturation time t_{sat} and the diffusivity D. Additional parameters are included to refine the observations for the two-step diffusion behaviour (abbreviated to Dual) individuals. Two parameters come from the Dual–Fick model: the water mass uptake at the first absorption step M_{inter}, and its time to be reached t_{inter} (Figure 1). The last three parameters, determined using the model developed by Carter and Kibler, that takes into account the functional groups of hydrophilicity, are: (1) γ the probability per time unit that free water molecules in the polymer will bond; (2) β the probability per time unit that bonded water molecules will liberate; and (3) K a parameter related to the material swelling. For the homogeneity purpose, all these parameters have been recalculated using the Carter and Kibler approximations, as a reminder [45]:

1- At the first step (intersection point between the two sorption stages):

$$\frac{M_{inter}}{M_{sat}} \cong \frac{\beta}{\gamma+\beta}; \quad (4)$$

2- At short times (curve linear domain):

$$\frac{M_{inter}}{M_{sat}} = \frac{4}{\pi^{3/2}}\left(\frac{\beta}{\gamma+\beta}\right)\sqrt{Kt}; \quad (5)$$

3- At longer times (beyond the first step):

$$\frac{M(t)}{M_{sat}} \cong \left\{1 - \frac{\gamma}{\gamma + \beta}(e^{-\beta t})\right\}; \qquad (6)$$

where: $K = \frac{\pi^2 D}{2h^2}$, D is the diffusivity, and h is the sample thickness.

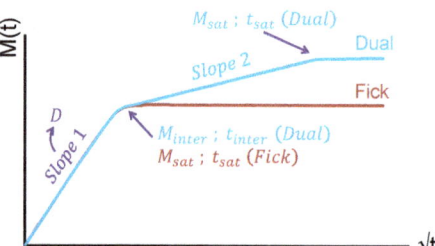

Figure 1. Fick and two-step (Dual) diffusion curves, positions of M_{sat}, t_{sat}, M_{inter} and t_{inter}.

All these parameters are related to the material parameters: prepolymer type, hardener type, reinforcement type, fibre architecture, fibre volume fraction v_f, thickness h, process; and conditioning parameters: conditioning (immersion in water or exposure to humid air), relative humidity RH, and ageing temperature T_{ageing}. Some variables are qualitative while others are quantitative (Table 1).

Table 1. List of variables used for the dispersion study.

	Material Parameters	**Ageing Parameters**	**Diffusion Parameters**
Qualitative	Prepolymer type Hardener type Reinforcement type Fibre architecture Process	Conditioning	Diffusion behaviour
Quantitative	v_f h	RH T_{ageing}	D M_{sat} t_{sat}

3. Descriptive Statistics: Study of Dispersion

To observe the variables' dispersion and study their mutual influence, the whole dataset is represented using box plots. This allows the removal of outliers and the highlighting of the median and central quartile values. As a result, the dataset is separated into four equally sized portions. Each of the three values that divide the elements of a statistical distribution is called a quartile. The box contains 50% of the individuals: 25% between the first quartile and the second quartile-or median-and 25% between the median and the third quartile. The "whiskers" are the lines that run across the box. They indicate the variability outside the quartiles, and each includes 25% of the individuals. The points outside the whiskers are atypical values.

With this first scatter analysis, the entire dataset is observed. The dispersion of the quantitative values saturation mass uptake M_{sat}, diffusivity D and saturation time scaled to thickness t_{sat}/h are studied, as a function of the categorical variables that describe the dataset: diffusion behaviour, prepolymer type, hardener type, reinforcement type, fibre architecture, process, and conditioning (Table 1). Time was reduced to thickness to normalize the results and obtain comparable durations. These three quantitative variables were also observed as a function of other quantitative variables: fibre volume fraction, relative humidity rate RH, and ageing temperature T_{ageing}. For this purpose, scatter plots were generated.

The R programming language and software, its packages *corrplot* and *factoextra*, as well as the *boxplot* function are used [114–116].

3.1. Diffusion Behaviour

The two types of moisture diffusion behaviour, Fick and Dual diffusions, are studied. Figure 2 shows the data distribution within quartiles for the variables t_{sat}/h, D and M_{sat}. All characteristic values of the box plots are shown. Outliers are not shown for the sake of clarity, but their percentage is indicated. One can observe that the distributions are not centered, and all the variables and behaviours have a higher skewness, which shows that they are not normal (Gaussian). This is reflected in the fact that some individuals have higher mass absorption, longer saturation time, and greater diffusivity, i.e., greater sensitivity to moisture. The box plot shows that the same median is obtained regardless of the moisture diffusion behaviour. In contrast, the third quartile and the upper whisker boundary are larger for specimens with two diffusion steps. As many individuals are located on either side of the median, but beyond this median, they are more widely spaced, and some have a large diffusivity that stretches the graph. Diffusivity and saturation time reduced to thickness reveal a very wide dispersion from one behaviour to another. For t_{sat}/h, the gap between the two behaviour types is around 10^2 for the median, which is very large.

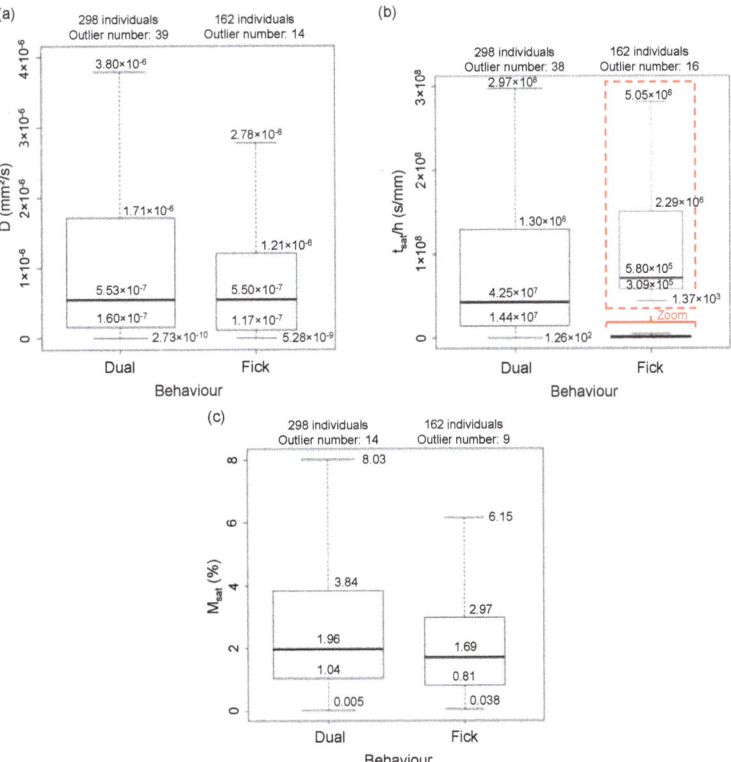

Figure 2. Box plots and characteristic values of (**a**) D, (**b**) t_{sat}/h and (**c**) M_{sat} as a function of the diffusion behaviour.

For the saturation mass uptake, the median is slightly lower for Fick individuals than for individuals with two diffusion steps. For Fick individuals, the distance between quartiles is similar. A low dispersion of epoxies and epoxy-based composites is observed.

Indeed, in this study, 75% of the individuals have a moisture absorption lower than 4%. The diffusivity and the saturation time show strong dispersion. While saturation can take considerable time, the saturation mass uptake varies only slightly after passing the first diffusion step. These differences in diffusion behaviour depend on the different parameters given in the Table 1. Indeed, the diffusion parameters differ according to the material parameters and the ageing parameters, detailed in the following.

3.2. Epoxy Prepolymer Type

Some polymers are more likely to absorb water due to their hydrophilicity [102,117–119], or the free volume presence [120–122]. Among the epoxy family, differences were found in the prepolymers functional groups and in their structure. The dataset is sorted according to the prepolymer type in order to observe the variation in the dispersion of M_{sat}, t_{sat}/h and D. For the sake of simplicity, only medians are shown (Table 2) and the entire box plots are available as Supplementary Materials (Figures S1–S3). Medians divide the studied populations into two sets comprising the same number of individuals. One can notice that these individuals show high skewness similarly to the box plots based on diffusion behaviour. Looking at the saturation mass uptake, we find values about 1.05% for the DGEBF resin and 6.37% for the DGEBA + novolac resin. In contrast, the DGEBA and novolac resins have weaker saturation mass uptake when unblended. The other blends (DGEBA + mTGAP, DGEBA + TGDDM) also appear to have lower moisture absorption than the neat resins. Furthermore, the mTGAP resins also show strong moisture absorption, which can be explained by the fact that they are two aromatic epoxy prepolymers with three oxirane groups and a nitrogen atom, which are very hydrophilic [25,117,118,123,124].

Table 2. Medians of M_{sat}, t_{sat}/h et D as a function of epoxy prepolymer type. List of pre-polymers and entire box plots are available as Supplementary Materials (Figures S1 and S3–S5).

Name (Individuals Number)	M_{sat} (%)	t_{sat}/h (s/mm $\times 10^7$)	D (mm²/s $\times 10^{-7}$)
DGEBA based (217)	1.61	0.80	5.23
DGEBA + mTGAP (3)	4.44	0.16	106
DGEBA + novolac (3)	6.37	22.30	0.50
DGEBA + TGDDM (2)	4.31	4.96	13.00
DGEBD (5)	6.15	0.02	17.40
DGEBF based (15)	1.05	7.02	3.49
ESO (11)	1.75	0.18	4.67
mTGAP (7)	5.35	0.16	56.61
Novolac (11)	2.26	0.36	15.50
Rubber-modified epoxy (2)	3.08	8.02	0.47
TCDAM (1)	3.84	8.02	0.47
TGDDM (80)	1.50	2.26	4.36
TGDDM + mTGAP (14)	2.02	1.71	6.62
Unknown epoxy (89)	2.64	1.69	9.07

Indeed, the hydrophilicity of the polymer is linked to the nature of the chemical groups composing the macromolecules. Water molecules are attracted to polar functions, such as hydroxyls or nitrogens. According to Van Krevelen, water absorption is an additive molar function. For a representative structural unit, independent of its environment, it is written (Equation (7)) [117,118]:

$$H = \frac{M(t) \cdot M}{1800}, \quad (7)$$

where H is the number of water moles, M is the molar mass and $M(t)$ is the percentage of mass uptake. There are universal values of H. For non-polar hydrocarbons (–CH–, –CH$_2$–, –CH$_3$), fluorinated groups or aromatic rings: $H = 0$. Moderately polar groups, such as esters and ethers, have $H < 0.3$. Hydrogen bonding groups, e.g., acids, alcohols, amides, amines and hydroxyls, are very hydrophilic and have: $H = 1$–2 [25,118,123–125]. Some structures combine hydrophilic and hydrophobic groups, such as polyamides and epoxies. The hydrophilicity then depends on the respective proportions of these groups. These predictions apply in the case where the contributions of the different groups are independent. For epoxies, where internal hydrogen bonds can interact with hydrogen bonds caused by moisture uptake, it is necessary to consider structural units that include groups interacting with each other [101,102,119,126].

However, the TGDDM prepolymers, composed of four oxirane groups and two nitrogen atoms, do not experience such large moisture absorption. Another phenomenon that affects moisture uptake is the gelation rate. Indeed, faster gelling traps more voids and has a lower cross-linking density. For example, Frank et al. report this to be the case for mTGAP/DDS, which has a greater saturation mass uptake than DGEBA/DDS. Nevertheless, the size of the free volumes decreases with the macromolecule functionality because of the increase of the cross-linking density. So although TGDDM/DDS has four functionalities against three for mTGAP/DDS, its longer and more flexible skeleton allows it to have smaller free volumes and lower absorption [41,44,119].

Different profiles are observed in Table 2. The highest absorption rates are not always associated with large diffusivities or long saturation times. The more polar functions the epoxy has, the greater the diffusion temperature dependence and the more hydrophilic it is. In this study, a pretty low saturation mass uptake is associated with a high saturation time and a low diffusivity, and vice versa: for example, DGEBF, ESO, TGDDM or DGEBA resins. On the other hand, not all specimens respect this behaviour as the DGEBA + Novolac mixtures and mTGAP combine low diffusivities, long saturation time and high saturation mass uptake. The diffusion depends on the absorption curve shape. According to [25], polar sites in hydrophilic systems could act as a barrier to diffusion. Polar groups act as "bottlenecks" and trap incoming water molecules, named bonded water, which inhibits their free diffusion within the free volumes. Considering the data dispersion, the curing effect on moisture sensitivity should also be taken into account. Indeed, Abdelkader and White find that imperfect resin cross-linking leaves uncrosslinked epoxy areas that are more sensitive to water and enable faster penetration into the material. In addition, cross-linking at excessive temperatures could reduce the polymer density through the porosities formation. These porosities allow water molecules to have more free volumes in which to settle [26,102].

3.3. Hardener Type

Due to the large number of hardener types, they are classified according to their chemistry: amine, amidoamine, dicyandiamide, anhydrid acid and phenol novolac. Since 276 individuals are amine, or 62%, this category of hardeners was divided into four, according to their structure: aliphatic, cycloaliphatic, aromatic and unknown when not given in the publication (Table 3).

Table 3. Medians of M_{sat}, t_{sat}/h and D as a function of the hardener type. List of hardeners and entire box plots are available as Supplementary Materials (Figures S2 and S6–S8).

Name (Individuals Number)	M_{sat} (%)	t_{sat}/h (s/mm $\times 10^7$)	D (mm^2/s $\times 10^{-7}$)
Aliphatic amine (51)	2.61	1.23	8.51
Aromatic amine (139)	2.63	1.62	5.79
Cycloaliphatic amine (14)	1.81	4.32	3.15
Amidoamine (16)	1.47	0.44	9.18
Dicyandiamide (61)	1.59	5.24	3.46
Anhydride acid (42)	1.01	0.87	4.42
Phenol Novolac (9)	2.10	0.36	5.56
Unknown amine (71)	0.79	0.28	7.33
Unknown hardener (57)	3.20	2.92	8.57

The classification by the type of hardener in this paper shows a lower dispersion than the classification by the type of prepolymer, for example, medians of the saturation mass uptake range from 0.79% to 3.20%. It should be taken into account that the crosslinking sites of thermoset materials are generally considered hydrophilic due to the high presence of hydrogen bridges, which tend to absorb more water [127–129]. For example, polyvinyl alcohol, polyacrylamides, amine and amide hardeners are very hydrophilic, which may explain this lower dispersion.

In the box plots, both aliphatic and aromatic amines show similar saturation mass uptake and saturation time medians. The diffusivity is lower for the aromatics. Cycloaliphatic amines have slightly lower saturation mass uptake, significantly lower diffusivity medians and longer saturation times. The unknown amines have much lower M_{sat} and t_{sat}/h, but D close to those of aliphatic amines. This result conflicts with some observations in the literature, where aliphatic amines have higher saturation masses than aromatic amines [26,41]. Amidoamines have much lower M_{sat} and t_{sat}/h than aliphatic, aromatic and cycloaliphatic amines. On the other hand, they show the highest D medians of the study. The dicyandiamides have the highest t_{sat}/h medians. The anhydrid acids have low M_{sat} and t_{sat}/h. Phenol novolacs exhibit low t_{sat}/h, with M_{sat} and D close to those of aromatic amines.

Hardeners with a bulky structure, such as amidoamines, enable us to obtain a flexible network because of their long macromolecules with a high functionality. The steric hindrance is very important, which causes significant free volume, allowing the rapid entrance of water molecules [26]. On the other hand, their chemistry does not present more hydrophilic sites than other types of hardeners, such as amines, which can explain the rather low M_{sat} medians. For aromatic amines, such as dianilines, the variation in their sensitivity to moisture is related to their reactivity and polarity. Furthermore, the hydroxyl and tertiary amine groups play a concerted role in the water bonding to the network, making their group contributions indistinguishable. The formation of a water-amine hydrogen bond competes directly or indirectly with an internal hydroxylamine hydrogen bond whose strength increases with the amine nucleophilicity. The indirect effect would occur if the water placement required a site of a particular steric configuration which, in its turn, depends on the amine-hydroxyl interaction [26,119].

Among the cycloaliphatic amines, many are IPDA, which are quite hydrophobic due to their CH_3 groups. These results have already been observed in the literature. The aliphatic and aromatic amines are all moderately hydrophilic due to their chemical composition. The most important saturation M_{sat} are linked to relatively low t_{sat}/h compared to the others. This may also apply to anhydride acid. In this hardener family, we observe hydrophobic specimens due to the CH_3 presence, which reduces their hydrophilicity. This may be the case with acetic anhydride. However, they are also more sensitive to humidity anhydrides, such as methylhexahydrophthalic anhydride (MHHPA). Epoxy anhydrides subjected to absorption followed by hydrolysis-related mass losses have been reported in the literature [22]. Hydrolysis-sensitive hardeners, such as some anhydrides or dicyandiamides, could then have lower mass uptakes.

3.4. Fibres Presence

While epoxy resin matrices are subject to wet ageing, carbon fibres can be considered impervious to moisture. Similarly, glass fibres are considered to have a low sensitivity to moisture over the material lifetime. It is important to note that despite their impermeability, these inorganic fibres impact moisture diffusion as they modify the behaviour of the polymer at the fibre-matrix interface. Differential swelling occurs and develops high stresses. Decohesion occurs if the adhesion is not sufficient. A void then appears at the interface, and the water uses it to propagate more rapidly through the material, like a moisture carrier [8,39,118]. In the study, composites based on carbon and glass fibres have the lowest median absorbed water mass uptake, with 1.39% for carbon and 0.80% for glass (Table 4). Their median diffusivities and saturation times are also the lowest among all the reinforcement types. As glass and carbon fibres are not very sensitive to water, M_{sat} should be reduced to the resin mass fraction in order to take into consideration only the matrix part in the composites. We also note that the M_{sat} and D box plots for the glass fibres are centred, i.e., the specimens are homogeneously distributed on either side of the medians, which is equal to the mean, despite the differences in resin type and conditioning. This low dispersion of values may be due to the diffusion inertia associated with the presence of fibres.

Table 4. Medians of M_{sat}, t_{sat}/h and D as a function of the reinforcement type. "None" refers to a neat resin, without reinforcement. Entire box plots are available as Supplementary Materials (Figures S9–S11).

Name (Individuals Number)	M_{sat} (%)	t_{sat}/h (s/mm $\times 10^7$)	D (mm^2/s $\times 10^{-7}$)
Aramid (30)	3.54	1.16	6.90
Carbon (127)	1.39	1.53	3.64
Flax (7)	9.82	0.60	14.20
Glass (88)	0.80	0.69	3.63
Hemp (6)	13.0	4.14	9.18
Hybrid carbon aramid (1)	1.62	1.30	3.52
Hybrid carbon glass (7)	3.33	0.04	25.92
Regenerated cellulose (2)	7.10	6.98	13.91
None (192)	2.87	1.59	6.54

The box plots highlight the flax, hemp and regenerated cellulose fibre composites, which have median saturation masses of 9.82, 13.0 and 7.10%, respectively. Their median diffusivity is slightly higher than the other materials. Thus, for the hemp and regenerated cellulose fibre individuals, the median saturation times is also higher. It is recognized in the literature that these organic vegetal fibres are sensitive to water. This sensitivity depends on their chemical composition, in particular on their lignin and hemicellulose content [130]. In the composite, in addition to the matrix, the fibres also absorb moisture. Indeed, organic fibres are subject to swelling, which can cause matrix cracking and accelerated water diffusion, resulting in a higher mass of absorbed water than in the neat resin [3,9,89]. Surface treatments can then be used on organic fibres to reduce their hydrophilicity and the moisture absorption they induce. For example, potassium hydroxide and sodium hydroxide are used to reduce the ability to create hydrogen bonds between natural fibres and water. For cellulose fibres, they will remove open hydroxyl groups. Silane can also be used to stabilize the fibres and make them resistant to leaching by masking hydroxyl groups, by creating silanol, and reducing the number of porosities [131–133]. Aramid fibres are also organic fibres, composed of very hydrophilic amide bonds, but they do not seem to cause disproportionate moisture absorption, with a median saturation mass uptake of 3.54%.

A composite material can have different behaviours on which the diffusion properties depend. The fibres' presence initially leads to a change in the flow path of water molecules, creating anisotropic diffusion at the macroscopic scale, which is the composite scale. Zhou and Lucas studied the dimensional changes that can be caused by the presence of fibres on carbon/epoxy composites immersed in distilled water at different temperatures. The diffusion along the fibre shows extreme stability, and no dimensional changes were measured. This stability is due to the carbon fibres high longitudinal stiffness. As they are impermeable to water absorption, the longitudinal dimension of the fibre is considered invariant. In width, the fibres hold and block the matrix, which prevents peeling. However, in the thickness, there are layers of neat epoxy that are not blocked by the fibres, which can peel under severe ageing. If saturation is reached, there is no longer any dimensional change [7]. The water diffusion in the composite material is strongly dependent on the fibre volume fraction v_f, and their arrangement, which affects both the gap size between two fibres and the diffusion path length through the matrix [35,110,112,134]. The effect of fibre volume fraction on moisture diffusion is complex, as the statistical study shows in Figure 3. M_{sat}, t_{sat}/h and D they show an increase as a function of v_f up to a certain threshold beyond which they fall. The fewer the gaps between the fibres, the shorter the diffusion paths through the matrix. As the fibre volume fraction increases, the diffusivity increases. However, if it is too important, the fibres are grouped and can touch each other.

In this situation, the local diffusion is blocked and therefore is very low [33,112,134]. In Figure 3, the evolution of the variables seems to be in accordance with these conclusions.

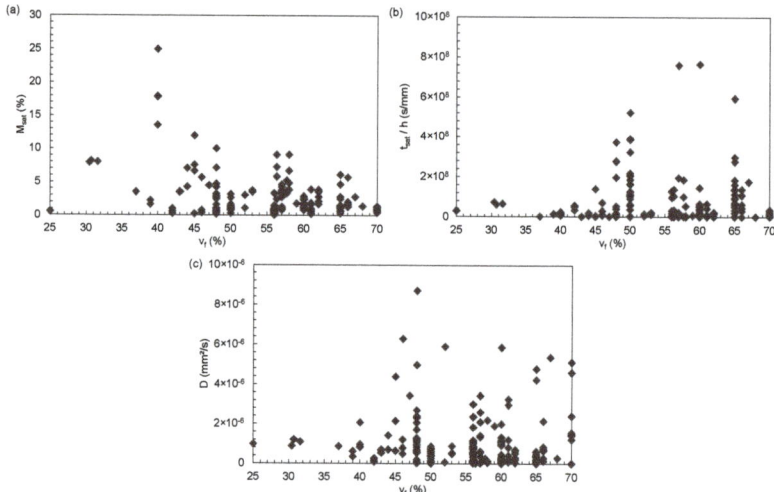

Figure 3. Evolution of (**a**) M_{sat}, (**b**) t_{sat}/h and (**c**) D as a function of the fibre volume fraction v_f.

Observing the reinforcement architectures used in the dataset, the composites composed of long fibre mats have a very important median absorbed mass uptake, which can be explained by the large voids created during its manufacture process (Table 5). This median absorbed mass uptake is quite similar for reinforcements arranged in one direction (UD), in two directions (2D), balanced, or in three dimensions (3D). It is slightly higher for UD and balanced reinforcements than for 3D. The median diffusivities are the greatest for the balanced and 3D composites, and the median saturation times are also the shortest for these two reinforcement configurations. The water molecules may have more directions to propagate, and and saturation is reached more rapidly. The literature is not uniform on this issue. Tang et al. found by modelling that their balanced woven fabrics diffused faster than their UD, particularly when the fibre waviness was greater [134]. For Yuan et al., the experimental data and modelling reveal a greater water mass absorbed in their 3D than in their UD. They also report a higher bound water proportion throughout the material [95]. On the other hand, Almudaihesh et al. observe that their UD composites absorb more water than their woven composites [135]. Finally, Wan et al. obtained a lower saturation mass uptake and diffusivity for their 3D than their UD. Their suggestion is that the moisture diffusion path is distorted in the 3D, which exerts a greater hindrance than for the UD [136].

Table 5. Medians of M_{sat}, t_{sat}/h and D as a function of architecture reinforcement. Entire box plots are available as Supplementary Materials (Figures S12–S14). UD: one direction, 2D: two directions, 3D: three dimensions.

Name (Individuals Number)	M_{sat} (%)	t_{sat}/h (s/mm $\times 10^7$)	D (mm^2/s $\times 10^{-7}$)
UD (93)	1.50	1.73	3.49
2D (94)	0.95	1.51	3.99
Balanced (47)	1.80	0.12	8.68
3D (5)	0.99	1.30	6.25
Long Fibre mat (4)	17.87	1.88	9.05
Unknown (2)	2.64	3.21	4.69

3.5. Manufacturing Process

Many processes have been developed to manufacture polymer or organic matrix composite materials. The process choice depends on various criteria: production speed, cost, desired performance, size and shape, resin and fibres nature. The quality of the process affects its cross-linking rate and the porosity percentage. Manufacturing processes applying pressure, such as resin transfer moulding (RTM), thermopressing or prepreg curing in the autoclave, result in a high fibre volume fraction and low porosity [136–138]. With these processes, the median saturation mass uptakes, saturation times and diffusivities appear relatively low. The dispersion of these values is also small (Table 6).

Table 6. Medians of M_{sat}, t_{sat}/h and D as a function of the process. Entire box plots are available as Supplementary Materials (Figures S15–S17).

Name (Individuals Number)	M_{sat} (%)	t_{sat}/h (s/mm $\times 10^7$)	D (mm^2/s $\times 10^{-7}$)
Autoclave (98)	1.52	6.36	3.48
Contact moulding (9)	3.41	2.07	36.60
Filament winding (2)	2.28	66.20	5.52
Heating table (9)	3.33	0.03	21.9
Infusion (28)	1.81	1.48	8.65
Moulding (195)	2.81	1.13	7.00
Pultrusion (1)	2.80	17.5	53.5
RTM (31)	0.89	0.89	7.90
Thermopressing (24)	2.70	2.86	9.48
Unknown (63)	0.94	0.32	1.71

Contact moulding and heating table processes show high median diffusivity and mass uptake. These processes have the common feature that they do not use pressure. They do not allow for high volume fractions of fibres. In addition, the fibres distribution and the resin content are not uniform, which can lead to voids formation. They show a noticeable data dispersion, probably because the dispersion and the quantity of porosities are variable, depending on the manufacturing parameters used.

Infusion, which has intermediate water absorption properties, also shows quite a large data dispersion. The vacuum bag low compaction is the cause of a greater void formation compared to other pressurized processes such as autoclave, which is responsible for greater diffusion and moisture absorption. Nevertheless, it is possible to improve the polymer quality by influencing its viscosity. A high viscosity may be responsible for a higher porosity. This porosity, therefore, varies according to the chosen resin type and cross-linking parameters, which may explain the wide dispersion of the data [139,140].

Concerning the unknown manufacturing process materials, it may be interesting to specify that these are generally aeronautical composites for which the information has not been revealed in the source publication. They may therefore be materials with high properties, which explains their low saturation mass uptakes and diffusivities.

3.6. Ageing Conditions
3.6.1. Conditioning Environment

Wet ageing can occur at various locations: in water, whether distilled water, seawater or deionised water, or in the humid air. In the literature, these environments have been widely studied. In addition, deionised water and humid inert atmosphere are also analyzed. By examining the individuals immersed in water, a similar median saturation mass uptake appears for distilled and deionised water, while it is much lower for seawater. Although the box plots for distilled water are almost centred, indicating good results homogeneity, the box plots for deionised water are eccentric towards the top (Table 7). The lowest saturation mass uptakes are obtained under humid air, while the highest are obtained in immersion in distilled or deionised water. Mass uptakes in seawater is also low. The mineral presence in the water slows down the water molecule diffusion through the material and leads to lower saturation masses, although these differences are mainly perceived over long ageing times [12,47,64]. In contrast, diffusivity under humid air is much higher and allows saturation to be reached most quickly. It is followed by the diffusivity under distilled water, which is very close. Finally, the diffusivity in deionised water appears to be the lowest, despite the absence of minerals that could slow down the water molecule diffusion. Nevertheless, in the database, the number of individuals immersed in deionised water or seawater is very low compared to the number of individuals immersed in distilled water, which could be linked to this divergence of results. It is possible that other parameters have a greater influence on the values of M_{sat}, t_{sat}/h and D than the water type.

Table 7. Medians of M_{sat}, t_{sat}/h and D as a function of the conditioning environment. Entire box plots are available as Supplementary Materials (Figures S18–S20).

Name (Individuals Number)	M_{sat} (%)	t_{sat}/h (s/mm $\times 10^7$)	D (mm^2/s $\times 10^{-7}$)
Humid air (215)	1.08	0.57	5.90
Inert atmosphere (2)	2.03	3.88	24.4
Deionised water (20)	3.09	0.89	0.95
Distilled water (203)	3.12	1.99	5.79
Sea water (20)	1.43	2.77	3.50

3.6.2. Relative Humidity

The relative humidity percentage influence on the absorbed water mass by epoxy-based materials has been demonstrated many times in the literature and is no longer in doubt [18,45,51,53–55]. This statistical study confirms that the higher the relative humidity, the greater the saturation mass uptake M_{sat} (Figure 4). Immersion of this material type in distilled water further increases its absorbed water mass [22].

This dataset effectively confirms the positive correlation between the absorbed water mass uptake and the relative humidity. Henry's law allows this correlation to be represented for humid air conditioning (Equation (8)) [20,124].

$$C_{sat} = SP_s, \tag{8}$$

where C_{sat} is the saturation water concentration and P_s is the water partial pressure, linked with RH (Equation (9)).

$$RH(\%) = 100 \frac{P_s}{P_{sat}}. \tag{9}$$

Several versions of this law adapted to Dual sorption were subsequently developed: power law (Equation (10)) [37]; Langmuir (Equation (11)) [141]; Dual sorption (Equation (12)) [142]; and Flory–Huggins (Equation (13)) [143].

$$C_{sat} = a\left(\frac{P_s}{P_{sat}}\right)^b \tag{10}$$

$$C_{sat} = \frac{cP_s}{1 + dP_s} \tag{11}$$

$$C_{sat} = SP_s + \frac{cP_s}{1 + dP_s} \tag{12}$$

$$\ln a_s = \ln \frac{P}{P_s} = \ln v + (1 - v) + \chi(1 - v)^2, \tag{13}$$

with a, b, c and d as coefficients, a_s the solvent activity, v the volume, and χ the polymer-solvent interaction coefficient. For pure resins, b is between 1.3 and 1.8 while it is close to 1 for composite materials. c and d are given approximately by the statistical thermodynamic treatment of Langmuir's.

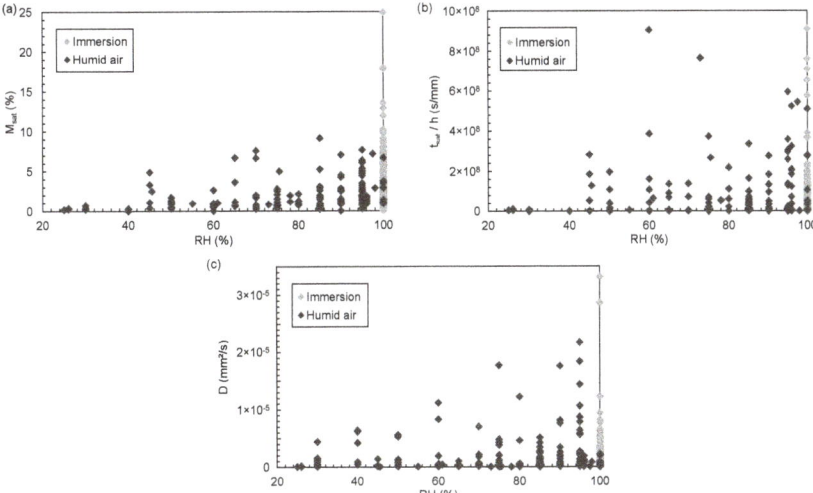

Figure 4. Evolution of (**a**) M_{sat}, (**b**) t_{sat}/h and (**c**) D as a function of relative humidity RH.

In immersion in a liquid, C_{sat} is related to the chemical potential of the water, i.e., it tends to decrease when the concentration of solutes increases [118]. The solvent concentration $[S]$ in an environment is in equilibrium with its partial pressure P_s in the atmosphere. There is a maximum concentration $[S]_{sat}$, which corresponds to the saturation pressure P_{sat} (Equation (14)).

$$\frac{P}{P_s} = \frac{[S]}{[S]_{sat}}. \tag{14}$$

If there is no extraction of soluble species, the polymer behaves in the same way in the liquid as in the saturated vapour. Indeed, equilibrium corresponds to the equality of the solvent chemical potentials in the polymer and in the environment. If the material is damaged, has pores or cracks, the solvent can also flow into it, as part of irreversible ageing. The presence of solutes in water, such as salt in seawater, causes the decrease of chemical potential, of saturation vapour pressure, and of solvent equilibrium concentration. Pure water causes greater moisture uptake than mineral water or seawater [20].

The diffusivity D also increases with the relative humidity in the environment. The calculation of this, according to the Fick model, is directly linked to M_{sat}, associated with the RH, and the thickness h (Equation (16)) [109,144]. For its part, the Carter and Kibler model links D to a number of free water molecules n and a number of bound water molecules N as a function of time t (Equation (16)) [45].

$$D = \frac{\pi}{16} \frac{h^2}{t} \left(\frac{M(t)}{M_{sat}}\right)^2 \tag{15}$$

$$D\frac{\delta^2 n}{\delta x^2} = \frac{\delta n}{\delta t} + \frac{\delta N}{\delta t}. \tag{16}$$

While M_{sat} and D increase with RH in the statistical study, t_{sat}/h evolution with RH is less evident. The high dispersion of this parameter does not allow us to conclude and is not suitable for a simple statistical dispersion study.

3.6.3. Temperature

The temperature T_{ageing} increase in the humid environment leads to accelerated ageing, as shown in the statistical study in Figure 5. The more important diffusion leads to an absorbed water mass stabilisation in a shorter time [33,39,54,55]. D, either Fick or Dual, is related to temperature by an Arrhenius law [37,145,146].

$$D = D_0 \exp\left(\frac{-E_a}{RT}\right). \tag{17}$$

T_{ageing} seems to increase D and to decrease t_{sat}/h. However, the strong dispersion of the data does not allow any conclusion.

In contrast, M_{sat} does not seem to be affected by the temperature. The non-relationship of saturation mass uptake with ageing temperature is under sorption isotherm laws such as Henry's. The saturation mass uptake does not depend on the temperature for the vast majority of studies, whether they are epoxy resins, epoxy-based composites, with varying reinforcements and architectures, Fick sorption kinetics or not. At 25 °C, epoxies have a solubility parameter δ close to 22 MPa$^{1/2}$, which is very far from the water solubility parameter, of 47.8 MPa$^{1/2}$ [117,147,148]. Water and epoxies are then hardly miscible. However, a higher temperature induces a decrease in the water solubility parameter while that of polymers remains reasonably constant or increases slightly. Substances that were not solvent can then become so. From an overall study point of view, and although there are exceptions, the variations in these two solubility parameter do not seem to be sufficient for the epoxies and the water to be miscible enough to induce a correlation between M_{sat} and T_{ageing}. For example, at 100 °C, the water solubility parameter is 44.4 MPa$^{1/2}$. That of the Epikote 828™/Epikure™ epoxy system increases from 21.9 at 25 °C to 22.5 MPa$^{1/2}$ at 100 °C [149].

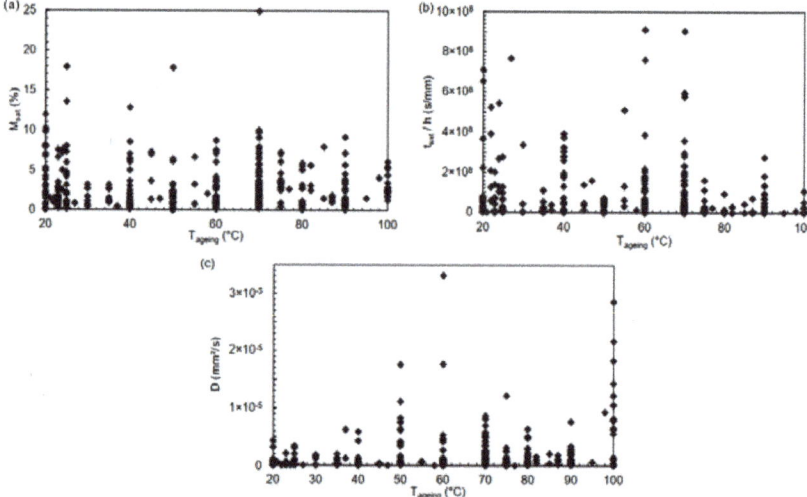

Figure 5. Evolution of (**a**) M_{sat}, (**b**) t_{sat}/h et (**c**) D as a function of the ageing temperature T_{ageing}.

However, as with the relative humidity, to observe the temperature correlations and be able to conclude, it is necessary to use another type of statistical analysis.

The wide dispersion of the data makes it difficult to conclude by simply observing the evolution of the variables in pairs. In addition, the data in the study show great heterogeneity due to a large number of prepolymers, hardeners and reinforcements, the diversity of processing methods, and the different conditioning parameters. All these parameters vary from one individual to another, which leads to other absorption behaviours, although irreversible behaviours have already been ruled out. A simple comparison of the variables in pairs is not sufficient to draw conclusions. To complete the statistical study, principal component analysis (PCA) is used in the rest of this paper.

4. Principal Component Analysis

4.1. Principal Component Analysis Introduction

The dataset used in this paper is associated with a large number of quantitative variables. Each of these variables can be related to a dimension. When the number of variables exceeds three, it is challenging to visualize a multidimensional space. Principal component analysis (PCA) can then be used. This is a multivariate statistical analysis method used when individuals are described by several quantitative variables, and it is used to determine the relationships between these different variables. PCA is based on the projection of a quantitative dataset belonging to a multidimensional space into several two-dimensional spaces. It simplifies the study by reducing the dimensions, i.e., the variable number, by highlighting principal components, which are linear combinations of the original variables. The n individuals are projected into a subspace of dimension q, which is the principal components space. To represent a data cloud S_i in a reduced space, we use a system of q linear combinations CP_q and p quantitative variables X^p. These linear combinations CP_q are the principal components [150,151].

PCA is particularly effective when the variables are highly correlated. In statistics and probability, studying the correlation between two variables is equivalent to studying the intensity of their links [114,116]. The correlations between variables and principal components can be read from a correlation circle of radius 1 where a vector of a given length represents each variable. The vector end coordinate corresponding to the variable on a principal component makes it possible to quantify its correlation. The closer the vector length is to 1, the better the variable is correlated in the principal component. It is also possible to observe correlations between variables if their vectors are long enough, so they are sufficiently well represented in the two-dimensional space. The correlation circle is interpreted as follows [152,153] :

- An acute angle (<90°) between the vectors of individual variables indicates a positive correlation between them;
- A 180° angle between the vectors of individual variables indicates a negative correlation between them;
- Variables whose vectors are orthogonal are not correlated with each other, and are therefore independent.

Therefore, principal component analysis enables, in addition to quantifying the correlation between a variable and a principal component, suggesting correlations between variables. The objective is to identify the dimensions or principal components along which the data variation is maximal. These data are represented in a system of X-Y coordinates. It is then possible to highlight relationships between variables that cannot be visualized in a space with more than two dimensions. Each individual in the study, characterized by these variables, is represented by a point. All these points form a data cloud described in a two-dimensional space. For the principal component analyses, the 448 individuals from the 90 publications are used. The set of variables is given in Table 8.

Table 8. List of variables used in the principal component analysis. The 8 qualitative variables are in italics and the 4 quantitative variables in Roman.

Material Parameters	Ageing Parameters	Diffusion Parameters
Prepolymer type	Conditioning	D
Hardener type	RH	M_{sat}
Reinforcement type	T_{ageing}	t_{sat}
h		M_{inter}
Process		t_{inter}
		β

4.2. Data Standardisation

A large dataset may contain heterogeneous individuals. This is particularly the case for some variables of our study, such as t_{sat} and D. It is then essential to center and reduce (i.e., to standardize) the dataset, which allows giving the same importance to the all the variables. The standardization is carried out so that the variables have a standard deviation equal to 1 and a mean value equal to 0 [116].

The PCA function of the package *FactoMineR* from R automatically normalises the data as explained above [116,154–156]. This function generates various indicators for each dimension or principal component: eigenvalues, variances, and cumulative variances. The eigenvalues can be used to determine the number of dimensions to keep for the study of the dataset. For this purpose, the data being centred-reduced, an eigenvalue of >1 is required. This means that the component concerned represents more variance than the original variable alone [157]. The eigenvalue quantifies the variance explained by each dimension. It is large for the first dimensions and small for the following ones (Table 9). Therefore, the first dimensions correspond to the directions that carry the maximum amount of variation contained in the dataset. PCA significantly reduces the number of principal components and compresses the variance into a smaller number of axes.

Table 9. Eigenvalues, variances and cumulative variances of the study dimensions.

Dimension	Eigenvalue	Variance	Cumulative Variance
Dim 1	2.18	27.19	27.19
Dim 2	1.55	19.39	46.58
Dim 3	1.13	14.09	60.68
Dim 4	0.88	11.10	71.78
Dim 5	0.86	10.73	82.52
Dim 6	0.79	9.84	92.37
Dim 7	0.58	7.19	99.56
Dim 8	0.03	0.43	100.00

Another method used to determine the number of the dimensions to choose from is to consider the graph or "scree plot" of the eigenvalues and stop at the level of the eigenvalue drop-off, beyond which they are relatively small (Figure 6) [150,158].

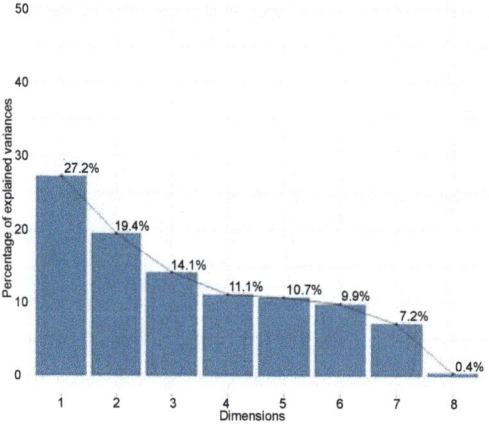

Figure 6. Eigenvalue scree plot.

If the whole dataset is considered, it could be possible to stop at the fourth principal component because its eigenvalue is less than 1. However, this only represents a cumulative variance of 60.7%. Looking at the eigenvalue graph, we do indeed see a drop-off at the fourth principal component, but components with relatively large variance follow it. We can choose to stop at this 4th principal component or to stop at the 6th before the second drop-out, that is to say, a cumulative variance of 92.4%. The latter solution is chosen. In addition, by observing the representation quality of the variables, we note that some of them are adequately represented on the principal components 4 and 6. This is the case for D and RH. This large number of different variables explains this large number of components to be studied. There is no universal method to decide how many principal components to choose for PCA. It depends on the dataset, the variances of each component, but also on how well the variables are represented in the different components.

4.3. Results

4.3.1. Overview

By using the PCA method on the 448 individuals in this study, we obtain individual projection graphs and correlation circles that provide information on the relationship (or non-relationship) between the variables. For each principal component, an individual graph and a correlation circle are obtained. The first two principal components represent respectively 27.2% and 19.4% of the variance of the individuals, i.e., 46.6%. It is necessary to observe the first 6 components so that 92.4% of the variance of the individuals is represented, which makes it possible to have an overall vision of the latter. The number of variables to be studied is therefore reduced from 8 to 6 principal components. The variables which constitute them are determined by analysis of the "coordinate" of their vector on the component, which is the cosine of the angle formed between the vector and the principal component: a cosine close to 1 is desired to have an accurate representation of the component. The 6 Equations (18)–(23) below describe the principal components and their compositions, the coefficients being the coordinates of the variables in the principal component:

$$CP_1 = M_{sat} \times 0.96 + M_{inter} \times 0.95 + RH \times 0.55 + h \times 0.18 - T_{ageing} \times 0.08 + D \times 0.08 + t_{sat} \times 0.04 + t_{inter} \times 0.004 \quad (18)$$

$$CP_2 = t_{inter} \times 0.79 - T_{ageing} \times 0.67 + t_{sat} \times 0.59 - D \times 0.34 - M_{inter} \times 0.05 - M_{sat} \times 0.02 + h \times 0.01 + RH \times 0.01 \quad (19)$$

$$CP_3 = h \times 0.75 + D \times 0.53 + t_{sat} \times 0.35 + T_{ageing} \times 0.29 + t_{inter} \times 0.19 - M_{inter} \times 0.12 - M_{sat} \times 0.11 + RH \times 0.10 \quad (20)$$

$$CP_4 = - D \times 0.74 + T_{ageing} \times 0.40 + h \times 0.28 + t_{sat} \times 0.22 - t_{inter} \times 0.16 + RH \times 0.16 - M_{inter} \times 0.04 - M_{sat} \times 0.02 \quad (21)$$

$$CP_5 = t_{sat} \times 0.54 - h \times 0.53 + RH \times 0.40 + T_{ageing} \times 0.23 + D \times 0.20 - t_{inter} \times 0.13 - M_{inter} \times 0.10 - M_{sat} \times 0.05 \quad (22)$$

$$CP_6 = -RH \times 0.70 + t_{sat} \times 0.33 + T_{ageing} \times 0.27 + M_{sat} \times 0.23 + M_{inter} \times 0.21 - h \times 0.13 + D \times 0.03 + t_{inter} \times 0.02 \quad (23)$$

The first principal component is mainly composed of M_{sat}, M_{inter} and to a lower level RH: it is the material moisture absorption. The second main component is dominated by t_{inter}, t_{sat} and $-T_{ageing}$. The third component is represented by h and D: it may illustrate the diffusion through the thickness. The fourth principal component is $-D$. The fifth principal component is also represented by t_{sat} and h, but their coordinates are not very large, so their representation is not very good. Finally, the sixth principal component consists mainly of $-RH$, whose representation is not good either. For a specific variable, the sum of the squared cosines on all the principal components equals 1. As they add up according to the different principal components, the representation of the squared cosines of the coordinates of the variables makes it possible to quickly visualize which principal components they are associated with, their representation quality, and therefore which correlation circles to study subsequently. However, in the literature, no threshold value of squared cosines or coordinates is formally declared (Figure 7).

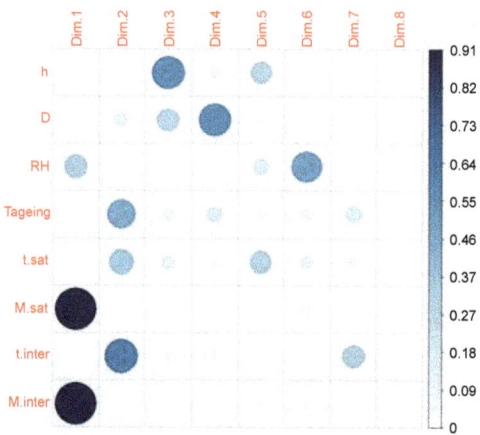

Figure 7. Representation quality of the study variables on the principal components of the overall PCA.

The correlation circles of the PCA, formed on the first 6 principal components, are then studied (Figure 8). The reduction by principal components allows us to limit the study to 6 and observe the variables by groups. It is already known that the variables of the following groups are correlated with each other because they are very well represented on their principal component:

- M_{sat} and M_{inter} are correlated with each other (CP_1);
- t_{sat} and t_{inter} are correlated with each other, and are anti-correlated with T_{ageing}, which means they decrease when T_{ageing} increases (CP_2);
- h and D are correlated with each other (CP_3);

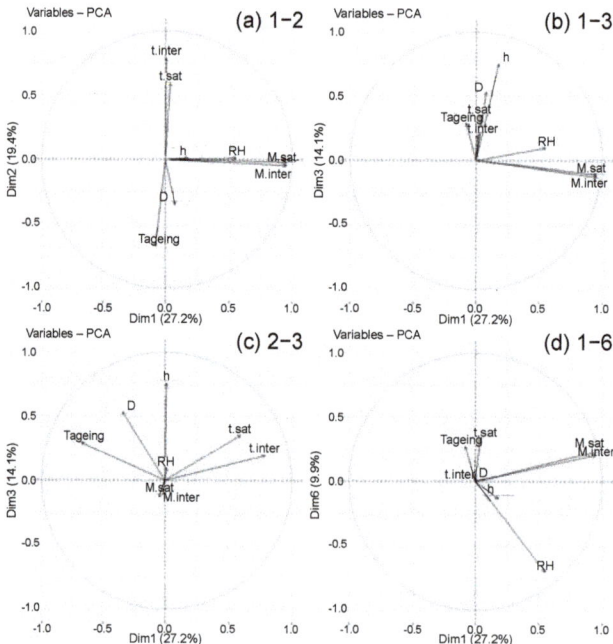

Figure 8. Correlation circles on planes (**a**) 1–2, (**b**) 1–3, (**c**) 2–3 and (**d**) 1–6 of the global PCA.

The vectors' placement representing the variables on the correlation circles is analyzed by taking into account their length. Figure 8 comprises the correlation circles of the 1–2, 1–3, 1–6 and 2–3 planes, which allow observation of all the analyses made in the following. The other circles, which do not provide more information, are not shown for the sake of clarity. RH, which was moderately well represented on principal component 1, is very well represented on component 6. The observation of the 1–6 plane confirms the correlation of M_{sat}, M_{inter} and RH. On the other hand, M_{sat} and M_{inter} are never correlated with the groups t_{sat}, t_{inter} and T_{ageing} or D and h, so their evolution is independent. A positive correlation is observed between D and T_{ageing} on the 2–3 plane, which is in agreement with the literature where Arrhenius laws are established between these two variables. Still on the 2–3 plane, h, t_{sat} and t_{inter} are correlated. The greater the h, the longer it takes to reach saturation time. Although t_{sat} and t_{inter} are anti-correlated with T_{ageing}, it is not possible to conclude on the correlation between the t_{sat} and t_{inter} group and D, their vectors not being of sufficient size. These observations are rather coherent with the conclusions made in the literature.

The points disposition on the individual graph is observed according to the diffusion behaviour type (Figure 9). Whether they follow a Fick law or a diffusion two-step derivative, most individuals are mixed in the graph centre. Nevertheless, there are only two-step specimens that extend along principal component 1, which represents M_{sat}, M_{inter} and RH. Furthermore, many samples in first part of the axis are Fick. Therefore, the strongest absorptions are associated with two-step materials, while the weakest are obtained with Fick materials, as we have observed with the box plots analysis. Points corresponding to the two types of behaviour studied stretch along component 2, which represents t_{sat}, t_{inter} and $-T_{ageing}$. Their intermediate and saturation times are therefore greater. Finally, the points that extend over component 3, corresponding to the thickness and the diffusivity, are associated with two-step diffusion materials. This material type, therefore, achieves the highest diffusivity.

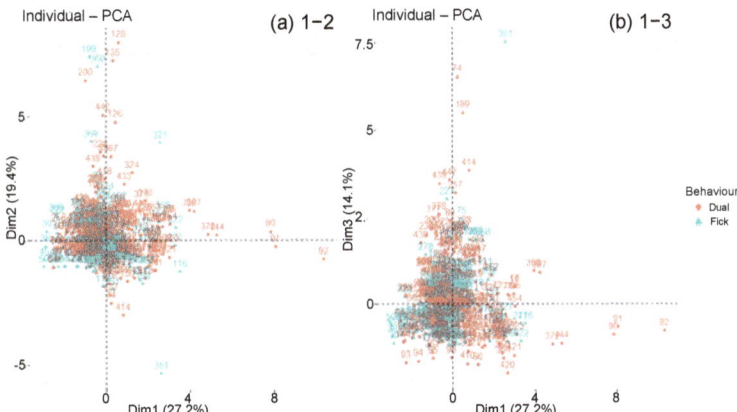

Figure 9. Graphs of individuals on planes (**a**) 1–2 and (**b**) 1–3 of the global PCA, according to diffusion behaviour.

The graphs of individuals along the principal components are now studied with respect to the qualitative variables presented previously. The objective is to highlight groupings according to one or more qualitative variables. However, the sorting according to the type of prepolymer or hardener does not clearly show any groups of individuals (Figures 10 and 11). All types of prepolymers and hardeners are mixed in the large cluster of graphs. Biobased epoxies, such as epoxidised soybean oil (ESO), blend into the cluster with typical M_{sat} and t_{sat} values [74]. The points that move away from the main cluster are not due to a difference in prepolymer or hardener.

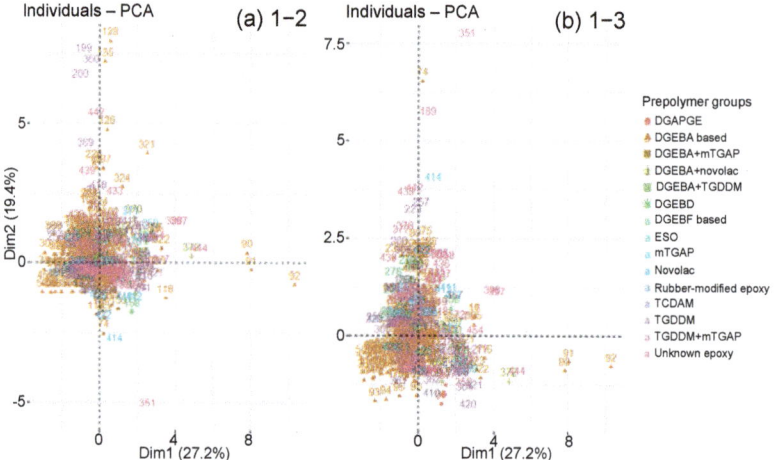

Figure 10. Graphs of individuals on planes (**a**) 1–2 and (**b**) 1–3 of the global PCA, according to prepolymer type.

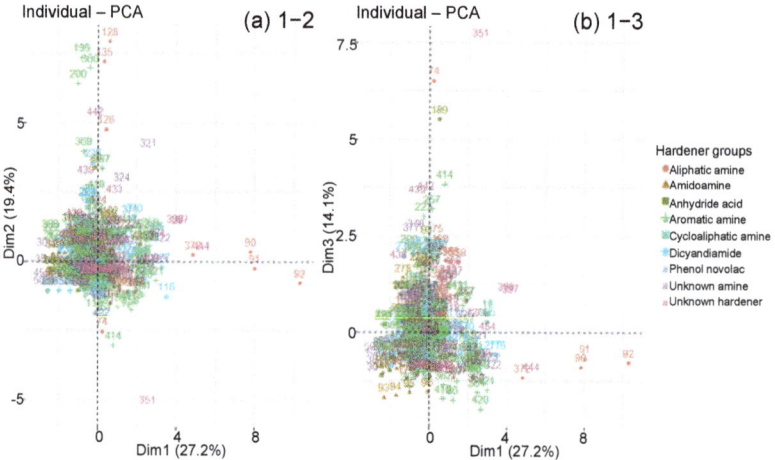

Figure 11. Graphs of individuals on planes (**a**) 1–2 and (**b**) 1–3 of the global PCA, according to hardener type.

Nevertheless, according to the reinforcement type, the graph makes it possible to distinguish several groupings on the 1–2 plane (Figure 12). In fact, towards the graph centre, the glass or carbon fibre composites are located. Then come the neat resins and finally the composites with flax, hemp, aramid and regenerated cellulose fibres. This group stretches along the principal component 1, which represents M_{sat}, M_{inter} and RH, which confirms the dispersion analysis as well as the literature: organic fibre composites absorb more water than inorganic fibre composites or even neat resins. Some points also stretch along component 2 which represents t_{sat}, t_{inter} and $-T_{ageing}$. Their intermediate and saturation times are therefore larger. These points represent carbon fibre composites and pure resins. There are no glass fibre composites.

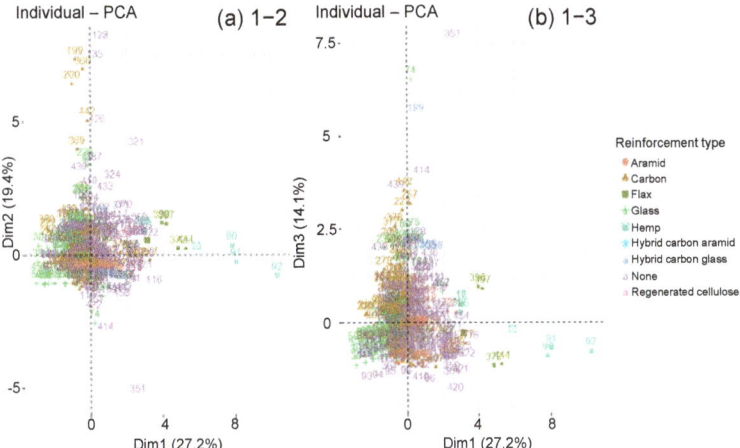

Figure 12. Graphs of individuals on planes (**a**) 1–2 and (**b**) 1–3 of the global PCA, according to reinforcement type.

By making the same observations with the type of conditioning, a cluster representing humid air and a cluster representing water are noticeable, whether distilled, seawater or deionised water (Figure 13). The humid air cluster is located in the left part of the first principal component. The individuals in this cluster, due to their smaller RH, show below-average M_{sat} and M_{inter} values. The water cluster extends along with component 1. High water mass uptakes distinguish several individuals immersed in distilled water. The two clusters are stretched along with components 2 and 3, which involve individuals with very long intermediate and saturation times for low temperatures. However, the individuals immersed in water with longer times are still more important than those under air.

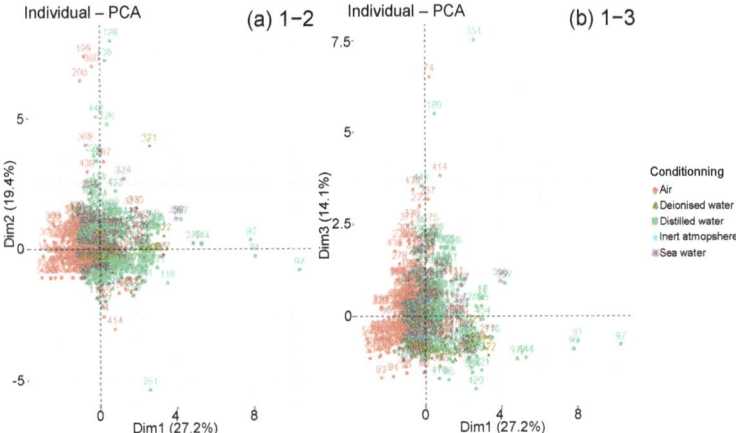

Figure 13. Graphs of individuals on planes (**a**) 1–2 and (**b**) 1–3 of the global PCA, according to conditioning environment.

About the manufacturing process, no data clustering is distinguished, except for the infused individuals that extend along with the first principal component (Figure 14).

The current dataset mixes neat resins and composite materials, but also materials with different absorption behaviour. The graph of individuals shows us clusters according to the diffusion behaviour and the presence or not of reinforcement as well as its nature.

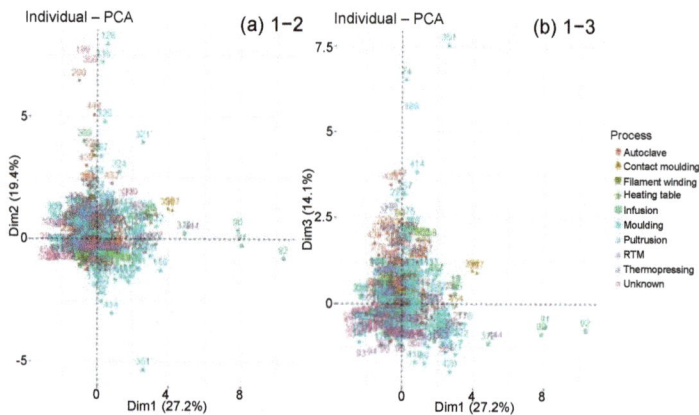

Figure 14. Graphs of individuals on planes (**a**) 1–2 and (**b**) 1–3 of the global PCA, according to manufacturing process.

For the continuation of the PCA study, the individuals are separated into four classified PCA according to their nature-neat resins or composites-and according to their moisture diffusion behaviour—Fick or Dual (Table 10). The objective is to study more precisely the correlations between variables, to compare the different behaviours, to further group the variables into principal components and to observe if new groupings of individuals are made. This separation of individuals also allows the introduction of variables not studied up to this point of the document. For composite materials, these variables are the fibre volume fraction and the fibre architecture. For Dual behaviour, they are M_{inter}, t_{inter}, β, γ and K, which are parameters linked to the Dual–Fick, and Carter *and* Kibler models, but have no meaning for the Fick model. As a reminder, M_{inter} and t_{inter} are respectively the values of the mass of water absorbed and the time at the intermediate stage. β is the probability that bound water molecules will be liberated, and γ is the probability that free water molecules will be bound. β and γ, therefore, illustrate the changes in the state of the water molecules that modify the diffusion behaviour of the materials. Finally, K is a parameter related to the swelling of the polymer (Table 11).

Table 10. Separation of the study individuals into four classified PCA groups.

	All	Resins, Fick	Composites, Fick	Resins, Dual	Composites, Dual
Individual number	448	56	98	147	147
Initial variables number	8	6	7	11	12
Individual variance of each initial variable	12.5%	16.7%	14.3%	9.1%	8.3%
Min. number of PC for an overview	6 (92.4%)	4 (85%)	4 (79.6%)	5 (76.3%)	6 (79.4%)

Table 11. List of variables used in the four classified principal component analysis. The 12 qualitative variables are in italics and the 7 quantitative variables in roman.

Material Parameters	Ageing Parameters	Diffusion Parameters
Prepolymer type	*Conditioning*	D
Hardener type	RH	M_{sat}
Reinforcement type	T_{ageing}	t_{sat}
Architecture		M_{inter}
Fibre orientation		t_{inter}
v_f		β
h		γ
Process		K

4.3.2. Classified PCA

On correlation circles, it is possible to distinguish four main groups of variables present in all the four classified PCA (Figure 15). M_{sat}, M_{inter} and RH represent moisture absorption. The T_{ageing}, D and h relate to the water diffusion speed. β, γ, K and D are associated with the water molecule diffusion and their placement. t_{sat}, t_{inter} and v_f are linked to the saturation time. This allows reducing the variable number. For example, the variables β, γ and K of the Carter *and* Kibler model are related to D. We also see the relationships between the different variables, confirming the experimental studies and models reported in the literature.

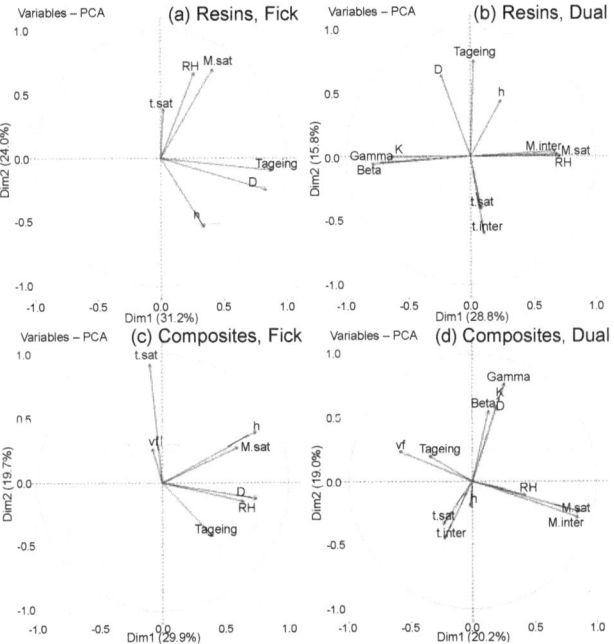

Figure 15. Correlation circles on the 1–2 planes of classified PCA: (**a**) Fick resins, (**b**) Dual resins, (**c**) Fick composites and (**d**) Dual composites. Further planes are given as Supplementary Materials: Fick resins in Figure S21, Fick composites in Figure S22, Dual resins in Figure S23 and Dual composites in Figure S24.

The classified PCA also showed differences. Depending on the material type (neat resin or composite) and the behaviour (Fick or Dual), further conclusions of correlation, anti-correlation or non-correlation can be made. In contrast, limited variations are observed depending on the nature of the epoxy prepolymer or hardener. In the case of neat resins and Fick composites, D depends on T_{ageing}, M_{sat} on RH and t_{sat} on T_{ageing}. In case of the fibre presence, t_{sat} is also influenced by v_f. All these parameters also depend on h, as Fick diffusion occurs in the material thickness, without chemical interaction.

For Dual diffusion materials, some correlations are modified in comparison with Fick diffusion materials. For neat resins, M_{sat} is dependent on h as for Fick materials, while for composites, it becomes independent. Moreover, t_{sat}, which is independent of D for Fick resins, becomes anti-correlated for Dual behaviour. M_{sat} changes from being correlated with D for Fick diffusion behaviour to being uncorrelated for Dual diffusion behaviour.

In neat resins, β, γ and K are correlated with D, anti-correlated with M_{sat} and uncorrelated with t_{sat}. For composites, β, γ and K are always correlated with D and T_{ageing}. T_{ageing} accelerates diffusion. It indirectly accelerates the propagation and the bonding and unbonding processes of water molecules. However, β, γ and K become uncorrelated with

M_{sat} and anti-correlated with t_{sat}. From the graphs of the individuals, these three variables are more important when the material is exposed to humid air rather than immersed, which may suggest that moisture diffusion is more complex in the air than in water. It should also be noted that v_f become anti-correlated with M_{sat} and correlated with t_{sat}, whereas, for the Fick cases, these two last parameters were not related to the mass uptake. The presence or absence of fibres has a strong impact on diffusion behaviour. Indeed, the increase of v_f decreases the mass water uptake. When there are too many fibres, they act as a barrier for the free water molecules, which diffuse in smaller numbers and more slowly. Different interactions may take place between the fibres and the water molecules, as differences are observed in the graphs of individuals representing the type of reinforcement used.

5. Discussion

With box plots, natural organic fibres appear to be the most sensitive to water, followed by aramid fibres which are synthetic organic fibres. In the study, composites based on organic fibres have a Dual behaviour. The presence of hydrophilic functional groups on their macromolecular skeleton amplifies the diffusion and moisture absorption, which can be higher than that of neat resins. Inorganic fibres such as glass and carbon have lower absorption percentages due to their impermeability. However, they still impact diffusion as they create differential swellings at the fibre/matrix interface leading to concentration gradients and possible degradations.

Another major difference highlighted by the correlation circles is the anti-correlation between T_{ageing} and M_{sat} in the case of Dual composites. M_{sat} is then influenced by T_{ageing} whereas it should not be in reversible diffusion models. Moreover, β, γ and K are parameters associated with the binding of water molecules to macromolecules. They change from correlated to M_{sat} for Dual Resins to uncorrelated for Dual Composites.

It is known that when the behaviour derives from the Fick model, the diffusion is not only due to the propagation of water molecules in the free volumes. Bonding interactions can take place between these water molecules and the material. These changes can lead to chemical degradation such as hydrolysis, chain breaks or oxidation. Hygroscopic swelling, cracking, osmotic damage and loss of admixtures or particles may occur [6,8,103,159–163]. In some cases, which were not studied in this paper, the gravimetry curves enable to observe these degradations, which are expressed by significant mass losses [7,22,105,112]. In other cases, weak irreversible alterations can take place in the material during hygrothermal ageing while having a two-stage diffusion. This is the case for many under-crosslinked industrial epoxy systems, which contain unreacted oxirane groups or polar compounds such as amines. These groups, which are very hydrophilic, become preferential sites for the formation of hydrogen bonds, for the initiation of hydrolysis and for the creation of concentration gradients [10,72]. As well, the sizing used on the fibres, the additives added to the matrix and the organic fibres can be hydrophilic, degrade with water and create osmotic degradation [29,133,160,164]. These irreversible modifications, which are hardly perceptible on the absorption gravimetry curve, can be observed by desorption. The mass variation $M(t)$ does not return to its initial value. Mass losses therefore occur during the absorption phase while the mass gain curve continues to increase slowly. Studying the desorption kinetics is then important to verify the presence of hidden mass losses or to highlight a chemical evolution [12,72,165–168]. As with the Fick equation, the Carter *and* Kibler and Dual–Fick equations model purely physical diffusion kinetics. When chemical degradation takes place, these models may no longer be representative of the material absorption behaviour. If irreversible phenomena enter the diffusion kinetics, it is possible that this will change the correlations between the different parameters. Because of T_{ageing} and M_{sat} anti-correlation, it is then conceivable that in a non-negligible number of cases, the behaviour of the individuals in the study is not purely related to physical diffusion. Degradations, although slight, may be present, making the chosen model inaccurate. Although the gravity curves fit well, the associated parameters may no longer represent the mechanisms for which they were chosen, such as *beta, gamma* and *K*. The consequences of an inappropriate model choice are the premature termination of gravimetric monitoring of materials, the non-detection of degradation phenomena that may appear at longer absorption times, inaccurate estimates of M_{sat} whose true value is masked by mass losses, and the performance of material characterisation at inappropriate M_{sat} [169,170].

6. Conclusions

In this paper, statistical tools were used to study how the parameters related to hygrothermal ageing influence each other. For this purpose, data were extracted presenting scientific publications from experimental tests carried out on neat epoxy resins and epoxy-based composites. The analysis started with a scattering study of the three most characteristic variables of hygrothermal ageing: saturation mass uptake M_{sat}, saturation time t_{sat} and diffusivity D, using box plots and scatter plots. The box plots enable us to differentiate these three parameters as a function of some qualitative parameters, such as pre-polymers type, reinforcement type, or process. The results obtained are in accordance with the observations made in the literature.

On the other hand, parameters such as hardener type or architecture reinforcement are more difficult to differentiate, as they are strongly linked to other parameters. For the quantitative parameters v_f, RH and T_{ageing}, the analyses are also more complex due to the high dispersion of the data coming from very diverse individuals.

These results led to combining the dispersion analysis with another type of statistical analysis: principal component analysis (PCA). This analysis, which centres and reduces the variables when they are highly dispersed, simplifies the study despite the variation of many parameters and allows the deduction of correlations between variables. PCA was then performed on different groups of data, classified according to the material nature (neat resin or composite) and the diffusion behaviour (Fick diffusion, or Dual diffusion), to observe if any differences are noticeable. Some connections between variables, identified in the correlations circles and graphs of individuals of the PCA, have been demonstrated many times in the literature. The study becomes more complex with the Dual diffusion behaviour composites. Some correlations between parameters change. In particular, D and T_{ageing}, correlated in the global PCA, become independents in the PCA of Dual composites. These changes do not correspond to the reversible diffusion models. It seems that, despite a two-step diffusion curve, without apparent mass losses in gravimetry, the Carter *and* Kibler or Dual–Fick models and their parameter do not seem to fit all individuals.

However, the study carried out in this paper has some limitations, due to the quality of representation of some variables, especially in the dual sorption composites PCA. Global PCA does not result in a significant reduction in the number of principal components (-2), although the classified PCA results in better reductions (from -2 to -6). PCA does not allow any remarkable differentiation according to the nature of the epoxies and their hardeners, as this type of polymer is hydrophilic and hygrothermal ageing depends on many parameters. In these two cases, the box plots allow a better visualisation of the data.

Other variables can be added to extend the study, such as mechanical properties, but this requires that these data have been reported in the literature, which is not always the case.

Supplementary Materials: The following supporting information can be downloaded at: https://www.mdpi.com/article/10.3390/polym14142832/s1, Figure S1: List of epoxy prepolymers, Figure S2: List of hardeners, Figures S3–S20: Box plots of D, t_{sat}/h and M_{sat} as a function of the study quantitative variables; Figure S21: Fick resins' PCA; Figure S22: Fick composites' PCA; Figure S23: Dual resins' PCA; Figure S24: Dual composites' PCA.

Author Contributions: Conceptualization, C.G., B.H.-R. and V.N.; methodology, C.G., F.T., B.H.-R. and V.N.; software, C.G. and F.T.; validation, C.G., F.T. and V.N.; formal analysis, C.G., F.T. and V.N.; investigation, C.G.; resources, B.H.-R., T.T. and V.N.; data curation, C.G.; writing—original draft preparation, C.G., F.T. and V.N.; writing—review and editing, C.G., F.T. and V.N.; supervision, V.N.; project administration, T.T. and V.N.; funding acquisition, T.T. and V.N. All authors have read and agreed to the published version of the manuscript.

Funding: This research was funded by Association Nationale de la Recherche et de la Technologie (ANRT) and SAFRAN Aircraft Engines through a CIFRE PhD fellowship: grant number n°2018/0355. The APC was funded by ANRT.

Institutional Review Board Statement: Not applicable.

Informed Consent Statement: Not applicable.

Data Availability Statement: The data presented in this study are available on request from the corresponding author.

Conflicts of Interest: The authors declare no conflict of interest.

References

1. Henry, L.; Kris, N. *Handbook of Epoxy Resins*; McGraw-Hill: New York, NY, USA, 1967.
2. Peters, S.T. (Ed.) *Handbook of Composites*, 2nd ed.; Springer Science and Business Media: Dordrecht, The Netherlands, 1998.
3. Jones, C.; Dickson, R.; Adam, T.; Reiter, H.; Harris, B. The environmental fatigue behaviour of reinforced plastics. *Proc. R. Soc. Lond. A. Math. Phys. Sci.* **1984**, *396*, 315–338.
4. Cavasin, M.; Sangermano, M.; Thomson, B.; Giannis, S. Exposure of Glass Fiber Reinforced Polymer Composites in Seawater and the Effect on Their Physical Performance. *Materials* **2019**, *12*, 807. [CrossRef]
5. Tsai, Y.; Bosze, E.; Barjasteh, E.; Nutt, S. Influence of hygrothermal environment on thermal and mechanical properties of carbon fiber/fiberglass hybrid composites. *Compos. Sci. Technol.* **2009**, *69*, 432–437. [CrossRef]
6. Lee, M.C.; Peppas, N.A. Water transport in graphite/epoxy composites. *J. Appl. Polym. Sci.* **1993**, *47*, 1349–1359. [CrossRef]
7. Zhou, J.; Lucas, J. The effects of a water environment on anomalous absorption behavior in graphite/epoxy composites. *Compos. Sci. Technol.* **1995**, *53*, 57–64. [CrossRef]
8. Weitsman, Y. Anomalous fluid sorption in polymeric composites and its relation to fluid-induced damage. *Compos. Part A Appl. Sci. Manuf.* **2006**, *37*, 617–623. [CrossRef]
9. Tsotsis, T. Considerations of failure mechanisms in polymer matrix composites in the design of aerospace structures. In *Failure Mechanisms in Polymer Matrix Composites*; Elsevier: Amsterdam, The Netherlands, 2012; pp. 227–278. [CrossRef]
10. Tcharkhtchi, A.; Bronnec, P.; Verdu, J. Water absorption characteristics of diglycidylether of butane diol–3,5-diethyl-2,4-diaminotoluene networks. *Polymer* **2000**, *41*, 5777–5785. [CrossRef]
11. Colin, X.; Verdu, J. Humid Ageing of Organic Matrix Composites. In *Durability of Composites in a Marine Environment. Solid Mechanics and Its Applications*; Davies, P., Rajapakse, Y.D.S., Eds.; Springer: Dordrecht, The Netherlands, 2014; pp. 47–114.
12. Deroiné, M.; Le Duigou, A.; Corre, Y.M.; Le Gac, P.Y.; Davies, P.; César, G.; Bruzaud, S. Accelerated ageing of polylactide in aqueous environments: Comparative study between distilled water and seawater. *Polym. Degrad. Stabil.* **2014**, *108*, 319–329. [CrossRef]
13. Boukhoulda, B.; Adda-Bedia, E.; Madani, K. The effect of fiber orientation angle in composite materials on moisture absorption and material degradation after hygrothermal ageing. *Compos. Struct.* **2006**, *74*, 406–418. [CrossRef]
14. Selzer, R.; Friedrich, K. Mechanical properties and failure behaviour of carbon fibre-reinforced polymer composites under the influence of moisture. *Compos. Part A Appl. Sci. Manuf.* **1997**, *28*, 595–604. [CrossRef]
15. De'Nève, B.; Shanahan, M. Water absorption by an epoxy resin and its effect on the mechanical properties and infra-red spectra. *Polymer* **1993**, *34*, 5099–5105. [CrossRef]
16. Ramirez, F.A.; Carlsson, L.A.; Acha, B.A. Evaluation of water degradation of vinylester and epoxy matrix composites by single fiber and composite tests. *J. Mater. Sci.* **2008**, *43*, 5230–5242. [CrossRef]
17. Halpin, J.C. *Effects of Environmental Factors on Composite Materials*; Technical report AFML-TR-67-423; Air Force Materials Laboratory, Wright-Patterson Air Force Base: Dayton, OH, USA, 1969; p. 45433.
18. Chen, C. Contribution à la Prise en Compte des Effets de l'Environnement sur la Tolérance aux Dommages d'Impact de Stratifiés Composites. Ph.D. Thesis, Institut Supérieur d'Aéronautique et de l'Espace, Paris, France, 2015.
19. Popineau, S.; Rondeau-Mouro, C.; Sulpice-Gaillet, C.; Shanahan, M.E. Free/bound water absorption in an epoxy adhesive. *Polymer* **2005**, *46*, 10733–10740. [CrossRef]
20. Fayolle, B.; Verdu, J. *Vieillissement Physique des Matériaux Polymères*; Techniques de l'Ingénieur: Paris, France, 2005; p. 22.
21. Grangeat, R. Durabilité des Assemblages Collés en Environnement Humide—Instrumentation par Capteurs à Fibre Optique. Ph.D. Thesis, Université de Nantes, Institut de Recherche en Génie Civil et Mécanique (GeM), Nantes, France, 2019.
22. Bonniau, P.; Bunsell, A.R. A comparative study of water absorption theories applied to glass epoxy composites. *J. Compos. Mater.* **1981**, *15*, 272–293. [CrossRef]
23. Morel, E.; Bellenger, V.; Verdu, J. Structure-water absorption relationships for amine-cured epoxy resins. *Polymer* **1985**, *26*, 1719–1724. [CrossRef]
24. Wong, K.J. Moisture Absorption Characteristics and Effects on Mechanical Behaviour of Carbon/Epoxy Composite: Application to Bonded Patch Repairs of Composite Structures. Ph.D. Thesis, Université de Bourgogne, Laboratoire DRIVE ISAT, Nevers, France, 2014.

25. Soles, C.L.; Yee, A.F. A Discussion of the Molecular Mechanisms of Moisture Transport in Epoxy Resins. *J. Polym. Sci. Part B Polym. Phys.* **2000**, *38*, 792–802. [CrossRef]
26. Abdelkader, A.F.; White, J.R. Water absorption in epoxy resins: The effects of the crosslinking agent and curing temperature. *J. Appl. Polym. Sci.* **2005**, *98*, 2544–2549. [CrossRef]
27. Sala, G. Composite degradation due to fluid absorption. *Compos. Part B Eng.* **2000**, *31*, 357–373. [CrossRef]
28. Doğan, A.; Arman, Y. The Effect of Hygrothermal Aging on the Glass and Carbon Reinforced Epoxy Composites for different Stacking Sequences. *Mechanics* **2018**, *24*, 19–25. [CrossRef]
29. Ray, B.C. Temperature effect during humid ageing on interfaces of glass and carbon fibers reinforced epoxy composites. *J. Colloid Interface Sci.* **2006**, *298*, 111–117. [CrossRef]
30. Augl, J.M.; Berger, A.E. *The Effect of Moisture on Carbon Fiber Reinforced Epoxy Composites. I. Diffusion*; Technical Report 76-7; White Oak Laboratory, Naval Surface Weapons Center: Silver Spring, MA, USA, 1976.
31. Clark, G.; Saunders, D.; van Blaricum, T.; Richmond, M. Moisture absorption in graphite/epoxy laminates. *Compos. Sci. Technol.* **1990**, *39*, 355–375. [CrossRef]
32. Gupta, V.B.; Drzal, L.T.; Rich, M.J. The physical basis of moisture transport in a cured epoxy resin system. *J. Appl. Polym. Sci.* **1985**, *30*, 4467–4493. [CrossRef]
33. Arnold, C.; Korkees, F.; Alston, S. The long-term water absorption and desorption behaviour of carbon-fibre/epoxy composites. In Proceedings of the ECCM15 Proceedings, ESCM: 15th European Conference on Composite Materials, Venice, Italy, 24–28 June 2012; Volume 15.
34. Simar, A. Impact du Vieillissement Humide sur le Comportement d'un Composite à Matrice Organique Tissé Fabriqué par Injection RTM: Mise en Evidence d'un Couplage entre Absorption d'Eau et Thermo-Oxydation de la Matrice. Ph.D. Thesis, École Nationale Supérieure de Mécanique et d'Aérotechnique, Institut Prime, Poitiers, France, 2014.
35. Kondo, K.; Taki, T. Moisture Diffusivity of Unidirectional Composites. *J. Compos. Mater.* **1982**, *16*, 82–93. [CrossRef]
36. Grangeat, R.; Girard, M.; Lupi, C.; Leduc, D.; Jacquemin, F. Measurement of the local water content of an epoxy adhesive by fiber optic sensor based on Fresnel reflection. *Mech. Syst. Sign. Process.* **2020**, *141*, 106439. [CrossRef]
37. Shen, C.H.; Springer, G.S. Moisture absorption and desorption of composite materials. *J. Compos. Mater.* **1976**, *10*, 2–20. [CrossRef]
38. Perreux, D.; Suri, C. A study of the coupling between the phenomena of water absorption and damage in glass/epoxy composite pipes. *Compos. Sci. Technol.* **1997**, *57*, 1403–1413. [CrossRef]
39. Pierron, F.; Poirette, Y.; Vautrin, A. A novel procedure for identification of 3D moisture diffusion parameters on thick composites: theory, validation and experimental results. *J. Compos. Mater.* **2002**, *36*, 2219–2243. [CrossRef]
40. Zainuddin, S.; Hosur, M.; Zhou, Y.; Kumar, A.; Jeelani, S. Durability studies of montmorillonite clay filled epoxy composites under different environmental conditions. *Mater. Sci. Eng. A* **2009**, *507*, 117–123. [CrossRef]
41. Frank, K.; Childers, C.; Dutta, D.; Gidley, D.; Jackson, M.; Ward, S.; Maskell, R.; Wiggins, J. Fluid uptake behavior of multifunctional epoxy blends. *Polymer* **2013**, *54*, 403–410. [CrossRef]
42. Sugita, Y.; Winkelmann, C.; La Saponara, V. Environmental and chemical degradation of carbon/epoxy lap joints for aerospace applications, and effects on their mechanical performance. *Compos. Sci. Technol.* **2010**, *70*, 829–839. [CrossRef]
43. Gaussens, C. Solutions Adhésives et Durabilité d'une Liaison Structurale d'un Capteur Céramique sur un Roulement Acier. Ph.D. Thesis, Institut National Polytechnique de Toulouse, Laboratoire Génie de Production (INP-ENIT), Tarbes, France, 2010.
44. Billy, F. Vieillissement et Propriétés Résiduelles de Matériaux Issus du Démantèlement d'Avions en Fin de Vie. Ph.D. Thesis, École nationale Supérieure de Mécanique et d'Aérotechnique, Institut Prime, Poitiers, France, 2013.
45. Carter, H.G.; Kibler, K.G. Langmuir-Type Model for Anomalous Moisture Diffusion In Composite Resins. *J. Compos. Mater.* **1978**, *12*, 118–131. [CrossRef]
46. Selzer, R.; Friedrich, K. Influence of water up-take on interlaminar fracture properties of carbon fibre-reinforced polymer composites. *J. Mater. Sci.* **1995**, *30*, 334–338. [CrossRef]
47. Zafar, A.; Bertocco, F.; Schjødt-Thomsen, J.; Rauhe, J. Investigation of the long term effects of moisture on carbon fibre and epoxy matrix composites. *Compos. Sci.Technol.* **2012**, *72*, 656–666. [CrossRef]
48. Choi, H.S.; Ahn, K.J. Hygroscopic aspects of epoxy/carbon fiber composite laminates in aircraft environments. *Compos. Part A Appl. Sci. Manuf.* **2001**, *32*, 709–720. [CrossRef]
49. Nguyen, T.H. Vieillissement Artificiel et Vieillissement Naturel en Ambiance Tropicale de Composites Modèles Epoxy/Verre: Approche Nanoscopique de l'Etude des Interphases. Ph.D. Thesis, Université du Sud-Toulon-Var, Laboratoire Matériaux Polymères-Interfaces-Environnement Marin, La Garde, France, 2013.
50. Zhou, J.; Lucas, J.P. Hygrothermal effects of epoxy resin. Part I: The nature of water in epoxy. *Polymer* **1999**, *40*, 5505–5512. [CrossRef]
51. Barton, S.J.; Pritchard, G. The moisture absorption characteristics of crosslinked vinyl terminated polyethers compared with epoxies. *Polym. Adv. Technol.* **1994**, *5*, 245–252. [CrossRef]
52. Weitsman, Y.J.; Guo, Y.J. A correlation between fluid-induced damage and anomalous fluid sorption in polymeric composites. *Compos. Sci. Technol.* **2002**, *62*, 889–908. [CrossRef]
53. Pérez-Pacheco, E.; Cauich-Cupul, J.I.; Valadez-González, A.; Herrera-Franco, P.J. Effect of moisture absorption on the mechanical behavior of carbon fiber/epoxy matrix composites. *J. Mater. Sci.* **2013**, *48*, 1873–1882. [CrossRef]
54. Dao, B.; Hodgkin, J.H.; Krstina, J.; Mardel, J.; Tian, W. Accelerated aging versus realistic ageing in aerospace composite materials. IV. Hot/wet ageing effects in a low temperature cure epoxycomposite. *J. Appl. Polym. Sci.* **2007**, *106*, 4264–4276. [CrossRef]

55. Dao, B.; Hodgkin, J.; Krstina, J.; Mardel, J.; Tian, W. Accelerated aging versus realistic aging in aerospace composite materials. V. The effects of hot/wet aging in a structural epoxy composite. *J. Appl. Polym. Sci.* **2010**, *115*, 901–910. [CrossRef]
56. Verghese, K.; Haramis, J.; Patel, S.; Senne, J.; Case, S.; Lesko, J. Enviro-mechanical durability of polymer composites. In Proceedings of the Long Term Durability of Structural Materials Workshop, Berkeley, CA, USA, 26–27 October 2001; pp. 121–132. [CrossRef]
57. Delozanne, J.; Desgardin, N.; Coulaud, M.; Cuvillier, N.; Richaud, E. Failure of epoxies bonded assemblies: Comparison of thermal and humid ageing. *J. Adhes.* **2018**, 1–24. [CrossRef]
58. Guermazi, N.; Haddar, N.; Elleuch, K.; Ayedi, H. Investigations on the fabrication and the characterization of glass/epoxy, carbon/epoxy and hybrid composites used in the reinforcement and the repair of aeronautic structures. *Mater. Des.* **2014**, *56*, 714–724. [CrossRef]
59. Dexter, H.B.; Baker, D.J. Flight service environmental effects on composite materials and structures. *Adv. Perform. Mater.* **1994**, *1*, 51–85. [CrossRef]
60. Vodicka, R.; Nelson, B.; van der Berg, J.; Chester, R. *Long-Term Environmental Durability of F/A-18 Composite Material*; Technical Report DSTO-TR-0826; DSTO Aeronautical and Maritime Research Laboratory: Melbourne, Australia, 1999.
61. McKague, E.; Halkias, J.; Reynolds, J. Moisture in composites: The effect of supersonic service on diffusion. *J. Compos. Mater.* **1975**, *9*, 2–9. [CrossRef]
62. Aktas, L.; Hamidi, Y.; Altan, M.C. Effect of Moisture Absorption on Mechanical Properties of Resin Transfer Molded Composites. In *Materials: Processing, Characterization and Modeling of Novel Nano-Engineered and Surface Engineered Materials, Proceedings of the ASME 2002 International Mechanical Engineering Congress and Exposition, New Orleans, LA, USA, 17–22 November 2002*; ASME: New York, NY, USA, 2002; Volume 2002, pp. 173–181. [CrossRef]
63. Sugiman, S.; Salman, S. Hygrothermal effects on tensile and fracture properties of epoxy filled with inorganic fillers having different reactivity to water. *J. Adhes. Sci. Technol.* **2019**, *33*, 691–714. [CrossRef]
64. Bordes, M.; Davies, P.; Cognard, J.Y.; Sohier, L.; Sauvant-Moynot, V.; Galy, J. Prediction of long term strength of adhesively bonded steel/epoxy joints in sea water. *Int. J. Adhes. Adhes.* **2009**, *29*, 595–608. [CrossRef]
65. Loh, W.K.; Crocombe, A.D.; Abdel Wahab, M.M.; Ashcroft, I.A. Modelling anomalous moisture uptake, swelling and thermalcharacteristics of a rubber toughened epoxy adhesive. *Int. J. Adhes. Adhes.* **2005**, *25*, 1–12. [CrossRef]
66. Scida, D.; Aboura, Z.; Benzeggagh, M. The effect of ageing on the damage events in woven-fibre composite materials under different loading conditions. *Compos. Sci. Technol.* **2002**, *62*, 551–557. [CrossRef]
67. Cândido, G.M.; Costa, M.L.; Rezende, M.C.; Almeida, S.F.M. Hygrothermal effects on quasi-isotropic carbon epoxy laminates with machined and molded edges. *Compos. Part B Eng.* **2008**, *39*, 490–496. [CrossRef]
68. Cao, Y.; Qian, X.; Liu, C.; Yang, J.; Xie, K.; Zhang, C. Controllable preparation of a novel epoxy/anhydride system with polyether-Polyester semi-interpenetrating structure and the excellent hydrothermal aging resistance properties. *Polym. Degrad. Stabil.* **2019**, *168*, 1–10. [CrossRef]
69. Wang, M.; Xu, X.; Ji, J.; Yang, Y.; Shen, J.; Ye, M. The hygrothermal aging process and mechanism of the novolac epoxy resin. *Compos. Part B Eng.* **2016**, *107*, 1–8. [CrossRef]
70. Guo, B.; Jia, D.; Fu, W.; Qiu, Q. Hygrothermal stability of dicyanate-novolac epoxy resin blends. *Polym. Degrad. Stabil.* **2003**, *79*, 521–528. [CrossRef]
71. Capiel, G.; Miccio, L.A.; Montemartini, P.E.; Schwartz, G.A. Water diffusion and hydrolysis effect on the structure and dynamics of epoxy-anhydride networks. *Polym. Degrad. Stabil.* **2017**, *143*, 57–63. [CrossRef]
72. El Yagoubi, J.; Lubineau, G.; Roger, F.; Verdu, J. A fully coupled diffusion-reaction scheme for moisture sorption-desorption in an anhydride-cured epoxy resin. *Polymer* **2012**, *53*, 5582–5595. [CrossRef]
73. Bréthous, R.; Colin, X.; Fayolle, B.; Gervais, M. Non-Fickian behavior of water absorption in an epoxy-amidoamine network. In Proceedings of the VIII International Conference on Times of Polymers and Composites, Naples, Italy, 19–23 June 2016; AIP Publishing: Melville, NY, USA, 2016; p. 020070. [CrossRef]
74. Pupure, L.; Doroudgarian, N.; Joffe, R. Moisture uptake and resulting mechanical response of biobased composites. I. constituents. *Polym. Compos.* **2013**, *35*, 1150–1159. [CrossRef]
75. Verge, P.; Toniazzo, V.; Ruch, D.; Bomfim, J.A. Unconventional plasticization threshold for a biobased bisphenol-A epoxy substitution candidate displaying improved adhesion and water-resistance. *Ind. Crops Prod.* **2014**, *55*, 180–186. [CrossRef]
76. Mannberg, P.; Nyström, B.; Joffe, R. Service life assessment and moisture influence on bio-based thermosetting resins. *J. Mater. Sci.* **2014**, *49*, 3687–3693. [CrossRef]
77. Núñez, L.; Villanueva, M.; Fraga, F.; Núñez, M.R. Influence of Water Absorption on the Mechanical Properties of DGEBA (n=0)/1, 2 DCH Epoxy System. *J. Appl. Polym. Sci.* **1999**, *74*, 353–358. [CrossRef]
78. Apicella, A.; Nicolais, L. Role of processing on the durability of epoxy composites in humid environments. *Ind. Eng. Chem. Prod. Res. Dev.* **1984**, *23*, 288–297. [CrossRef]
79. Suzuki, T.; Oki, Y.; Numajiri, M.; Miura, T.; Kondo, K. Free-volume characteristics and water absorption of novolac epoxy resins investigated by positron annihilation. *Polymer* **1996**, *37*, 3025–3030. [CrossRef]
80. Cadu, T.; Van Schoors, L.; Sicot, O.; Moscardelli, S.; Divet, L.; Fontaine, S. Cyclic hygrothermal ageing of flax fibers' bundles and unidirectional flax/epoxy composite. Are bio-based reinforced composites so sensitive? *Ind. Crops Prod.* **2019**, *141*, 1–12. [CrossRef]

81. Assarar, M.; Scida, D.; El Mahi, A.; Poilâne, C.; Ayad, R. Influence of water ageing on mechanical properties and damage events of two reinforced composite materials: Flax–fibres and glass–fibres. *Mater. Des.* **2011**, *32*, 788–795. [CrossRef]
82. Newman, R.H. Auto-accelerative water damage in an epoxy composite reinforced with plain-weave flax fabric. *Compos. Part A Appl. Sci. Manuf.* **2009**, *40*, 1615–1620. [CrossRef]
83. Scida, D.; Assarar, M.; Poilâne, C.; Ayad, R. Influence of hygrothermal ageing on the damage mechanisms of flax-fibre reinforced epoxy composite. *Compos. Part B Eng.* **2013**, *48*, 51–58. [CrossRef]
84. Yan, L.; Chouw, N. Effect of water, seawater and alkaline solution ageing on mechanical properties of flax fabric/epoxy composites used for civil engineering applications. *Construct. Build. Mater.* **2015**, *99*, 118–127. [CrossRef]
85. Perrier, A.; Touchard, F.; Chocinski-Arnault, L.; Mellier, D. Quantitative analysis by micro-CT of damage during tensile test in a woven hemp/epoxy composite after water ageing. *Compos. Part A Appl. Sci. Manuf.* **2017**, *102*, 18–27. [CrossRef]
86. Doroudgarian, N.; Pupure, L.; Joffe, R. Moisture uptake and resulting mechanical response of bio-based composites. II. Composites. *Polym. Compos.* **2015**, *36*, 1510–1519. [CrossRef]
87. Islam, M.S. The Influence of Fibre Processing and Treatments on Hemp Fibre/Epoxy and Hemp Fibre/PLA Composites. Ph.D. Thesis, University of Waikato, Hamilton, New Zealand, 2008.
88. Wan, Y.; Wang, Y.; Huang, Y.; Luo, H.; He, F.; Chen, G. Moisture absorption in a three-dimensional braided carbon/Kevlar/epoxy hybrid composite for orthopaedic usage and its influence on mechanical performance. *Compos. Part A Appl. Sci. Manuf.* **2006**, *37*, 1480–1484. [CrossRef]
89. Aronhime, M.T.; Neumann, S.; Marom, G. The anisotropic diffusion of water in Kevlar-epoxy composites. *J. Mater. Sci.* **1987**, *22*, 2435–2446. [CrossRef]
90. Piasecki, F. Résines Polyépoxydes Nanostructurées aux Propriétés d'Adhésion et à la Tenue au Vieillissement Améliorées. Ph.D. Thesis, Université de Bordeaux-I, Laboratoire de Chimie des Polymères Organiques, Bordeaux, France, 2013.
91. Akay, M.; Ah Mun, S.K.; Stanley, A. Influence of moisture on the thermal and mechanical properties of autoclaved and oven-cured Kevlar-49/epoxy laminates. *Compos. Sci. Technol.* **1997**, *57*, 565–571. [CrossRef]
92. Hahn, H.T.; Kim, K.S. Hygroscopic Effects in Aramid Fiber/Epoxy Composite. *J. Eng. Mater. Technol.* **1988**, *110*, 153–157. [CrossRef]
93. Sinchuk, Y.; Pannier, Y.; Antoranz-Gonzalez, R.; Gigliotti, M. Analysis of moisture diffusion induced stress in carbon/epoxy 3D textile composite materials with voids by μ-CT based Finite Element Models. *Compos. Struct.* **2019**, *212*, 561–570. [CrossRef]
94. Wan, Y.Z.; Wang, Y.L.; Cheng, G.X.; Han, K.Y. Three-dimensionally braided carbon fiber-epoxy composites, a new type of material for osteosynthesis devices. I. Mechanical properties and moisture absorption behavior. *J. Appl. Polym. Sci.* **2002**, *85*, 1031–1039. [CrossRef]
95. Yuan, Y.; Zhou, C.w. Meso-Scale Modeling to Characterize Moisture Absorption of 3D Woven Composite. *Appl. Compos. Mater.* **2016**, *23*, 719–738. [CrossRef]
96. Abacha, N.; Kubouchi, M.; Sakai, T.; Tsuda, K. Diffusion behavior of water and sulfuric acid in epoxy/organoclay nanocomposites. *J. Appl. Polym. Sci.* **2009**, *112*, 1021–1029. [CrossRef]
97. Dell'Anno, G.; Lees, R. Effect of water immersion on the interlaminar and flexural performance of lowcost liquid resin infused carbon fabric composites. *Compos. Part Eng.* **2012**, *43*, 1368–1372. [CrossRef]
98. Larbi, S.; Bensaada, R.; Bilek, A.; Djebali, S. Hygrothermal ageing effect on mechanical properties of FRP laminates. In Proceedings of the AIP 4th International Congress in Advances in Applied Physics and Materials Science, Fethiye, Turkey, 24–24 April 2015; AIP Publishing: Melville, NY, USA, 2015. [CrossRef]
99. Collings, T.; Stone, D. Hygrothermal effects in CFC laminates: Damaging effects of temperature, moisture and thermal spiking. *Compos. Struct.* **1985**, *3*, 341–378. [CrossRef]
100. Vanlandingham, M.R.; Eduljee, R.F.; Gillespie, J.W. Moisture diffusion in epoxy systems. *J. Appl. Polym. Sci.* **1999**, *71*, 787–798. [CrossRef]
101. Johncok, P.; Tudgey, G.F. Some Effects of Structure, Composition and Cure on the Water Absorption and Glass Transition Temperature of Amine-cured Epoxies. *Br. Polym. J.* **1986**, *18*, 292–302. [CrossRef]
102. Bellenger, V.; Verdu, J.; Morel, E. Structure-properties relationships for densely cross-linked epoxide-amine systems based on epoxide or amine mixtures. Part 2: Water absorption and diffusion. *J. Mater. Sci.* **1989**, *24*, 63–68. [CrossRef]
103. Xiao, G.; Shanahan, M. Swelling of DGEBA/DDA epoxy resin during hygrothermal ageing. *Polymer* **1998**, *39*, 3253–3260. [CrossRef]
104. Heman, M.B. Contribution à l'Etude des Interphases et de Leur Comportement au Vieillissement Hygrothermique dans les Systèmes à Matrice Thermodurcissable Renforcés de Fibres de Verre. Ph.D. Thesis, Université du Sud-Toulon-Var, Laboratoire Matériaux à Finalités Spécifiques, La Garde, France, 2008.
105. Xiao, G.; Shanahan, M. Irreversible effects of hygrothermal aging on DGEBA/DDA epoxy resin. *J. Appl. Polym. Sci.* **1998**, *69*, 363–369. [CrossRef]
106. Shirrell, C.D.; Halpin, J. Moisture absorption and desorption in epoxy composite laminates. In *Proceedings of the Composite Materials: Testing and Design*; Davis West, J., Ed.; ASTM International: West Conshohocken, PA, USA, 1977; pp. 514–528.
107. Martin, R. *Ageing of Composites. Composites Science and Engineering*; Woodhead Publishing Series: Cambridge, UK, 2008.
108. Lefebvre, D.; Dillard, D.; Ward, T. A model for the diffusion of moisture in adhesive joints. Part I: Equations governing diffusion. *J. Adhes.* **1989**, *27*, 1–18. [CrossRef]
109. Crank, J. *The Mathematics of Diffusion*; Clarendon Press: Oxford, UK, 1975.

110. Weitsman, Y.J. Diffusion models. In *Fluid Effects in Polymers and Polymeric Composites*; Springer: New York, NY, USA, 2012; pp. 69–94. [CrossRef]
111. Placette, M.D.; Fan, X. A Dual Stage Model of Anomalous Moisture Diffusion and Desorption in Epoxy Mold Compounds. In Proceedings of the 12th International Conference on Thermal, Mechanical and Multiphysics Simulation and Experiments in Micro-Electronics and Micro-Systems EuroSimE, Linz, Austria, 18–20 April 2011; IEEE: Linz, Autriche, 2011; pp. 1–8.
112. Alston, S.; Korkees, F.; Arnold, C. Finite element modelling of moisture uptake in carbon fibre/epoxy composites: A multi-scale approach. In Proceedings of the ECCM15 Proceedings; ESCM: 15th European Conference on Composite Materials, Venice, Italy, 24–28 June 2012.
113. Gurtin, M.E.; Yatomi, C. On a Model for Two Phase Diffusion in Composite Materials. *J. Compos. Mater.* **1978**, *13*, 126–130. [CrossRef]
114. Chambers, J.M.; Cleveland, W.S.; Kleiner, B.; Tukey, P.A. *Graphical Methods for Data Analysis*; Chapman and Hall/CRC: Boca Raton, FL, USA, 2018.
115. Ugarte, M.D.; Militino, A.F.; Arnholt, A.T. *Probability and Statistics with R*; CRC Press: Boca Raton, FL, USA, 2008.
116. Cornillon, P.A.; Husson, F.; Jégou, N.; Matzner-Lober, E.; Josse, J.; Guyader, A.; Rouvière, L.; Kloareg, J. *Statistiques avec R; Pratique de la Statistique*; Presses Universitaires de Rennes: Paris, France, 2012.
117. Van Krevelen, D.W. *Properties of Polymers: Their Correlation with Chemical Structure Their Numerical Estimation and Prediction from Additive Group Contributions*, 4th ed.; Elsevier: Amsterdam, The Netherlands, 2009.
118. Verdu, J. *Action de l'Eau sur les Plastiques*; Techniques de l'Ingénieur: Paris, France, 2000; pp. 1–11.
119. Morel, E.; Bellenger, V.; Verdu, J. *Relations Structure-Hydrophilie des Réticultats Epoxyde-Amine*; Pluralis Ed: Paris, France, 1984; pp. 597–614.
120. McKague, E.L.; Reynolds, J.D.; Halkias, J.E. Swelling and glass transition relations for epoxy matrix material in humid environments. *J. Appl. Polym. Sci.* **1978**, *22*, 1643–1654. [CrossRef]
121. Cotugno, S.; Larobina, D.; Mensitieri, G.; Musto, P.; Ragosta, G. A novel spectroscopic approach to investigate transport processes in polymers: The case of water-epoxy system. *Polymer* **2001**, *42*, 6431–6438. [CrossRef]
122. Karad, S.K.; Jones, F.R. Mechanisms of moisture absorption by cyanate ester modified epoxy resin matrices: The clustering of water molecules. *Polymer* **2005**, *46*, 2732–2738. [CrossRef]
123. Fuller, R.T.; Fornes, R.E.; Memory, J.D. NMR study of water absorbed by epoxy resin. *J. Appl. Polym. Sci.* **1979**, *23*, 1871–1874. [CrossRef]
124. Pascault, J.P.; Sautereau, H.; Verdu, J.; Williams, R.J.J. *Thermosetting Polymers*; Marcel Dekker: New York, Ny, USA, 2002.
125. Bouvet, G. Relations Entre Microstructure et Propriétés Physico-Chimiques et Mécaniques de Revêtements Epoxy Modèles. Ph.D. Thesis, Université de La Rochelle, Laboratoire des Sciences de l'Ingénieur pour l'Environnement, Paris, France, 2014.
126. Apicella, A.; Nicolais, L.; de Cataldis, C. Characterization of the morphological fine structure of commercial thermosetting resins through hygrothermal experiments. *Adv. Polym. Sci.* **2005**, *66*, 189–207.
127. Bellenger, V.; Decelle, J.; Huet, N. Ageing of a carbon epoxy composite for aeronautic applications. *Compos. Part Eng.* **2005**, *36*, 189–194. [CrossRef]
128. Alam, P.; Robert, C.; Ó Brádaigh, C.M. Tidal turbine blade composites—A review on the effects of hygrothermal aging on the properties of CFRP. *Compos. Part Eng.* **2018**, *149*, 248–259. [CrossRef]
129. Adamson, M.J. Thermal expansion and swelling of cured epoxy resin used in graphite/epoxy composite materials. *J. Mater. Sci.* **1980**, *15*, 1736–1745. [CrossRef]
130. Al-Maharma, A.Y.; Al-Huniti, N. Critical review of the parameters affecting the effectiveness of moisture absorption treatments used for natural composites. *J. Compos. Sci.* **2019**, *3*, 27. [CrossRef]
131. Azwa, Z.; Yousif, B.; Manalo, A.; Karunasena, W. A review on the degradability of polymeric composites based on natural fibres. *Mater. Des.* **2013**, *47*, 424–442. [CrossRef]
132. Hamid, M.R.Y.; Ab Ghani, M.H.; Ahmad, S. Effect of antioxidants and fire retardants as mineral fillers on the physical and mechanical properties of high loading hybrid biocomposites reinforced with rice husks and sawdust. *Ind. Crop. Prod.* **2012**, *40*, 96–102. [CrossRef]
133. Xie, Y.; Hill, C.A.; Xiao, Z.; Militz, H.; Mai, C. Silane coupling agents used for natural fiber/polymer composites: A review. *Compos. Part Appl. Sci. Manuf.* **2010**, *41*, 806–819. [CrossRef]
134. Tang, X.; Whitcomb, J.D.; Li, Y.; Sue, H.J. Micromechanics modeling of moisture diffusion in woven composites. *Compos. Sci. Technol.* **2005**, *65*, 817–826. [CrossRef]
135. Almudaihesh, F.; Holford, K.; Pullin, R.; Eaton, M. The influence of water absorption on unidirectional and 2D woven CFRP composites and their mechanical performance. *Compos. Part Eng.* **2020**, *182*, 107626. [CrossRef]
136. Wan, Y.; Wang, Y.; Huang, Y.; He, B.; Han, K. Hygrothermal aging behaviour of VARTMed three-dimensional braided carbon-epoxy composites under external stresses. *Compos. Part Appl. Sci. Manuf.* **2005**, *36*, 1102–1109. [CrossRef]
137. Gollins, K.; Chiu, J.; Delale, F.; Liaw, B.; Gursel, A. Comparison of Manufacturing Techniques Subject to High Speed Impact. In Proceedings of the ASME International Mechanical Engineering Congress and Exposition, Montreal, QC, Canada, 14–20 November 2014; Volume 46583, p. V009T12A019.
138. Bhatt, A.T.; Gohil, P.P.; Chaudhary, V. Primary manufacturing processes for fiber reinforced composites: History, development & future research trends. In *Proceedings of the IOP Conference Series: Materials Science and Engineering*; IOP Publishing: Bristol, UK, 2018; Volume 330, p. 012107.

139. Park, S.Y.; Choi, C.H.; Choi, W.J.; Hwang, S.S. A Comparison of the Properties of Carbon Fiber Epoxy Composites Produced by Non-autoclave with Vacuum Bag Only Prepreg and Autoclave Process. *Appl. Compos. Mater.* **2019**, *26*, 187–204. [CrossRef]
140. Wolter, N.; Beber, V.C.; Yokan, C.M.; Storz, C.; Mayer, B.; Koschek, K. The effects of manufacturing processes on the physical and mechanical properties of basalt fibre reinforced polybenzoxazine. *Compos. Commun.* **2021**, *24*, 100646. [CrossRef]
141. Hopfenberg, H.; Stannett, V. The diffusion and sorption of gases and vapours in glassy polymers. In *The Physics of Glassy Polymers*; Springer: Berlin/Heidelberg, Germany, 1973; pp. 504–547.
142. Barrer, R.M.; Barrie, J.A.; Slater, J. Sorption and diffusion in ethyl cellulose. Part III. Comparison between ethyl cellulose and rubber. *J. Polym. Sci.* **1958**, *27*, 177–197. [CrossRef]
143. Flory, P.J. *Principles of Polymer Chemistry*; Cornell University Press: Cornell, UK, 1953.
144. Fick, A. Über diffusion. *Ann. Phys. Hig. Poggendroff* **1855**, *170*, 59–86. [CrossRef]
145. Obeid, H. Durabilité de Composites à Matrice Thermoplastique Sous Chargement hygromécanique: Etude Multi-Physique et Multiéchelle des Relations Microstructure-Propriétés—états Mécaniques. Ph.D. Thesis, Université Bretagne-Loire, Institut de Recherche en Génie Civil et Mécanique (GeM), Nantes, France, 2016.
146. Loos, A.C.; Springer, G.S. Moisture Absorption of Graphite-Epoxy Composites Immersed in Liquids and in Humid Air. *J. Compos. Mater.* **1979**, *13*, 131–147. [CrossRef]
147. Zeng, W.; Du, Y.; Xue, Y.; Frisch, H.L. Solubility parameters. In *Physical Properties of Polymers Handbook*, 2nd ed.; Mark, J.E., Ed.; Springer: New York, NY, USA, 2006; pp. 289–305.
148. Charlas, M. Étude et Durabilité de Solutions de Packaging Polymère de Module d'Electronique de Puissance à Application Aéronautique. Ph.D. Thesis, Université de Pau et des Pays de l'Adour, Institut des Sciences Analytiques et de Physico-Chimie pour l'Environnement et les Matériaux, Pau, France, 2009.
149. Hansen, C.M. *Hansen Solubility Parameters: A User's Handbook*, 2nd ed.; CRC Press: Boca Raton, FL, USA, 2007.
150. Jolliffe, I.T. *Principal Component Analysis*; Springer: New York, NY, USA, 1986.
151. Meglen, R.R. Examining large databases: A chemometric approach using principal component analysis. *J. Chemom.* **1991**, *5*, 163–179. [CrossRef]
152. Jolliffe, I.T.; Cadima, J. Principal component analysis: A review and recent developments. *Philos. Trans. R. Soc. Math. Phys. Eng. Sci.* **2016**, *374*, 1–16. [CrossRef] [PubMed]
153. Villar Montoya, M. Procédé de Soudage Laser de Polymères Haute Performance: Etablissement des Relations entre les Paramètres du Procédé, la Structure et la Morphologie du Polymère et les Propriétés Mécaniques de l'Assemblage. Ph.D. Thesis, Institut National Polytechnique de Toulouse, Laboratoire Génie de Production (INP-ENIT), Paris, France, 2018.
154. Husson, F.; Lê, S.; Pagès, J. *Exploratory Multivariate Analysis by Example Using R. Computer Science & Data Analysis*; Chapman Hall/CRC Press: Boca Raton, FL, USA, 2011.
155. Lê, S.; Josse, J.; Husson, F. FactoMineR: An R Package for Multivariate Analysis. *J. Stat. Softw.* **2008**, *25*, 1–18. [CrossRef]
156. Dunteman, G.H. *Principal Components Analysis*; Sage Publications: Thousand Oaks, CA, USA, 1989; p. 69
157. Kassambara, A. *Practical Guide To Principal Component Methods in R; Vol. Multivariate Analysis II, Statistical Tools for High-throughput Data Analysis (STHDA)*; Sage Publications: Thousand Oaks, CA, USA, 2017.
158. Peres-Neto, P.R.; Jackson, D.A.; Somers, K.M. How many principal components? Stopping rules for determining the number of non-trivial axes revisited. *Comput. Stat. Data Anal.* **2005**, *49*, 974–997. [CrossRef]
159. Simar, A.; Gigliotti, M.; Grandidier, J.C.; Ammar-Khodja, I. Decoupling of water and oxygen diffusion phenomena in order to prove the occurrence of thermo-oxidation during hygrothermal aging of thermosetting resins for RTM composite applications. *J. Mater. Sci.* **2018**, *53*, 11855–11872. [CrossRef]
160. Farrar, N.R.; Ashbee, K.H.G. Destruction of epoxy resins and of glass-fibre-reinforced epoxy resins by diffused water. *J. Phys. Appl. Phys.* **1978**, *11*, 1009–1015. [CrossRef]
161. Fedors, R.F. Osmotic effects in water absorption by polymers. *Polymer* **1980**, *21*, 207–212. [CrossRef]
162. Gautier, L.; Mortaigne, B.; Bellenger, V.; Verdu, J. Osmotic cracking nucleation in hydrothermal-aged polyester matrix. *Polymer* **2000**, *41*, 2481–2490. [CrossRef]
163. Colin, X.; Verdu, J.; Rabaud, B. Stabilizer Thickness Profiles in Polyethylene Pipes Transporting Drinking Water Disinfected by Bleach. *Polym. Eng. Sci.* **2011**, *51*, 1539–1549. [CrossRef]
164. Moisan, J.Y. Diffusion des additifs du polyéthylène I: Influence de la nature du diffusant. *Eur. Polym. J.* **1980**, *16*, 979–987. [CrossRef]
165. Abdessalem, A.; Tamboura, S.; Fitoussi, J.; Ben Daly, H.; Tcharkhtchi, A. Bi-phasic water diffusion in sheet molding compound composite. *J. Appl. Polym. Sci.* **2020**, *137*, 1–12. [CrossRef]
166. Berthé, V.; Ferry, L.; Bénézet, J.; Bergeret, A. Ageing of different biodegradable polyesters blends mechanical and hygrothermal behavior. *Polym. Degrad. Stab.* **2010**, *95*, 262–269. [CrossRef]
167. Gillet, C.; Nassiet, V.; Poncin-Epaillard, F.; Hassoune-Rhabbour, B.; Tchalla, T. Chemical behaviour of water absorption in a carbon/epoxy 3D woven composite. In *Macromolecular Symposia, Proceedings of the10th International Conference on Times of Polymers and Composites, Ischia, Italy, 30 April 2022*; Wiley Online Library: Hoboken, NJ, USA, 2022.
168. Gillet, C.; Hassoune-Rhabbour, B.; Poncin-Epaillard, F.; Tchalla, T.; Nassiet, V. Contributions of atmospheric plasma treatment on a hygrothermal aged carbon/epoxy 3D woven composite material. *Polym. Degrad. Stab.* **2022**, *202*, 110023. [CrossRef]

169. Guloglu, G.E.; Hamidi, Y.K.; Altan, M.C. Fast recovery of non-fickian moisture absorption parameters for polymers and polymer composites. *Polym. Eng. Sci.* **2017**, *57*, 921–931. [CrossRef]
170. Cocaud, J.; Célino, A.; Fréour, S.; Jacquemin, F. What about the relevance of the diffusion parameters identified in the case of incomplete Fickian and non-Fickian kinetics? *J. Compos. Mater.* **2019**, *53*, 1555–1565. [CrossRef]

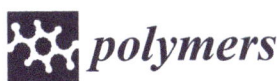

Review

Biopolymeric Nanoparticles–Multifunctional Materials of the Future

Andrey A. Vodyashkin [1,*], Parfait Kezimana [1,2], Alexandre A. Vetcher [1,3,*] and Yaroslav M. Stanishevskiy [1]

[1] Institute of Biochemical Technology and Nanotechnology, Peoples Friendship University of Russia (RUDN University), 6 Miklukho-Maklaya Str., 117198 Moscow, Russia; kezimana-p@rudn.ru (P.K.); stanyar@yandex.ru (Y.M.S.)

[2] Department of Agrobiotechnology, Peoples Friendship University of Russia (RUDN University), 6 Miklukho-Maklaya Str., 117198 Moscow, Russia

[3] Complementary and Integrative Health Clinic of Dr. Shishonin, 5 Yasnogorskaya Str., 117588 Moscow, Russia

* Correspondence: vodyashkin_aa@pfur.ru (A.A.V.); avetcher@gmail.com (A.A.V.)

Abstract: Nanotechnology plays an important role in biological research, especially in the development of delivery systems with lower toxicity and greater efficiency. These include not only metallic nanoparticles, but also biopolymeric nanoparticles. Biopolymeric nanoparticles (BPNs) are mainly developed for their provision of several advantages, such as biocompatibility, biodegradability, and minimal toxicity, in addition to the general advantages of nanoparticles. Therefore, given that biopolymers are biodegradable, natural, and environmentally friendly, they have attracted great attention due to their multiple applications in biomedicine, such as drug delivery, antibacterial activity, etc. This review on biopolymeric nanoparticles highlights their various synthesis methods, such as the ionic gelation method, nanoprecipitation method, and microemulsion method. In addition, the review also covers the applications of biodegradable polymeric nanoparticles in different areas—especially in the pharmaceutical, biomedical, and agricultural domains. In conclusion, the present review highlights recent advances in the synthesis and applications of biopolymeric nanoparticles and presents both fundamental and applied aspects that can be used for further development in the field of biopolymeric nanoparticles.

Keywords: biopolymeric nanoparticles; biodegradable; synthesis; applications; medicine; agriculture

Citation: Vodyashkin, A.A.; Kezimana, P.; Vetcher, A.A.; Stanishevskiy, Y.M. Biopolymeric Nanoparticles–Multifunctional Materials of the Future. *Polymers* **2022**, *14*, 2287. https://doi.org/10.3390/polym14112287

Academic Editor: Agnieszka Tercjak

Received: 29 April 2022
Accepted: 31 May 2022
Published: 4 June 2022

Publisher's Note: MDPI stays neutral with regard to jurisdictional claims in published maps and institutional affiliations.

Copyright: © 2022 by the authors. Licensee MDPI, Basel, Switzerland. This article is an open access article distributed under the terms and conditions of the Creative Commons Attribution (CC BY) license (https:// creativecommons.org/licenses/by/ 4.0/).

1. Introduction

Nanopar ticles are very promising materials with applications in various industries. Their active use is mainly due to their ultra-small size, easy functionalization, simple production methods, economical production, and easy scalability of their synthesis. They are used in industries such as the paper industry [1], water purification [2], catalysis [3], drug delivery [4], antibacterial products [5], and the food industry [6].

Materials with specific physical and chemical properties at the nano scale can potentially penetrate tissues, cells, and organelles, and interact with functional biomolecular structures, causing toxicity [7,8]. The toxicity of nanoparticles is affected by several parameters, such as their size, nature, surface functionalization, etc. [9,10].

Metal-based nanoparticles are the most popular nanoparticles found in commercially available products [11], but many of them are highly toxic to the human body [12–14], so more stringent quality and safety controls are needed for their use. Moreover, it is also necessary to strictly regulate the use of these particles in various applications [15,16]. In addition, metal nanoparticles can harm the environment (e.g., soil, water) and have a negative impact on the life of organisms therein [17,18]. It is worth mentioning that the use of metal nanoparticles for biomedical purposes is severely limited by the inability to excrete or metabolize them in the human body [19].

In this regard, biodegradable nanoparticles are needed not only for biomedical purposes, but also to improve the safety of various industries [20,21]. Biological polymeric compounds present several properties, such as biocompatibility, antioxidant, and antibacterial properties, photoprotection, and active surface functionality [22–24], making them good candidates for the development of nanomaterials with biological activities.

Among the biodegradable nanoparticles, the most common are biopolymeric nanoparticles (BPNs), which are nanoparticles that are constructed from the natural polymers found in biological species such as proteins (e.g., collagen, gelatin, β-casein, zein, and albumin), protein-mimicking polypeptides (e.g., cationic polypeptides such as polylysine and polyornithine [25]), and polysaccharides (e.g., hyaluronic acid, chitosan, alginate, pullulan, starch, and heparin). These nanoparticles from various biopolymers are currently being created to solve problems related to toxicity, biocompatibility, and biodegradability. Biopolymeric nanoparticles (BPNs) are also applied in various industries, as are metallic NPs, but the former has increased safety and are environmentally friendly [26–28]. Currently, BPNs are actively used in the food industry, for biomedical purposes, in the development of diagnostic tools, and in the household chemical industry [29–31].

Although this manuscript concentrates on polysaccharide-based nanoparticles (e.g., chitosan nanoparticles, cellulose nanoparticles, etc.), we should also note that polypeptide-based nanoparticles have been reported as ideal materials for gene and drug delivery, due to their versatile traits, including excellent biocompatibility, biodegradability, low immunogenicity [32], precise secondary structure conformations, and self-assembling properties [33]. Moreover, with their bio-mimicked nature and surface functionalization, the use of polypeptide-based nanomaterials can help minimize adverse nano–bio interactions. For example, smooth excretion of PEG-polypeptides from the body via the biliary route within 3 h after their intended biomedical use further empowers their biocompatible nature [34].

The simplicity of carbohydrate-based BPNs' methods of preparation, along with their wide application in various industries, has attracted several researchers, and there has been an increasing number of papers reporting on their synthesis, parameters, and applications. In this review, we discuss the main methods of their preparation, and their potential use in life sciences—from the delivery of biomolecules in organisms, diagnostic materials, and antibacterial agents, to improving agricultural production. Thus, this review presents a systematized analysis of the synthesis and application of BPNs.

2. Main Methods of Synthesis of Biodegradable Polymeric Nanoparticles

To date, several methods for the preparation of BPNs with different shapes and surface charges have been reported. The properties and, consequently, the application of BPNs depend on their shape, as spherical nanoparticles have been reported to have a high potential for cell penetration [35]. Therefore, there are requirements related to the shape of nanoparticles prepared from biodegradable polymers—especially those used as delivery agents in the human body. This requirement is of particular importance when choosing the preparation method and is discussed in the section on the synthesis of biodegradable polymeric nanoparticles.

2.1. The Ionic Gelation Method

Gelation is the process of a polymer solution's transition from liquid to solid form under the influence of polymer crosslinking reactions taking place in the solution. This process has been used since ancient times, and in recent years has been actively used to prepare BPNs [36]. Ionic gelation is a process that takes place under the influence of electrostatic interactions between ionic polymers and crosslinking agents, as illustrated in Figure 1. This method produces stable nanoparticles of chitosan, starch alginate, cellulose, and other biopolymers. Nanoparticles obtained via the ionic gelation method show good encapsulation efficiency and preservation of the bioactivity of the embedded molecules [37–40].

For the preparation of chitosan nanoparticles, Fan et al. demonstrated that the optimal ratio of chitosan to sodium tripolyphosphate is 3.3:1, the optimal pH of the solution is 4.7–4.8, and the optimal concentration of acetic acid solution is 0.2 mg/mL. These conditions make it possible to obtain nanoparticles of 138.1 nm in size, with the narrowest distribution [41].

Chitosan nanoparticles have also been obtained via ionic gelation, for which chitosan was dissolved in a 1% acetic acid solution to obtain a 0.1% solution (w/v and adjusted to a certain pH with a NaOH solution. A 0.1% solution of sodium tripolyphosphate in deionized water was prepared separately. While stirring, sodium tripolyphosphate solution was added to the chitosan solution at a rate of 0.1 mL/min, until a certain mass ratio of chitosan to sodium tripolyphosphate was reached. After adding the sodium tripolyphosphate solution, the resulting mixture was subjected to ultrasound for 10 min. It was found that factors such as the molecular weight of chitosan, the pH of the chitosan solution, and the mass ratio of chitosan to sodium tripolyphosphate affect the size of the resulting nanoparticles. The smallest particle size of 26 nm can be obtained by using low-molecular-weight chitosan, bringing the pH of the chitosan solution to 4.6 and the mass ratio of chitosan to sodium tripolyphosphate to 3:1. Moreover, the use of low-molecular-weight chitosan makes it possible to not use ultrasonic effects to obtain nanoparticles of a given size [42].

Anand et al. also obtained chitosan NPs via ionic gelation. In their method, they brought the pH of the chitosan solution to 6.0, and spherical nanoparticles with a wide distribution of 8–80 nm were obtained [43].

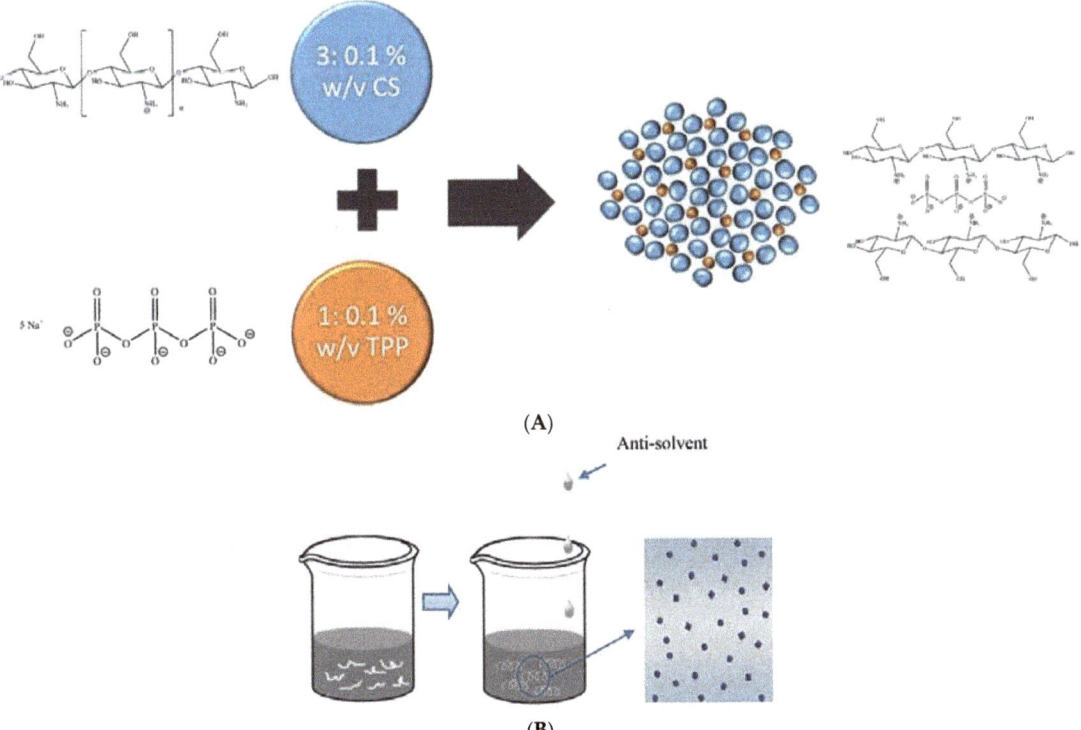

Figure 1. (A) Crosslinking of charged chitosan with sodium tripolyphosphate, reprinted with permission from [38], 2018, Elsevier. (B) Schematic illustration of the nanoprecipitation method, reprinted with permission from [44], 2019, Elsevier.

Ji et al. obtained chitosan nanoparticles with encapsulated amylase via ionic gelation [45]. A solution of carboxymethyl short-chain amylose was added to chitosan dissolved in an aqueous solution of acetic acid and stirred continuously for 6 h at room temperature. After that, the formed complex nanoparticles were isolated by centrifugation. The result was spherical nanoparticles with an average size of 262 nm. The work of Ji et al. demonstrates that the ionic gelation method allows loading of biodegradable polymer nanoparticles with active substances in the process of nanoparticle synthesis [45].

Anita et al. further demonstrated the possibility of preparing a complex of biodegradable polymer–metal nanoparticles and loading them with biomolecules using the ionic gelation method [46]. Copper nanoparticles were obtained by mixing copper sulfate, sodium hydroxide, and starch, which was used as a capping agent. To increase the functionality of the nanoparticles, microencapsulation was performed, which consisted of coating the nanoparticles with calcium alginate. This process was performed using the method of ionic gelation, the essence of which was to mix sodium alginate gel and copper nanoparticles, and to atomize calcium chloride into this system.

Qing et al. reported a study describing the preparation of negatively charged carboxymethylene branched starch nanoparticles via ionic gelation [37]. Starch and sodium hydroxide were added to ethanol, and the mixture was heated to 40 °C and incubated under stirring for one hour. Then, a mixture of monochloroacetic acid dissolved in ethanol was slowly added, after which the pH of the reaction mixture was adjusted to 12.0. The reaction mixture was incubated at 40 °C and under constant stirring for 4 h. At the end of the reaction, the pH of the mixture was adjusted to neutral with hydrochloric acid. The nanoparticles were isolated by centrifugation, and then washed with ethanol. As a result, starch nanoparticles with a spherical shape and a size of 50 to 100 nm were obtained.

Qing et al. also reported another way of preparing starch nanoparticles via ionic gelation, whereby a suspension of oxidized branched starch was gelatinized and slowly added to a solution containing calcium ions under stirring at room temperature, and then incubated for 2 h. The result was spherical starch nanoparticles with sizes ranging from 30 to 50 nm. It was noted that as the concentration of the calcium ion solution decreased, the size of the nanoparticles became smaller, opening the possibility of adjusting the size of the resulting particles [47].

The method of ionic gelation also makes it possible to obtain nanocomposites consisting of two biodegradable polymers. Subramanian et al. obtained composite nanoparticles consisting of chitosan and starch. Chitosan and starch were dissolved in an aqueous 2% acetic acid solution and, after complete dissolution, Tween-80 was added for uniform dispersion. A solution of a pre-synthesized anticancer agent and sodium tripolyphosphate was added to the mixture. The particles were isolated by centrifugation, and then lyophilized. As a result, nanoparticles of spherical shape, ranging in size from 173 to 297 nm, were obtained [48].

Nait Mohamed and Laraba-Djebari obtained alginate nanoparticles via ionic gelation [49]. Calcium chloride solution was slowly added to sodium alginate solution at room temperature and under constant stirring, and the mixture was incubated for one hour. The nanoparticles were isolated by centrifugation. As a result, nanoparticles of spherical shape, with a wide distribution of 150–350 nm, were obtained.

Stable nanoparticles of chitosan, alginate, starch, and other biopolymers can be obtained via the ionic gelation method. Nanoparticles obtained via this method show good encapsulation efficiency, small size, and high colloidal stability. Studies demonstrating the ability to adjust the size and distribution width of the resulting nanoparticles have been carried out.

2.2. The Nanoprecipitation Method

The nanoprecipitation method is a simple and fast process that differs from emulsion-based methods (e.g., emulsion–diffusion, emulsion–evaporation, and salting-out methods) in that no precursor emulsion is required. In practice, the hydrophobic solute (i.e., polymer or lipid molecules) is first dissolved in a polar organic solvent (usually ethanol, acetone, or THF). This solution is then added to a large amount of non-solvent (usually water) of the dissolved substance, with which the polar solvent is mixed in all ratios. The mixed binary solution becomes a non-solvent for the hydrophobic molecules, and the system evolves toward phase separation, resulting in the formation of hydrophobic solute particles (Figure 1B). The organic solvent can then be removed by evaporation. The formation of nanoparticles by nanoprecipitation ensures high drug content due to the coating of the nanoparticle surface with protective agents [44].

The resulting nanoparticles are separated by centrifugation or removal of the solvent with a rotary evaporator, and then washed with non-solvent (anti-solvent) or its mixture with distilled water, and subjected to lyophilic drying [50,51].

This method opens the possibility of obtaining nanodrug delivery agents. Moreover, instead of using multistep methods for obtaining nanoparticles and then incorporating the reagent into them, it is possible to introduce the drug into the nanoparticles directly during the nanoprecipitation process. El-Naggar et al. obtained starch nanoparticles with encapsulated diclofenac sodium via nanoprecipitation. Starch was dissolved in sodium hydroxide solution and incubated under stirring and at room temperature for 30 min, and then a mixture of diclofenac sodium and Tween-80 dissolved in distilled water was added to the resulting suspension. After 30 min, sodium tripolyphosphate solution was added, and the mixture was stirred continuously for 2 h. After 2 h, absolute ethanol was added to the mixture, causing the starch nanoparticles to precipitate. The formed nanoparticles were centrifuged and washed with a mixture of ethanol and water. As a result, spherical nanoparticles with an average diameter of 21.04 nm were obtained. The obtained nanoparticles were used as a delivery model for sodium diclofenac [52].

In the process of implementing the nanoprecipitation method, it is possible to regulate the size of the resulting nanoparticles. The solvent system used to dissolve the starch has a direct influence on the size of the resulting nanoparticles. Chin et al. obtained starch nanoparticles via nanoprecipitation. Starch was dissolved in a system of sodium hydroxide and urea solutions. The presence of sodium hydroxide in the solvent system is necessary for hydrolysis and for breaking the hydrogen bonds of the starch molecules. Urea plays an important role in preventing the self-association of starch molecules, leading to their increased solubility. The resulting mixture was slowly added to absolute ethanol and stirred at room temperature for 30 min. The resulting starch nanoparticles were isolated by centrifugation and washed with absolute ethanol to remove alkali and urea residues. The resulting nanoparticles were spherical in shape, ranging in size from 300 to 400 nm. The authors also demonstrated that adding the surfactant Tween-80 to the solvent system can reduce the size of the obtained particles to a range of 250–300 nm, and added hexadecyl(cetyl)trimethylammonium bromide to 150–200 nm [53].

Chin et al. explained the importance of the presence of a component such as urea in the starch solvent. However, it is worth highlighting a work that searched for the optimal concentrations of the components in the starch solvent. G. Gutierrez et al. obtained starch nanoparticles via nanoprecipitation. Their work studied the effects of the concentrations of the components of the starch solvent system, as well as the effect of the rate of addition of starch solution to absolute ethanol, on the size of the resulting particles. Starch was dissolved in an aqueous solution of sodium hydroxide and urea with different concentrations of components, and stirred at 80 °C for 30 min, after which the resulting mixture was slowly introduced into absolute ethanol. Spherical nanoparticles, 13.9 ± 9.03 nm in size, were obtained when concentrations of 8% (w/v) and 10% (w/v) urea were observed in the solvent system for dissolving the starch, and with an injection rate of the starch solution into absolute ethanol of 4 mL/h [54].

The nanoprecipitation method uses dimethyl sulfoxide (DMSO) as an alternative solvent for starch. DMSO is able to break the inter- and intramolecular hydrogen bonds of starch before replacing the hydroxy hydrogen bonds with DMSO–starch hydrogen bonds [55,56].

Pan et al. obtained starch nanoparticles via nanoprecipitation; in contrast to the previously listed works, the authors used DMSO as a starch solvent, and used ultrasound treatment to grind the resulting particles. Starch was added to DMSO and continuously stirred for 10 min, and then treated with ultrasound for 1 h. Then, deionized water was slowly added to the mixture while stirring, treated with ultrasound for another 1 h, and the obtained nanoparticles were isolated by centrifugation. In this work, the authors subjected the resulting nanoparticles to dialysis to remove impurities, and then lyophilized them. As a result, nanoparticles of spherical shape and a size of 68.63 ± 9.21 nm were obtained [55].

In addition to obtaining starch nanoparticles, the method of nanoprecipitation is also used for obtaining chitosan nanoparticles. Sotelo-Boyas et al. dissolved chitosan in acetic acid solution, and a mixture of lime and methanol was added to the solution while stirring. The formed nanoparticles were isolated by removing the solvent with a rotary evaporator. No additional purification was used for the obtained nanoparticles. As a result, the formed nanoparticles had a spherical shape and a size of 5.8 ± 1.6 nm. The obtained nanoparticles were used to encapsulate lime essential oil; as a result, the composite was studied for its antibacterial properties [57].

Chitosan-precipitated nanoparticles, as well as starch nanoparticles, have the potential to be drug carriers [58]. Thyme essential oils have antibacterial properties due to the volatile components they contain [59], but due to their volatile nature, thyme ester oils lose their antibacterial properties over time. The possibility of storing volatile components is opened by their encapsulation in nanocarriers. Sotelo-Boyas et al. used chitosan nanoparticles obtained via nanoprecipitation as nanocarriers for thyme essential oil components. The encapsulation process was performed directly during nanoparticle production. Chitosan dissolved in an aqueous solution of acetic acid was slowly added to a mixture of thyme essential oil and methanol under stirring. The resulting nanoparticles were isolated by removing the solvent using a rotary evaporator. As reported by the authors, 68% of the components of thyme essential oil were incorporated into the nanoparticles. As a result, the obtained particles had a spherical shape and an average size of 6.4 ± 0.5 nm [60].

Thus, the nanoprecipitation method opens the possibility of obtaining nanocarriers consisting of biodegradable polymers such as starch, chitosan, etc. There are possibilities for encapsulating drugs in nanoparticles directly during nanoparticle synthesis. The size of the resulting nanoparticles can be controlled by such parameters as the concentration of the reaction system components, the rate of adding "anti-solvent" to the polymer solution, and the use of ultrasound treatment.

2.3. The Microemulsion Method

Microemulsions are isotropic, thermodynamically stable, transparent (or translucent) colloidal dispersions consisting of at least three components: a non-polar phase, a polar phase, and a surfactant. The structural organization of microemulsions is usually of the oil-in-water (M/W) or water-in-oil (W/M) forms [61]. The microemulsion method is based on the addition of an aqueous polymer solution to an organic solvent including a surfactant during homogenization to form a finely dispersed water-in-oil (W/AM) microemulsion, in which nanoparticles are deposited [54]. The microemulsion method makes it possible to obtain nanoparticles of biodegradable polymers [62,63].

Trinh and Hung obtained starch nanoparticles via the microemulsion method. An alkaline starch solution was added to ethanol containing Tween-80 and soybean oil under stirring. The mixture was continuously stirred for 1 h. The formed nanoparticles were isolated by centrifugation and lyophilized. As a result, nanoparticles of spherical shape, with an average size of 70–200 nm, were obtained [64].

The microemulsion method makes it possible to obtain nanocomposite compounds such as two or more polymers crosslinked together, or metal nanoparticles encapsulated in polymers. Wang et al. obtained magnetic chitosan nanoparticles via the microemulsion method. Chitosan was dissolved in hydrochloric acid solution, and cyclohexane, hexanol-1, and divalent iron salt were added to the resulting solution while stirring. To the resulting mixture, TX-100 was slowly added until the mixture became transparent. After that, the reaction mixture was continuously stirred and heated at 60 °C for 2 h. Then, the particles were separated by centrifugation, washed with distilled water and ethanol, and lyophilized. As a result, chitosan nanoparticles with Fe_3O_4 nanoparticles encapsulated within them were obtained; the average diameter of the nanoparticles was 10–20 nm. The obtained nanoparticles were used as an adsorbent for albumin protein [65].

Dubey et al. obtained a chitosan–alginate nanocomposite via the microemulsion method. Chitosan was pre-dissolved in an aqueous solution of acetic acid and alginate in distilled water. Two prepared solutions and paraffin oil—which was used as an oil phase—were mixed in one flask, and then stirred continuously for 30 min, after which sodium tripolyphosphate and calcium salt solution were added. After that, they were stirred continuously at room temperature for another 3 h. The formed nanoparticles were isolated by centrifugation and washed with acetone. As a result, nanoparticles of spherical shape with an average size of 65 nm were obtained. The obtained chitosan–alginate nanocomposite was used as an adsorbent for mercury ions in aqueous solutions [66].

The use of ionic liquids in the microemulsion method allows the substitution of polar and non-polar phases and surfactants. The use of ionic liquids constitutes green chemistry because there is the possibility of their reuse, making them a profitable tool for use in reaction media. Ionic liquids have a melting point below 100 °C, which allows them to be used as solvents [67]. Zhou et al. obtained starch nanoparticles via microemulsion crosslinking using ionic liquid. Starch powder was added to 1-octyl-3-methylimidazolium acetate (an ionic liquid replacing the aqueous phase) and stirred under heating for 2.5 h. Next, cyclohexane was added to the mixture, after which a mixture of the surfactant TX-100 and butanol-1 was added while stirring until the mixture became transparent. Epichlorohydrin was added to the resulting mixture as a crosslinking agent. The mixture was then stirred at 50 °C for 3 h. After 3 h, the mixture was cooled to room temperature and the starch nanoparticles were precipitated with anhydrous ethanol under vigorous stirring, followed by centrifugation. The precipitate was washed thoroughly with absolute ethanol to remove unreacted epichlorohydrin, butanol-1, and TX-100. After that, the solid was centrifuged and dried in vacuo at 45 °C for 24 h. As a result, starch nanoparticles of spherical and oval shape, with an average size of 96.9 nm, were obtained [68].

Wang et al. obtained starch nanoparticles via microemulsion crosslinking, the technique of which was similar to that of Zhou et al., with the only difference being that the authors used 1-hexadecyl-3-methylimidazolium bromide as the ionic liquid instead of 1-octyl-3-methylimidazolium acetate. As a result, nanoparticles of indeterminate shape, with an average diameter of 80.5 nm, were obtained [69]. The use of ionic liquids makes it possible to obtain nanoparticles of biodegradable polymers up to 100 nm in size.

The microemulsion method is complex in terms of its implementation since the process is implemented in several steps. This method is suitable for obtaining nanodrugs based on biodegradable polymers. The use of ionic liquids in this method opens opportunities for technical simplification and cheaper process.

2.4. The Coacervation Method

Coacervation is the spontaneous formation of a superdense liquid phase from a homogeneous macromolecular solution with low solvent affinity. During coacervation, a homogeneous solution of charged macromolecules undergoes liquid–liquid phase separation, resulting in a polymer-rich dense phase coexisting with the supernatant. The two liquid phases do not mix, but strongly interact, allowing the formation of nanoparticles of different biopolymers [70].

Starch nanoparticles have been prepared via coacervation, with solutions of pre-modified starch of two kinds—positively charged (with an amino group) and negatively charged (with a carboxyl group)—being mixed at different molar ratios under stirring and at room temperature. Nanoparticles were formed within 15 min. The formed nanoparticles were centrifuged and lyophilized. With this method, spherical nanoparticles were obtained, and the optimal ratio for obtaining nanoparticles with the smallest size of 138.9 ± 0.3 nm was found to be 1:3 [71]. Studies have shown that with a similar methodology, composite nanoparticles of biodegradable polymers of chitosan and carboxymethyl starch can also be prepared [72].

Another method of coacervation is the saturation of the polymer solution followed by the formation of the coacervate phase. Tavares et al. obtained chitosan nanoparticles via the simple addition of sodium sulfate solution to a chitosan solution with fixed volumes and concentrations under magnetic stirring [73].

Patra et al. used a modified coacervation method, using a binary mixture of acetone and ethanol to initiate phase separation. This approach allowed the formation of a compact structure of the polymer chain, which contributed to the reduction in the particle size. Gelatin nanoparticles were obtained via the binary coacervation method. A mixture of acetone and alcohol was added to an aqueous gelatin solution heated to 40 °C, observing a 1:1:1 volume ratio of the components. Next, an ethanol solution of the crosslinking agent—glutaric aldehyde—was added to the mixture. Moreover, the aqueous gelatin solution was brought to different pH values—ranging from 3 to 11—to study the effect of pH on the size of the resulting particles. The resulting particles were washed several times with distilled water, treating the system with ultrasound for 5 min during each wash cycle, and then centrifuged at 2000 rpm. As a result, the minimum particle size was achieved at pH = 7, with a size of 55.67 ± 43.74 nm [74]. It is worth noting that this method can produce nanoparticles of biopolymers with a very small size, but a wide distribution, which may limit some of their applications.

2.5. The Electrospray Method

The electrospray method is based on forces such as electromechanical and hydrodynamic forces. The basic elements of this method are a syringe pump, a nozzle equipped with a cooling jacket, a high-voltage power supply, and a metal manifold. Important factors in this process are the distance from the needle to the manifold, the flow rate through the nozzle, and the applied voltage. In this method, a solution of the substance from which the nanomaterial is to be produced flows through the nozzle under voltage and is sprayed onto the collector at varying speeds. The collector, in turn, acts as a grounding agent. Special drums or solutions can also be used instead of the collector. This method is traditionally used to create drug nanocrystals (Figure 2) [75], but there has also been works on obtaining nanoparticles of biodegradable polymers via this method.

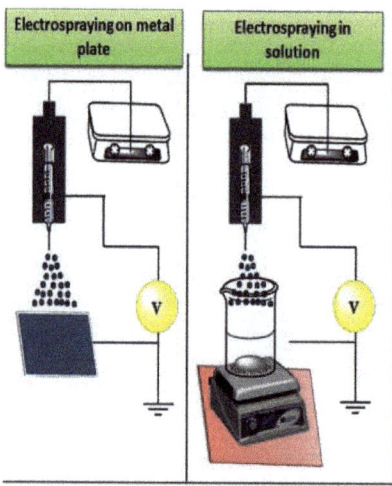

Figure 2. Illustration of a setup of the electrospray method, reprinted with permission from [75], 2018, Elsevier.

Applying an electric charge to the polymer solution plays an important role in electrospraying, and the use of a high electric voltage deforms the droplet interface and results in nanometer-sized droplets [76]. The proposed mechanism is that when an electric field is applied to a liquid droplet, an electrostatic force in the droplet—called the Coulomb force—is generated, which competes with the adhesion force in the droplet. When the applied Coulomb force overcomes the adhesion force of the droplet, the surface tension is released, and nanodroplet formation occurs [77].

Wang et al. obtained chitosan nanoparticles encapsulating tea polyphenols. A solution of chitosan in acetic acid was prepared, and then tea polyphenols were added at various mass ratios, and the pH of the mixture was maintained in the range of 4.6–4.7. A solution of sodium tripolyphosphate was prepared separately, to which the mixture of chitosan and tea polyphenols was fed through a nozzle with a diameter of 0.5 mm; the distance from the nozzle was 5 to 15 cm, the feed rate of the mixture was 0.25–1 mL/h, and the voltage applied to the nozzle was 10–20 kV. The formed nanoparticles were isolated by centrifugation and lyophilized. As a result, the smallest size of chitosan nanoparticles with encapsulated tea polyphenols was 128 ± 13 nm, using the following parameters: nozzle-to-solution distance of 5 cm, spray rate of 1 mL/h, and voltage of 15 kV [78].

Chitosan nanoparticles were also obtained by electrospraying using a special unit for voltage generation, with a syringe with an internal needle diameter of 0.65 mm and a working distance of 10 cm. Chitosan dissolved in acetic acid was fed at a rate of 0.2 mL/h to 1.1 mL/h. After that, it was dried in a vacuum. It is worth noting that solutions of different concentrations and molecular weights can be used to produce chitosan nanoparticles [79].

3. Main Applications of Biodegradable Polymeric Nanoparticles

Based on their diverse range of molecular and physicochemical properties, biopolymeric nanoparticles exhibit a broad range of functional attributes. They are therefore used in different areas, including medical, agricultural, and industrial applications. In the present review, we concentrate on the medical and agricultural uses.

3.1. Drug Delivery Using Nanoparticles of Biodegradable Polymers

Nanoparticles of biodegradable polymers are unique systems for the delivery of various drugs into the human body [80,81]. The small size and easy possibility of functionalization of nanoparticles allow them to be used as a universal system for the delivery of various drugs [82,83]. Nanoparticles of polymers can be loaded with a high concentration of a drug substance, but due to its inclusion in the nanoparticles it does not cause harm to the body [84,85]. The drug included in nanoparticles is protected from the body's environment, helping it to maintain its bioactivity [86–88].

Polysaccharide-based nanoparticles can reduce uptake by the mononuclear phagocyte system compared to other types of NPs, prolonging the in vivo NP residence time and increasing the likelihood of disease focus accumulation [89]. Polymer nanoparticles allow increased accessibility, as well as increased hydrophilicity of the therapeutic agent [90]. Moreover, by incorporating various drugs into polymer nanoparticles, targeted delivery systems can be created, allowing the drug to be delivered to the location of the tumor or thrombus and, thus, greatly increasing the therapeutic effect [91,92].

One example of biopolymer nanoparticles used for drug delivery is starch nanoparticles. Xie et al. prepared cassava starch nanoparticles via the microemulsion method, using $POCl_3$ as a crosslinking agent. In this work, the adsorption of the anticancer drug paclitaxel (PTX, $C_{47}H_{51}NO_{14}$) onto starch nanoparticles was studied, and the kinetics of the process were studied; 94.19% of free PTX was released from the dialysis bag in 0.9% saline (pH = 7) within 8 h. Compared with free PTX, the release rate of CA-CSNP-loaded PTX was 17.04% after 8 h and 37.61% after 96 h under the same conditions, showing the possibility of prolonging the therapeutic effect of the drug, which can improve the effectiveness of antitumor therapy [93]. All of this confirms the high prospects for the use of starch nanoparticles as drug carriers to provide a targeted effect on tumor cells.

Alp et al. proposed a method of producing starch nanoparticles synthesized using the emulsion–solvent diffusion method, used to deliver CG-1521 (7-phenyl-2,4,6-heptatrienoylhydroxamic acid), which is a histone deacetylase inhibitor (HDAC). The authors demonstrated the release kinetics of the drug, consisting of a release of approximately 40% within the first 10 h and 64% within 120 h. The authors demonstrated localization of NPs in the cytoplasm and perinuclear region of MCF-7 cells, indicating that they were effectively absorbed by the cells after 12 h. During the study, it was found that the cytotoxic activity of CG-1521 incorporated into starch nanoparticles was more than three times higher compared to the pure CG-1521 preparation [94]. This system demonstrated high cytotoxic activity against MCF-7 cells. In addition, it was shown that it is possible to create systems with prolonged action, which can increase the effectiveness of the therapy.

Another promising material for drug delivery is chitosan nanoparticles; de Oliveira Pedro et al. developed a method for producing chitosan nanoparticles based on the self-assembly process, with quercetin incorporated into them. Due to the non-polar molecules in the core of chitosan nanoparticles, it is possible to include hydrophobic molecules such as quercetin. The authors studied the kinetics of quercetin's release, and suggested mechanisms that affect the release of this drug. Chitosan nanoparticles with quercetin demonstrated an inhibitory effect against MCF-7 cells comparable to that of free quercetin. With a confocal laser microscopy, it was found that some of the nanoparticles accumulate on the cell membrane, while others are internalized. In addition, these chitosan nanoparticles were shown to have suitable compatibility with blood which, together with their biodegradability, allows them to be used as an effective delivery system for hydrophobic therapeutic molecules [95].

Theoretical calculations of real systems can help to create more efficient drug delivery systems, as well as increasing the therapeutic effects of existing drugs. Rahbar et al. performed quantum chemical calculations on the use of chitosan nanoparticles as carriers for the anticancer drug 5-fluorouracil. Negative values of the bonding energies showed that the functionalization of CrT PFs is energetically advantageous, and the solvation energies showed an increase in the solubility of 5-fluorouracil. Such a system would help increase the solubility of the hydrophobic anticancer drug, as well as reducing the burden on the whole body by incorporating 5-fluorouracil into chitosan nanoparticles [96].

Zhang et al. proposed functionalized alginate nanoparticles crosslinked with calcium ions and modified with mannose to deliver ovalbumin. Ovalbumin was attached via a pH-sensitive bond and used as a model antigen (Figure 3). The release of ovalbumin from the alginate nanoparticles was in the region of 30% after 43 h, at a pH of 5.5. Using confocal microscopy, alginate NFs were found to facilitate antigen transport, uptake, and delivery to the cytosol of dendritic cells. In vivo experiments on laboratory mice proved that alginate encapsulated in nanoparticles can be gradually transported from the injection site to the draining lymph nodes, and can remain there for several days, providing a targeted and prolonged therapeutic effect. The antitumor effect was evaluated in OVA-expressing E.G7-OVA cells of mice. The average tumor weight was at least four times less than in the control group and the group with free ovalbumin, showing the synergistic effect of the inclusion of ovalbumin in nanoparticles of modified alginate. This system allows a significant increase in antigen absorption, provides increased efficacy of antitumor therapy, and can also provide a prolonged therapeutic effect (Figure 3) [97].

Cellulose nanoparticles can also successfully serve for drug delivery to the human body. Suk Fun Chin et al. demonstrated a simple method for producing cellulose nanoparticles and optimized the loading efficiency and studied the release kinetics of the substance with a model hydrophilic drug [98].

Modified biopolymers are increasingly being used in the delivery of various therapeutic agents into the body. Various theranostic and therapeutic systems are particularly promising. An example of such a system could be amino-functionalized nanoparticles of nanocrystalline cellulose modified with 1-lysine [99]. In their work, Moghaddam et al. presented a method of obtaining nanocrystalline cellulose functionalized with aminoprapyl-3-methoxysilane and conjugated with 1-lysine. This nanocarrier was used for simultaneous delivery of two drugs: methotrexate (MTX) and curcumin (CUR). The average size of the nanoparticles was 240 nm, as confirmed by data from DLS and SEM. The critical parameter for any injectable drug is the hemocompatibility of the carrier, which allows its use in the form of injections. Modified cellulose nanoparticles had absolute safety when injected at a concentration of 1000 μg/mL. The cellulose nanoparticles showed high cytotoxicity against MCF-7 and MDA-MB-231 cells, with an IC$_{50}$ of 2.253 and 15.34, respectively. It is especially noteworthy that the modified cellulose nanoparticles have a significant effect on cell morphology and cell number and exceed these figures for free drugs (according to a DAPI staining study) (Figure 4) [99]. Nanoparticles proved to be a highly effective platform for drug delivery via the bloodstream to fight cancer. In addition, the authors studied the safety and kinetics of drug release. This study showed the high promise of cellulose nanoparticles for use in the delivery of cytotoxic and antitumor drugs.

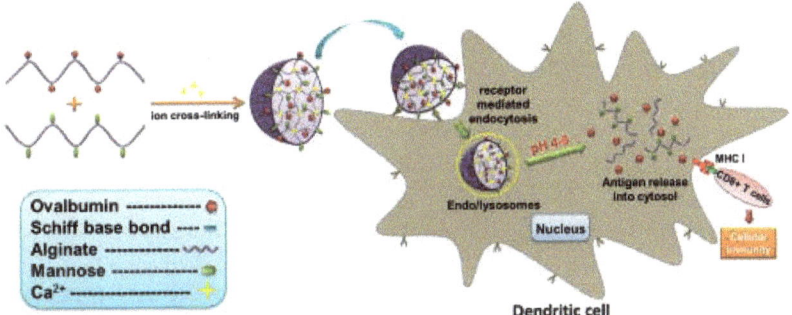

Figure 3. Illustration of the potential use of functionalized alginate nanoparticles as a nanovaccine for cancer immunotherapy. Reproduced with permission from [97], 2017, Elsevier.

Composite nanoparticles of biopolymers can also be actively used for the delivery of various drugs. Sorasitthivanukarn et al. demonstrated the possibility of creating alginate/chitosan nanoparticles obtained via ionotropic gelation and emulsification; three-level Box–Behnken statistical design was used to optimize the synthesis and loading parameters of curcumin di-glutaric acid as an antitumor agent. During exposure of curcumin di-glutaric acid immobilized in nanoparticles and in free form, it was proven that the nanocomposite material protects the antitumor agent from photodegradation. In an in vitro simulated experiment, it was found that the chitosan/alginate nanoparticles are not subject to degradation under conditions close to the gastrointestinal tract (data confirmed using DLS and SEM). In addition, these nanoparticles showed the ability to penetrate Caco-2 cells (used to evaluate the absorption of orally administered drugs across the gastrointestinal barrier) via the endocytic route. The chitosan/alginate nanoparticles showed a complete absence of cell toxicity, whereas the nanoparticles loaded with the model drug (curcumin di-glutaric acid) showed high cytotoxicity toward MDA-MB-23, HepG2, and Caco-2 cells [100].

The specifics of the chitosan/alginate nanosystems allow us to speak about their high potential for use as agents for the oral delivery of biologically active substances. This system shows absolute absence of toxicity to cells and high permeability inside GIT cells. Furthermore, nanoparticles can provide increased preservation of the therapeutic drug, as well as prolonged action of the drug substance.

Alkholief M. proposed a method to create self-assembled lecithin/chitosan nanoparticles used for the delivery of various therapeutic drugs. The authors developed and optimized the nanoparticle parameters using Box–Behnken design. The chitosan/lecithin nanoparticles had the ability to exert prolonged action through gradual release of the drug. This system, based on mathematical modeling, showed promise in the delivery and prolongation of drug action, but further in vivo experiments are needed to confirm the effectiveness of this system on living subjects, and a hemocompatibility study is needed to form a complete picture of the possibility of using lecithin/chitosan nanoparticles on living subjects [101].

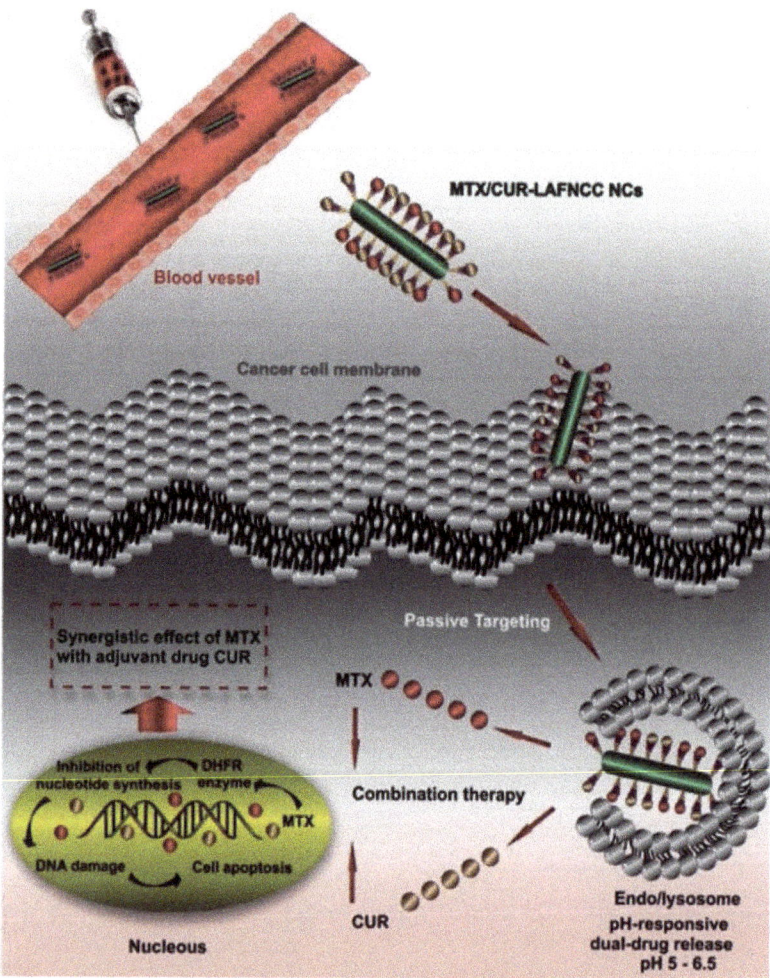

Figure 4. Schematic diagram for dual-drug delivery of a lysine-embedded cellulose-based nanosystems. Reproduced with permission from [99], 2020, Elsevier.

Biopolymer nanoparticles can be used to deliver a range of therapeutic agents into the human body. Biopolymer nanoparticles can protect the drug substance from the body's environment, as well as providing prolonged release of the drug, which can increase the effectiveness of the therapy many times over; in addition, biopolymer nanoparticles can be used not only for venous delivery, but also for oral and ophthalmic delivery of drugs [102,103]. In combination with their nano scale and biocompatibility, biopolymer nanoparticles can play a leading role in the delivery of drugs to the human body.

3.2. Diagnostic Systems Based on Nanoparticles of Biodegradable Polymers

Systems based on biodegradable polysaccharides have been successfully used as the basis for various diagnostic and theranostic systems. In recent years, the use of polysaccharide nanoparticles for the diagnosis of various diseases has become relevant [104,105]. Nanoparticles of biopolymers can be loaded with various substances, providing highly synergistic effects and diagnostic/therapeutic efficiency [106,107].

Nanoparticles of chitosan and its various modifications can successfully serve for theranostics of hepatocellular carcinoma. Chitosan nanoparticles can control drug release and increase drug stability. Additionally, due to the positive charge of the amino group, they can better bind to the negatively charged membranes of tumor cells, increasing the theranostic effect [108].

Kim et al. proposed a way to produce chitosan nanoparticles modified with Cy5.5 (IR dye) and encapsulating paclitaxel (an anticancer drug). Tumor-binding specificity was tested in vivo on squamous-cell carcinoma cells of SCC7 mice. One day after the injection of modified chitosan nanoparticles (CNPs), early-stage tumors were clearly separated from the surrounding normal tissue, demonstrating CNPs' specificity with respect to tumor targeting, with the NIRF signal being proportional to the tumor size. Chitosan nanoparticles accumulated throughout the tumor tissue, which could be used for improved detection of tumor localization. When studying biodistribution, it was found that the NIRF signal per gram of each organ in tumor tissues was 4–7 times higher than in other organs. In vivo experiments on mice showed a high survival rate of mice that received paclitaxel incorporated into chitosan nanoparticles, as opposed to free paclitaxel (Figure 5) [109]. All of the above shows the possibility of using a system based on chitosan nanoparticles and paclitaxel as an effective theranostic system to diagnose cancer at early stages, deliver the drug, and monitor tumor parameters during treatment.

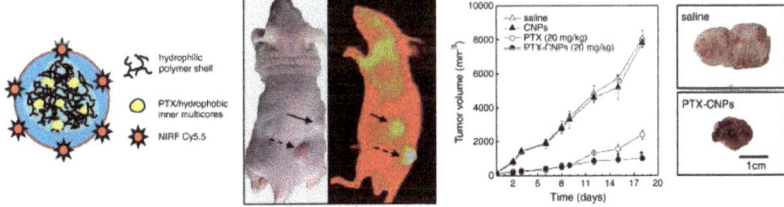

Figure 5. Schematic diagram of chitosan-based nanoparticles used in cancer diagnosis and therapy. Reproduced with permission from [109], 2010, Elsevier.

Nanoparticles (nanocrystals) of cellulose can also be used to create various diagnostic systems. Mahmoud et al. proposed a method for obtaining nanocrystalline cellulose derived from linen fibers, which had an average diameter of about 20 nm and a length of 120 nm. The authors studied the cellular uptake of cellulose nanoparticles incorporating the fluorescent dye rhodamine B isothiocyanate in HEK 293 and Sf9 cells. The nanoparticles were mainly dispersed throughout the cytoplasm, although some of them were identified by their vesicular structure. In the nuclei of both cell lines there was no significant amount of nanoparticles, indicating less interaction between these nanocrystals and the nuclear membrane. In addition, an MTT test was performed to determine the cytotoxicity of the nanoparticles with respect to the cells, and it was found that at a concentration of 100 µg/mL, the cell viability was 90% of that of the control sample. Nanocrystalline cellulose can be considered as a new generation of probes for bioimaging, given its large surface area of 150 m^2/g and the ability to adjust the surface charge to penetrate the cell without causing cell damage [110].

Photoacoustic imaging (PAI)—a biomedical analysis technique that provides actionable information about the molecular characteristics of tissue—is a recently developed hybrid biomedical imaging technique for monitoring tumor angiogenesis and detecting skin melanoma, as well as for monitoring and diagnosing various other neoplasms [111].

Hyaluronic acid nanoparticles modified with Cy5.5 IR dye and copper (II) sulfide were used as a contrast agent for photoacoustic and fluorescent imaging. Hyaluronic acid nanoparticles were injected intravenously, and fluorescence images were obtained at various time points after injection. After the injection of hyaluronic acid nanoparticles with Cy5.5 and copper (II) sulfide, hyaluronic acid was cleaved by hyaluronidase in the tumor area, and the quenching effect between CuS and Cy5.5 was reduced, resulting in the recovery of the fluorescence signal over time (Figure 6) [112]. Photoacoustic imaging allows detailed examination of tumor blood vessels and increases the effectiveness of therapy. PA contrast at the tumor site increased over time and reached a maximum after 6 h; it is also worth noting that using only copper sulfide did not significantly increase the PA signal. When studying the biodistribution, in addition to the tumor, the authors detected a signal in the kidneys, which was associated with the biodegradation of hyaluronic acid nanoparticles. The main mechanisms to which the high selectivity towards tumor cells can be attributed are the effects of increased permeability and retention, as well as receptor-mediated endocytosis. Hyaluronidase overexpression in tumor cells can effectively degrade hyaluronic acid nanoparticles and, consequently, enhance strong fluorescence signals for detecting tumor localization. In addition, hyaluronic acid nanoparticles have shown promise as carriers of signal-enhancing substances for PAI [112].

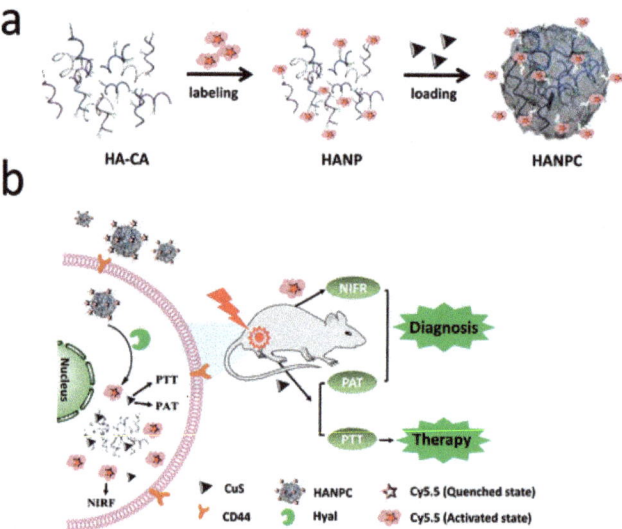

Figure 6. Illustration of (**a**) synthesis of activatable Cy5.5–HANP/CuS (HANPC) nanocomposites, and (**b**) in vivo applications. Reproduced with permission from [112], 2014, American Chemical Society.

3.3. Therapeutic and Tissue Engineering Applications

Chen et al. proposed a method to produce chitosan nanoparticles obtained via ionic crosslinking, which can be used for photothermal therapy. Then, the chitosan nanoparticles were modified with 5-aminolevulinic acid (5-ALxAl) and IR-780 iodide. When heating the modified nanoparticles in solution for 2 min, a temperature of 67.2 °C was recorded, which is a sufficient value for the application of these nanoparticles as an agent for photothermal (PTT) and photodynamic therapy (PDT). The authors performed a comparative analysis of the cytotoxicity of nanoparticles loaded alternately with 5-aminolevulinic acid (5-ALA) and IR-780 iodide and mixed with chitosan nanoparticles (Figure 7). Nanoparticles loaded with both drugs showed higher efficacy relative to other chitosan systems. Moreover, nanoparticles modified with two drugs showed greater areas of tumor necrosis, fragmentation, and tumor tissue damage compared to chitosan nanoparticles with the inclusion of one drug each. The biodistribution of chitosan nanoparticles and their safety for the human body were also studied in this work [113]. This work successfully demonstrated the high efficacy of chitosan nanoparticles in photothermal and photodynamic therapies and laid the foundation for the transition to medical practice with the use of nano-objects, due to the safety of this complex preparation of chitosan nanoparticles for anticancer therapy.

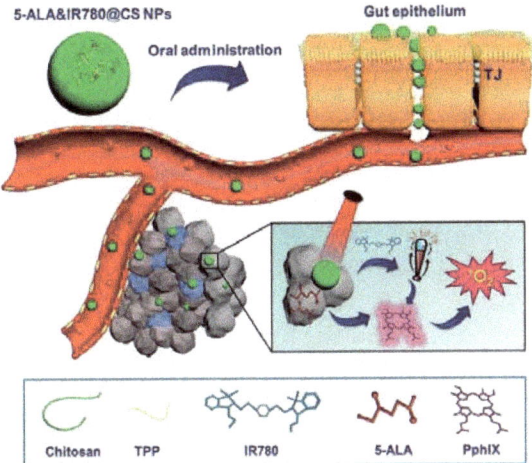

Figure 7. Schematic diagram of chitosan-nanoparticle-based 5-ALA and IR780 delivery for oral PTT and PDT in cancer treatment. Reproduced with permission from [113], 2020, Elsevier.

Biopolymeric nanoparticle composite materials can also be used for the treatment of tumors, and as agents in photothermal therapy. Wang et al. proposed a nanopolysaccharide composite consisting of chitosan nanoparticles loaded with polypyrrole and 5-fluorouracil nanoparticles, with an outer shell of carboxymethyl cellulose crosslinked with disulfide. The composite material had an average diameter of 254 nm. Using a laser of 808 nm and a power of 1.5 W/cm^2, the temperature of the nanoparticle solution increased from 20 to 42 °C in 300 s, while the buffer temperature increased by 2 °C. In addition, the high photothermal stability and reproducibility of the effect caused by the nanopolysaccharide complex were proved (Figure 8). The photothermal conversion efficiency was calculated from the descending part of the curve and was 21.6. The nanopolysaccharide complex also exhibited antitumor properties, as confirmed by in vivo experiments [114].

Biopolymer nanoparticles can be successfully used as a basis for drug delivery in photothermal therapy, as well as having antitumor effects.

In recent years, the development of biocompatible and non-toxic nanoparticles that can be used in bone engineering, implants, and other biomedical systems has been relevant.

Chitosan induces mineralization during the differentiation of mesenchymal stem cells' osteoblasts by enhancing genes related to mineralization and calcium binding, such as collagen type I alpha I, integrin-binding sialoprotein, osteopontin, osteonectin, and osteocalcin [115]. Chitosan nanoparticles, in contrast to chitosan, can also be used for the delivery of various drugs, as well as to provide stimulation of the immune system and antibacterial activity in this system. Jafary and Vaezifar proposed the use of chitosan nanoparticles obtained via ionic gelation to incorporate alkaline phosphatase enzymes. Overall, 71% of the alkaline phosphatase retained its native activity, with 55% protein–chitosan nanoparticle binding. The expression of the major bone markers osteocalcin, alkaline phosphatase, and collagen at day 7 increased 35-, 6-, and 13-fold, respectively, in the group with the addition of chitosan nanoparticles relative to the control group. At the same time, the formation of bone nodules and an increase in the size of the nodules during the differentiation of osteoblasts in the presence of immobilized ALP/chitosan nanoparticles were observed by days 7 and 14 of differentiation, as established by means of von Koss staining. At the same time, the number of bone nodules was many times higher than in untreated cells. This work proved that chitosan nanoparticles can be used as an effective component in bone tissue therapy. Simultaneous use with alkaline transferase increases the expression of genes that are associated with the mineralization and diffusion of osteoblasts [116].

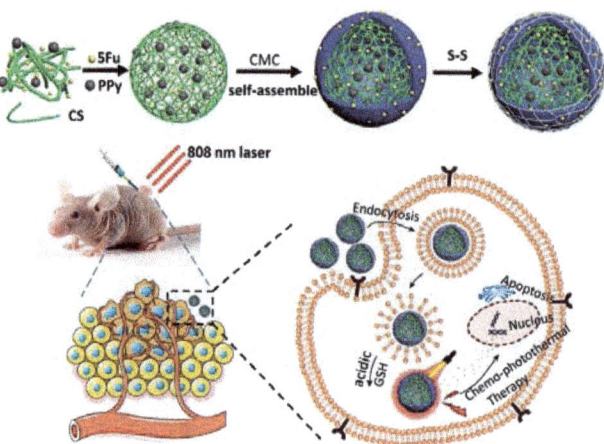

Figure 8. Schematic diagram of the synthesis of a drug-loaded complex of chitosan NPs and carboxymethyl cellulose, and its application in photothermal chemotherapy for tumor cells. Reproduced with permission from [114], 2022, Elsevier.

Herdocia-Lluberes et al. proposed a method of creating a composite material based on nanohydroxyapatite and nanocellulose using the gelation method. The average size of these nanoparticles was 300 nm. The toxicity of these nano-objects at a concentration of 5 mg/mL was studied in relation to osteoblast cells. It was shown that nanocellulose/nanohydroxyapatite nanoparticles modified with the BMP-2 protein induced the increased growth of osteoblasts compared to the control group. This study demonstrated the possibility of using cellulose nanocomposites for bone tissue engineering [117].

A similar system based on nanocellulose, hydroxyapatite, and silk fibroin has been proposed for the regeneration of a rat brain/skull defect (Figure 9) [118]. When investigating the effect of the nanocomposite on MC3T3-F1 cells' viability, it was found that the cell viability at 5 and 7 days was much higher than that of samples without the composite material. The hybrid materials increased alkaline phosphatase activity at days 7 and 14 of the cell studies. In this work, new murine skull-bone formation was measured after injury using micro-CT. Rat skulls treated with the composite material showed significant growth at 12 weeks, relative to silk fibroin, as well as to compositions of silk fibroin–nanocrystalline cellulose and silk fibroin–nanohydroxyapatite. With the help of histological analysis, it was proven that in the formation of the skull modified with the composite material based on nanocellulose, hydroxyapatite, and silk fibroin, the main role is played by the large amount of new bone. High biocompatibility and efficacy in the repair of rat skull vault defects confirmed the prospects of composite materials based on nanocrystalline cellulose in bone tissue engineering.

Nanocellulose has also been reported to improve the functional properties of chitosan-based nanocomposites, since chitosan has poor mechanical properties that limit its applications in bone tissue engineering [119]. Nanoparticles are usually used in tissue engineering, and they are able to mimic the natural extracellular matrix, thus increasing the possibility of controlling scaffold features, such as biomechanical and biological properties [120,121].

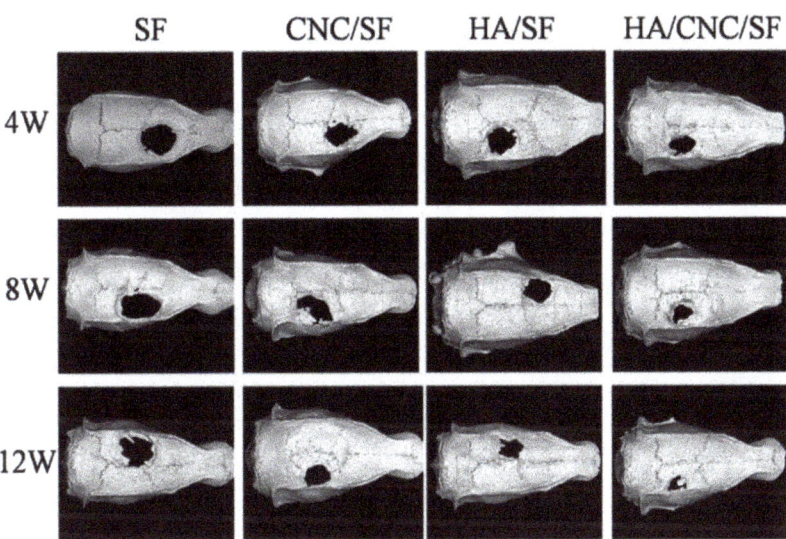

Figure 9. Illustration of the different repair results in rat calvarial defects assessed by micro-CT. Reproduced with permission from [118], 2011, Royal Society of Chemistry.

3.4. Antibacterial Properties of Nanoparticles of Biodegradable Polymers

Antibacterial properties against various bacteria are necessary for use in various areas of human life. The antibacterial properties of various metal nanoparticles are widely known [122–124]. Metal nanoparticles are highly toxic to human cells, and can be harmful not only to bacteria, but also to humans [125,126]. Biocompatible antibacterial systems have high potential for current industrial applications. Biopolymer nanoparticles can exhibit antibacterial activity against several bacteria [127,128]. The mechanisms of action and toxicity against different strains manifest individually for each nanoparticle. One of the biopolymers exhibiting antibacterial activity is chitosan. The antibacterial properties of chitosan depend on the transfer of amino groups' protons, which can react with the negative charge of the cell surface, leading to the destruction of the bacterial cell. Chitosan interacts with the bacterial membrane, changing the cell permeability [129,130].

Some authors have demonstrated that low-molecular-weight chitosan nanoparticles can penetrate the cell, bind to DNA, and inhibit the replication mechanism [131]. Chitosan nanoparticles exhibited higher antimicrobial activity than chitosan against a wide range of pathogens, including fungi as well as Gram-positive and Gram-negative bacteria. Chitosan nanoparticles obtained via ionic gelation were medium-sized and exhibited antibacterial activity against *N. gonorrhea*. Their MIC_{50} was 0.16 mg/mL at pH 5.5. Bacterial adhesion studies showed that chitosan nanoparticles reduced the adhesion of multidrug-resistant *N. gonorrhea* (WHO Q) to HELA cells at the lowest concentration of chitosan nanoparticles, and inhibited bacterial growth compared to cells not exposed to chitosan nanoparticles [132]. It is worth noting that when the pH was increased to 7.5, chitosan nanoparticle activity decreased, consistent with other reports in the literature [133]. Chitosan nanoparticles showed promise in their use as an antivaccine agent and were also cytocompatible and reduced bacterial cell adhesion.

The antibacterial properties of chitosan nanoparticles are widely used in industry. Mousavi et al. proposed a fabric conditioner using chitosan nanoparticles. This conditioner exhibited antibacterial properties against bacteria such as *Streptococcus mutans*, *Enterococcus faecalis*, and *Pseudomonas aeruginosa*, as well as the fungus *Candida albicans*. Complete inhibition of microbial growth was performed at a concentration of 2.5–10% of chitosan nanoparticles in the system [134].

Furthermore, various nanosystems based on chitosan in combination with various biopolymers have been used as a basis for antibacterial systems used in various industries [135,136].

3.5. Biodegradable Polymer Nanoparticles in Agriculture

The importance of nanotechnology has increased in all branches of science, including agriculture. With the goal of enhancing crop yield and providing food security, several researchers have turned to agricultural nanotechnology [137]. Studies have shown that the use of nanomaterials in agriculture can be oriented towards (a) plant nutrition by controlling the release of agrochemicals; (b) plant protection, e.g., release control of pesticides, detection of plant diseases; and (c) crop improvement, e.g., delivery of genetic materials [138]. Despite the advantages of nanomaterials, one of the parameters that usually hinders their use in agriculture is their toxicity, which has made the use of non-toxic, biocompatible, and biodegradable polymeric nanoparticles more advisable [137,138]. Therefore, several biopolymers—such as carbohydrates (e.g., chitosan, cellulose, starch)—have been used to prepare biopolymeric nanoparticles. In addition to the usual advantages of nanoparticles in agriculture, biopolymeric nanoparticles also play an important role in agriculture, as they have plant growth and plant protection properties [139]. Moreover, with their unique properties, biopolymeric nanoparticles can be used in agriculture to increase global food production and enhance food quality. They are suitable for application in different areas of agriculture—such as plant protection, plant growth, stress tolerance, smart delivery systems, crop improvement, biosensors, and precision farming—and in this part of the review, the application of biopolymeric nanomaterials in crop science is highlighted.

3.5.1. Biopolymeric NPs in Plant Growth and Productivity

One of the main reasons for low crop productivity is environmental factors, such as low temperature, pests, and weeds. Thus, it is important to find growth promotors and biosensors that can be used to effectively assess the biochemical and physiological changes in plant growth. Nanomaterials, including biopolymeric nanoparticles, can be used as delivery systems for plant growth promoters to increase their efficacy. In their work, Santo Pereira, A.E. et al. developed two systems composed of alginate/chitosan (ALG/CS) or chitosan/tripolyphosphate (CS/TPP) and used them to deliver the plant hormone gibberellic acid (GA). These systems showed effects on both morphological and biochemical parameters [140]. At the biochemical level, chitosan nanoparticle systems had different effects on the pigment contents, as the ALG/CS–GA nanoparticles increased the levels of chlorophyll "a", relative to the control, while the CS/TPP–GA nanoparticles increased the levels of chlorophyll "b". Regarding the content of carotenoids, only the ALG/CS–GA nanoparticles were found to increase it [140]. Morphological parameters, such as leaf area and root length, were also important. Roots are critical for plant growth, as they are responsible for the uptake of water and nutrients [141–144], and the CS/TPP–GA nanoparticles were found to increase the root length by 5.5 and 5.5 cm at concentrations of 0.012 and 0.025%, respectively, compared to the control. The free hormone showed an increase in root length of 4 cm at concentrations of 0.025% [140]. Regarding the leaf area, a large leaf area increases the photosynthetic activity of the plant and, therefore, results in an increase in the energy available for plant development. The experiment showed that the use of ALG/CS–GA nanoparticles at concentrations of 0.012 led to greater leaf area compared to other variants (Figure 10) [140].

The added advantage of the use of such chitosan-based nanosystems is that studies have shown improved antifungal activity of chitosan nanoparticles [145–147]; thus, they can help improve the plant growth while also protecting the crop.

Apart from the delivery of plant growth promoters, biopolymeric nanoparticles have also been shown to have growth-promoting properties. Several studies have shown that chitosan nanoparticles promote seed germination in crops and can therefore be used to produce healthy seedlings [148–150].

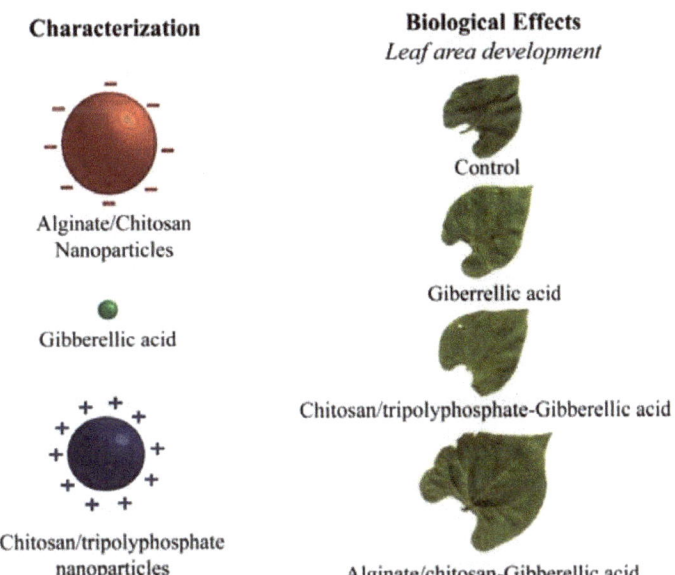

Figure 10. ALG/CS and CS/TPP nanoparticles, gibberellic acid, and the effects of treatment with GA, ALG/CS–GA, and CS/TPP–GA nanoparticles on the leaves' area. Reproduced with permission from [140], 2017, Elsevier.

Kadam et al. also showed the effect of chitosan-based nanoparticles on plant growth by developing salicylic acid–chitosan nanoparticles (SA–CS NPs). Their application increased the activities of seed reserve food-remobilizing enzymes during seedling growth, such as α-amylase and proteases, with higher enzyme activities observed from the 3rd to the 7th days. In addition, they also observed that the shoot–root length and fresh weight were significantly higher in nanoparticle treatments, compared to the control, bulk chitosan, and salicylic acid treatments [151].

Chitosan nanoparticles also effect the germination and seedling growth of wheat (*Triticum aestivum* L.), as shown by the work of Li et al., who showed that the chitosan nanoparticles had a positive impact on the seed germination and seedling growth of wheat at a lower concentration compared to chitosan treatment. Such efficacy was suggested to be due to the high adsorption of chitosan nanoparticles on the surface of wheat seeds observed via energy-dispersive spectroscopy and confocal laser scanning microscopy [149].

In addition to laboratory experiments on seedling growth, studies in greenhouses also showed the effects of chitosan nanoparticles, as their application increased the chlorophyll content and the photosynthesis intensity in coffee leaves in comparison to controls. They also improved the nutrient uptake and had a significant impact on the growth of the coffee seedlings, with the growth parameters of the treated coffee higher than those of the controls [152].

In addition to the treatment of seeds, foliar application of biopolymeric nanoparticles improves plant growth. Chitosan nanoparticles loaded with macronutrients—nitrogen, phosphorus, and potassium (NPK)—were easily applied to the leaf surfaces of wheat plants, and they entered the stomata via gas uptake. Transmission electron microscopy showed that the nanoparticles were taken up and transported through phloem tissues. The foliar application of the nanoparticles with NPK increased the wheat yield [153].

Loading or encapsulating nutrients in biopolymeric nanoparticles can also help minimize economical losses and control the release of nutrients. In addition to chitosan, other carbohydrate- and protein-based polymeric nanoparticles have been used in the preparation of such systems for the controlled release of nutrients [154–157].

Overall, biopolymeric nanoparticles can be used to promote plant growth and productivity, but their mode of application and efficacy may vary depending on the type of plant species and their growth stages.

3.5.2. Biopolymeric NPs in Plants' Defense against Pathogens and Abiotic Stress Factors

Nanoparticles in plant protection can be used either for the delivery of pesticides or for the rapid diagnosis of plant diseases [158], and the ecological safety of bio-based nanomaterials makes them more attractive for use in plant protection. In addition, several studies have shown that the use of biopolymeric nanomaterials upregulates defense enzymes/genes in plants and protects crops from pathogens [139,159,160]. Among them, chitosan and other carbohydrate-based nanoparticles have shown their efficacy in crop protection [154,161].

Among them, β-d-glucan nanoparticles showed antifungal activity against *P. aphanidermatum*—a devastating fungus that affects major crop plants [162–164]. Chitosan nanoparticles have also been found to be able to suppress blast disease of rice, caused by the fungus *Pyricularia grisea* [165].

The antifungal activity of chitosan nanoparticles was studied against rice sheath blight disease caused by the pathogenic fungus *Rhizoctonia solani*, and it was determined that the use of chitosan nanoparticles led to an increase in the levels of defense enzymes, such as peroxidase, phenylalanine ammonia-lyase, and chitinase enzymes, thus showing that these are potent plant immunity boosters that are able to help suppress up to 90% of disease in detached leaf assay and 75% under greenhouse conditions [166].

In their studies, Choudhary et al. showed that chitosan nanoparticles can inhibit the plant pathogen *Fusarium solani* under in vitro conditions [167], while Chen et al. showed that the antimicrobial activity of chitosan-based nanoparticles is due to their zeta potential, which plays a key role in binding with negatively charged microbial membranes [168].

In addition to chitosan nanoparticles, nanocomposites of other biopolymers have also shown antimicrobial properties. Pinto et al. showed the antifungal activity of nanocomposite thin films of pullulan and silver against *Aspergillus niger* [169].

Biopolymeric nanoparticles have also been used against plant nematodes, such as the pine wood nematode, which is devastating to pine trees. In their study, Liang et al. tried to improve the poor availability of the widely used bionematicide avermectin—due to its poor solubility in water and rapid photolysis—by encapsulating it within nanoparticles composed of poly-γ-glutamic acid and chitosan. Their results showed that the encapsulation of avermectin within biopolymeric nanoparticles reduced its losses through photolysis by more than 20.0%, and that the mortality rate of nematodes that were treated with 1 ppm of avermectin encapsulated in nanoparticles was 98.6% [170].

In addition to the direct antifungal activity of polymeric nanoparticles, they can also be used to control pathogens by controlling the fungicides, as shown by the work of Liu et al., who used nanoparticle systems based on the polymeric materials polyvinylpyridine and polyvinyl pyridine-co-styrene to slowly release the fungicides tebuconazole and chlorothalonil to control a common brown rot wood decay fungus [171]. Other than these polymers and inorganic nanoparticles used as delivery systems for agrochemicals [172], several reports have highlighted the preparation and application of chitosan-based nanoparticles for the delivery and controlled release of pesticides in agriculture [137]. In these studies, chitosan-based nanomaterials have been used for the delivery of insecticides such as rotenone [173] and imidacloprid [174], the biopesticide azadirachtin [175], and the essential oil of *Lippia sidoides*, with insecticidal properties [176].

Another interesting use of biopolymeric nanoparticles is in the controlled release of herbicides, since they can increase herbicide efficacy at reduced doses, thus decreasing the environmental impact of said herbicides. Several studies have shown the efficacy of chitosan-based nanomaterials in the delivery of herbicides. Silva et al. used alginate/chitosan nanoparticles for the delivery of paraquat herbicides [177], while Grillo et al. used chitosan/tripolyphosphate nanoparticles as carriers of the same herbicide [137,178].

The results of these studies showed that biopolymeric nanoparticles were able to decrease the herbicides' toxicity, thus reducing the negative impacts caused by paraquat [137].

Many of these studies show that biopolymer-based nanotechnologies should be considered as useful tools for environmental protection against the negative impacts of agrochemicals.

Along with plant pathogens, abiotic stress factors—such as salinity and drought—are another reason for low crop productivity. Chitosan nanoparticles have been reported to improve the salinity stress response in several plants [179–181]. In their work, Balusamy et al. discussed the beneficial effect of chitosans, chitosan nanoparticles (CsNPs), and modified chitosan biomaterials (CsBMs) under salt stress to improve crop performances by stimulating physiological changes in plants, including changes in primary metabolite production, the jasmonic acid signaling pathway, antioxidant activity, secondary metabolite production, and membrane permeability, leading to increased salt resistance (Figure 11) [182].

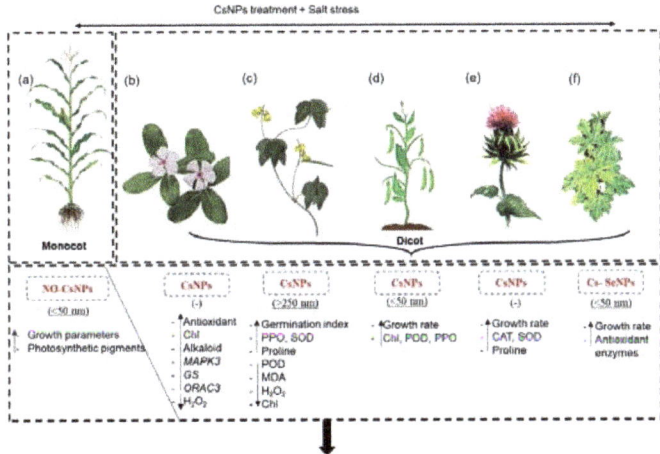

Figure 11. Treatment with CsNPs/modified CsBMs under salinity stress improved the plant performance in various plant species (**a**) maize, (**b**) periwinkle, (**c**) mung bean, (**d**) bean, (**e**) milk thistle, (**f**) bitter melon. Reproduced with permission from [182], 2022, Elsevier.

In addition to salt tolerance, chitosan nanoparticles have also shown beneficial effects for plants under drought conditions [183]. The results of the work of Behboudi et al. indicated that the application of chitosan NPs—especially at 90 ppm—can mitigate adverse effects of drought on wheat [184]. Ali et al. reported that the application of chitosan NPs mitigated the negative effects of drought stress, such as the reduction in relative water content, stomatal conductance, and total chlorophyll. NPs increased the activity of the enzymes catalase and ascorbate peroxidase, as well as the accumulation of proline. They also reduced the oxidative stress and the accumulation of H_2O_2 and malondialdehyde, thus preserving membrane integrity. Their results also showed an increase in the accumulation of alkaloids, which was associated with the expression of genes of the alkaloid biosynthetic pathway—strictosidine synthase, deacetylvindoline-4-O-acetyltransferase, peroxidase 1, and geissoschizine synthase [185].

From those works, we can see that chitosan nanoparticles help mitigate the negative effects of abiotic stress through biochemical changes and gene expression modulation.

3.5.3. Biopolymeric NPs in Crop Improvement

Crop improvement via genetic engineering and genomic editing can help speed up the development of improved crops. One of the challenges in plant engineering is gene delivery through plant cell walls. The traditional gene transfer methods—such as Agrobacterium-mediated gene transfer, electroporation, PEG-mediated gene transfer, particle gun bombardment, and others—are expensive [186], and can cause significant perturbation to the growth of cells [137]. Recent knowledge and technology have enabled the use of genome editing in crop improvement, making it simple and precise by increasing the precision of the correction or insertion, and preventing cell toxicity [187].

Several studies have reported the use of nanomaterials in genetic engineering and genome editing [188–193]. Nanomaterials have enabled efficient, targeted delivery into diverse plant species and tissues—both monocot and dicot plants—contributed to the controlled release of DNA and proteins and their protection from degradation, and in some cases even allowed imaging of the delivery and release processes in plants [193,194].

Biopolymeric nanoparticles, such as chitosan nanoparticles, have been used as nanocarriers for the delivery of DNA, RNA, and proteins in genetic engineering and the CRISPR (clustered regularly interspaced short palindromic repeats)/Cas9 (CRISPR-associated protein 9) system for genome editing [195,196]. Duceppe and Tabrizian showed the potential of chitosan NPs for DNA delivery, since they can effectively condense and complex DNA through the electrostatic interactions between the positively charged amino groups on the glucosamine units of chitosan and the negatively charged phosphates on DNA [197]. In addition, it was reported that chitosan-based delivery systems can protect DNA from nuclease degradation [198].

Kwak et al. used chitosan-complexed single-walled nanotubes for the transformation of chloroplasts. They demonstrated the potential of nanoparticle-mediated chloroplast transgene delivery tools, which can be used to transform mature plants. Their chitosan-nanoparticle-mediated chloroplast transgene delivery platform was noted as simple, easy to use, cost-effective, and applicable to mature plants across different species, and did not require specialized, expensive equipment, enabling its widespread application in plant bioengineering and plant biology studies [199]. The authors also hypothesized that their approach could be used in genome editing for the delivery of zinc-finger nucleases, transcription activator-like effector nucleases (TALENs), and clustered regularly interspaced short palindromic repeats (CRISPR)–CRISPR-associated protein 9 (Cas9) vectors.

Since several studies have suggested that the *Bbm* and *Wus2* genes can be used to improve the efficiency of plant engineering and promote plant regeneration [191,200,201], Lv et al. suggested that nanoparticles can be used to carry these genes—*Wus2* and *Bbm*—and CRISPR/Cas9 into intact plant cells, thus producing a transgenic plant without the plant tissue culture processes (Figure 12) [191].

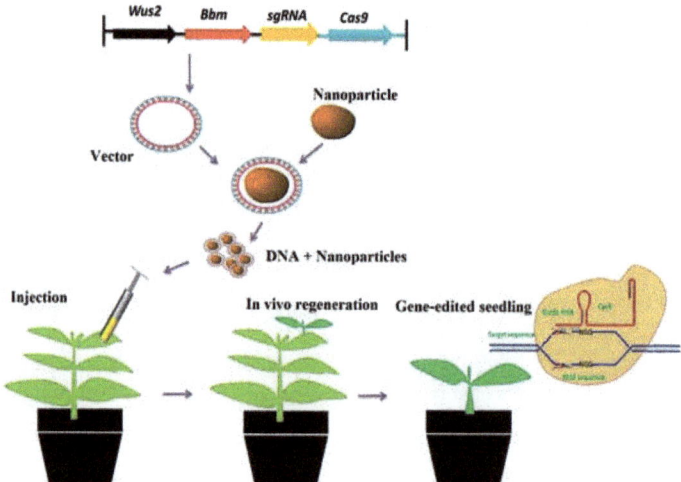

Figure 12. Illustration of prospective applications of nanoparticle-based delivery of CRISPR/Cas9 in plant engineering. Reproduced with permission from [191], 2020, John Wiley and Sons.

Overall, recent studies show the potential advantages of using nanoparticles, including biopolymeric nanoparticles, to deliver biomolecules (e.g., DNA/RNA/proteins/ribonucleoproteins) in plants to accelerate the process of genetic transformation and the development of improved plant genotypes. They can even help in the transformation of crops that are non-compatible with traditional transformation methods and tissue culture protocols. In addition to the applied research, biopolymeric nanoparticles can also help advance the fundamental understanding of plants' cellular and organellar genetics, and aid in the discovery of new genes and gene clusters, as well as epigenetic indicators of plant metabolism.

4. Conclusions

Nanoparticles have been developed from various materials, such as synthetic polymers and natural biological polymers. Biopolymeric nanoparticles offer several advantages, such as biodegradability, biocompatibility, and reduced toxicity. This review presents the main methods for the preparation of biopolymeric nanoparticles, such as ionic gelation, nanoprecipitation and microemulsion methods, spray-drying, etc. The promising applications of biopolymeric nanoparticles are also presented, with a special emphasis placed on biomedical and agricultural applications. For the biomedical applications, we present the potential use of biopolymeric NPs in drug delivery and diagnostic systems, as well as in tissue engineering, and photodynamic and photothermal therapy. Moreover, we also highlight other advantages, as well as the possibility of creating antibacterial systems based on biopolymeric nanoparticles against many microbial strains. The potential use of biopolymeric nanoparticles in agriculture is also presented, with several studies reporting on the growth-promoting properties of several NPs. In addition, biopolymeric NPs can also be used as delivery systems for agrochemicals, and in crop improvement.

The cytotoxicity, biodegradability, and biocompatibility of biopolymeric nanoparticles enable their use in various fields, and the present review was designed to present recent advances that could help researchers in the large-scale production of these nanoparticles and—by combining sciences such as bioengineering, chemical modification, and nanotechnology—aid in the development of nanoparticles with improved efficiency and effectiveness, thus increasing their applicability.

Author Contributions: Manuscript conception, A.A.V. (Andrey A. Vodyashkin); design of the article, A.A.V. (Andrey A. Vodyashkin) and P.K.; literature data analysis, A.A.V. (Andrey A. Vodyashkin); writing—original draft preparation, A.A.V. (Andrey A. Vodyashkin) and P.K.; writing—review and editing, A.A.V. (Andrey A. Vodyashkin), A.A.V. (Alexandre A. Vetcher), Y.M.S. and P.K. All authors have read and agreed to the published version of the manuscript.

Funding: This research received no external funding.

Data Availability Statement: The data presented in this study are available in the article.

Acknowledgments: This publication was prepared with the support of the RUDN University Strategic Academic Leadership Program.

Conflicts of Interest: The authors declare no conflict of interest.

References

1. Vasilev, S.; Vodyashkin, A.; Vasileva, D.; Zelenovskiy, P.; Chezganov, D.; Yuzhakov, V.; Shur, V.; O'Reilly, E.; Vinogradov, A. An Investigative Study on the Effect of Pre-Coating Polymer Solutions on the Fabrication of Low Cost Anti-Adhesive Release Paper. *Nanomaterials* **2020**, *10*, 1436. [CrossRef] [PubMed]
2. Vodyashkin, A.A.; Kezimana, P.; Prokonov, F.Y.; Vasilenko, I.A.; Stanishevskiy, Y.M. Current Methods for Synthesis and Potential Applications of Cobalt Nanoparticles: A Review. *Crystals* **2022**, *12*, 272. [CrossRef]
3. Li, Y.-H.; Li, J.-Y.; Xu, Y.-J. Bimetallic Nanoparticles as Cocatalysts for Versatile Photoredox Catalysis. *EnergyChem* **2021**, *3*, 100047. [CrossRef]
4. Mitchell, M.J.; Billingsley, M.M.; Haley, R.M.; Wechsler, M.E.; Peppas, N.A.; Langer, R. Engineering Precision Nanoparticles for Drug Delivery. *Nat. Rev. Drug Discov.* **2021**, *20*, 101–124. [CrossRef]

5. Cai, Y.; Yang, F.; Wu, L.; Shu, Y.; Qu, G.; Fakhri, A.; Kumar Gupta, V. Hydrothermal-Ultrasonic Synthesis of CuO Nanorods and CuWO4 Nanoparticles for Catalytic Reduction, Photocatalysis Activity, and Antibacterial Properties. *Mater. Chem. Phys.* **2021**, *258*, 123919. [CrossRef]
6. Hua, Z.; Yu, T.; Liu, D.; Xianyu, Y. Recent Advances in Gold Nanoparticles-Based Biosensors for Food Safety Detection. *Biosens. Bioelectron.* **2021**, *179*, 113076. [CrossRef] [PubMed]
7. Wang, S.; Guo, H.; Li, Y.; Li, X. Penetration of Nanoparticles across a Lipid Bilayer: Effects of Particle Stiffness and Surface Hydrophobicity. *Nanoscale* **2019**, *11*, 4025–4034. [CrossRef]
8. Zhang, Y.; Lin, R.; Li, H.; He, W.; Du, J.; Wang, J. Strategies to Improve Tumor Penetration of Nanomedicines through Nanoparticle Design. *WIREs Nanomed. Nanobiotechnology* **2019**, *11*, e1519. [CrossRef]
9. Ajdary, M.; Moosavi, M.; Rahmati, M.; Falahati, M.; Mahboubi, M.; Mandegary, A.; Jangjoo, S.; Mohammadinejad, R.; Varma, R. Health Concerns of Various Nanoparticles: A Review of Their In Vitro and In Vivo Toxicity. *Nanomaterials* **2018**, *8*, 634. [CrossRef] [PubMed]
10. Jorge de Souza, T.A.; Rosa Souza, L.R.; Franchi, L.P. Silver Nanoparticles: An Integrated View of Green Synthesis Methods, Transformation in the Environment, and Toxicity. *Ecotoxicol. Environ. Saf.* **2019**, *171*, 691–700. [CrossRef] [PubMed]
11. Vance, M.E.; Kuiken, T.; Vejerano, E.P.; McGinnis, S.P.; Hochella, M.F.; Rejeski, D.; Hull, M.S. Nanotechnology in the Real World: Redeveloping the Nanomaterial Consumer Products Inventory. *Beilstein J. Nanotechnol.* **2015**, *6*, 1769–1780. [CrossRef]
12. Buzea, C.; Pacheco, I. Toxicity of Nanoparticles. In *Nanotechnology in Eco-efficient Construction*; Elsevier: Amsterdam, the Netherlands, 2019; pp. 705–754, ISBN 978-0-08-102641-0.
13. Baranowska-Wójcik, E.; Szwajgier, D.; Oleszczuk, P.; Winiarska-Mieczan, A. Effects of Titanium Dioxide Nanoparticles Exposure on Human Health—a Review. *Biol. Trace Elem. Res.* **2020**, *193*, 118–129. [CrossRef] [PubMed]
14. Pacheco-Blandino, I.; Vanner, R.; Buzea, C. Toxicity of Nanoparticles. In *Toxicity of Building Materials*; Elsevier: Amsterdam, the Netherlands, 2012; pp. 427–475, ISBN 978-0-85709-122-2.
15. Buzea, C.; Pacheco, I.I.; Robbie, K. Nanomaterials and Nanoparticles: Sources and Toxicity. *Biointerphases* **2007**, *2*, MR17–MR71. [CrossRef] [PubMed]
16. Brohi, R.D.; Wang, L.; Talpur, H.S.; Wu, D.; Khan, F.A.; Bhattarai, D.; Rehman, Z.-U.; Farmanullah, F.; Huo, L.-J. Toxicity of Nanoparticles on the Reproductive System in Animal Models: A Review. *Front. Pharmacol.* **2017**, *8*, 606. [CrossRef] [PubMed]
17. Simonin, M.; Martins, J.M.F.; Le Roux, X.; Uzu, G.; Calas, A.; Richaume, A. Toxicity of TiO$_2$ Nanoparticles on Soil Nitrification at Environmentally Relevant Concentrations: Lack of Classical Dose–Response Relationships. *Nanotoxicology* **2017**, *11*, 247–255. [CrossRef]
18. Fajardo, C.; Saccà, M.L.; Costa, G.; Nande, M.; Martin, M. Impact of Ag and Al$_2$O$_3$ Nanoparticles on Soil Organisms: In Vitro and Soil Experiments. *Sci. Total Environ.* **2014**, *473–474*, 254–261. [CrossRef]
19. Sufian, M.M.; Khattak, J.Z.K.; Yousaf, S.; Rana, M.S. Safety Issues Associated with the Use of Nanoparticles in Human Body. *Photodiagn. Photodyn. Ther.* **2017**, *19*, 67–72. [CrossRef]
20. Arroyo-Maya, I.J.; McClements, D.J. Biopolymer Nanoparticles as Potential Delivery Systems for Anthocyanins: Fabrication and Properties. *Food Res. Int.* **2015**, *69*, 1–8. [CrossRef]
21. Joye, I.J.; McClements, D.J. Biopolymer-Based Nanoparticles and Microparticles: Fabrication, Characterization, and Application. *Curr. Opin. Colloid Interface Sci.* **2014**, *19*, 417–427. [CrossRef]
22. Zhang, X.; Li, Z.; Yang, P.; Duan, G.; Liu, X.; Gu, Z.; Li, Y. Polyphenol Scaffolds in Tissue Engineering. *Mater. Horiz.* **2021**, *8*, 145–167. [CrossRef]
23. Yang, P.; Zhu, F.; Zhang, Z.; Cheng, Y.; Wang, Z.; Li, Y. Stimuli-Responsive Polydopamine-Based Smart Materials. *Chem. Soc. Rev.* **2021**, *50*, 8319–8343. [CrossRef] [PubMed]
24. Hu, J.; Yang, L.; Yang, P.; Jiang, S.; Liu, X.; Li, Y. Polydopamine Free Radical Scavengers. *Biomater. Sci.* **2020**, *8*, 4940–4950. [CrossRef]
25. Dirisala, A.; Uchida, S.; Li, J.; Van Guyse, J.F.R.; Hayashi, K.; Vummaleti, S.V.C.; Kaur, S.; Mochida, Y.; Fukushima, S.; Kataoka, K. Effective MRNA Protection by Poly(L-ornithine) Synergizes with Endosomal Escape Functionality of a Charge-Conversion Polymer toward Maximizing MRNA Introduction Efficiency. *Macromol. Rapid Commun.* **2022**, *2022*, 2100754. [CrossRef] [PubMed]
26. Hu, K.; McClements, D.J. Fabrication of Biopolymer Nanoparticles by Antisolvent Precipitation and Electrostatic Deposition: Zein-Alginate Core/Shell Nanoparticles. *Food Hydrocoll.* **2015**, *44*, 101–108. [CrossRef]
27. Nitta, S.; Numata, K. Biopolymer-Based Nanoparticles for Drug/Gene Delivery and Tissue Engineering. *Int. J. Mol. Sci.* **2013**, *14*, 1629–1654. [CrossRef] [PubMed]
28. Jones, O.G.; Lesmes, U.; Dubin, P.; McClements, D.J. Effect of Polysaccharide Charge on Formation and Properties of Biopolymer Nanoparticles Created by Heat Treatment of β-Lactoglobulin–Pectin Complexes. *Food Hydrocoll.* **2010**, *24*, 374–383. [CrossRef]
29. Kumar, S.S.D.; Rajendran, N.K.; Houreld, N.N.; Abrahamse, H. Recent Advances on Silver Nanoparticle and Biopolymer-Based Biomaterials for Wound Healing Applications. *Int. J. Biol. Macromol.* **2018**, *115*, 165–175. [CrossRef] [PubMed]
30. Nakache, E.; Poulain, N.; Candau, F.; Orecchioni, A.; Irache, J. Biopolymer and Polymer Nanoparticles and Their Biomedical Applications. In *Handbook of Nanostructured Materials and Nanotechnology*; Elsevier: Amsterdam, the Netherlands, 2000; Volume 5, pp. 577–635, ISBN 978-0-12-513760-7.
31. Ltaief, S.; Jabli, M.; Ben Abdessalem, S. Immobilization of Copper Oxide Nanoparticles onto Chitosan Biopolymer: Application to the Oxidative Degradation of Naphthol Blue Black. *Carbohydr. Polym.* **2021**, *261*, 117908. [CrossRef]

32. Cabral, H.; Miyata, K.; Osada, K.; Kataoka, K. Block Copolymer Micelles in Nanomedicine Applications. *Chem. Rev.* **2018**, *118*, 6844–6892. [CrossRef] [PubMed]
33. Klinker, K.; Schäfer, O.; Huesmann, D.; Bauer, T.; Capelôa, L.; Braun, L.; Stergiou, N.; Schinnerer, M.; Dirisala, A.; Miyata, K.; et al. Secondary-Structure-Driven Self-Assembly of Reactive Polypept(o)Ides: Controlling Size, Shape, and Function of Core Cross-Linked Nanostructures. *Angew. Chem. Int. Ed.* **2017**, *56*, 9608–9613. [CrossRef] [PubMed]
34. Dirisala, A.; Uchida, S.; Toh, K.; Li, J.; Osawa, S.; Tockary, T.A.; Liu, X.; Abbasi, S.; Hayashi, K.; Mochida, Y.; et al. Transient Stealth Coating of Liver Sinusoidal Wall by Anchoring Two-Armed PEG for Retargeting Nanomedicines. *Sci. Adv.* **2020**, *6*, eabb8133. [CrossRef] [PubMed]
35. Zhang, X.-Q.; Xu, X.; Bertrand, N.; Pridgen, E.; Swami, A.; Farokhzad, O.C. Interactions of Nanomaterials and Biological Systems: Implications to Personalized Nanomedicine. *Adv. Drug Deliv. Rev.* **2012**, *64*, 1363–1384. [CrossRef]
36. van der Linden, E.; Foegeding, E.A. Gelation. In *Modern Biopolymer Science*; Elsevier: Amsterdam, The Netherlands, 2009; pp. 29–91, ISBN 978-0-12-374195-0.
37. Liu, Q.; Cai, W.; Zhen, T.; Ji, N.; Dai, L.; Xiong, L.; Sun, Q. Preparation of Debranched Starch Nanoparticles by Ionic Gelation for Encapsulation of Epigallocatechin Gallate. *Int. J. Biol. Macromol.* **2020**, *161*, 481–491. [CrossRef] [PubMed]
38. Sullivan, D.J.; Cruz-Romero, M.; Collins, T.; Cummins, E.; Kerry, J.P.; Morris, M.A. Synthesis of Monodisperse Chitosan Nanoparticles. *Food Hydrocoll.* **2018**, *83*, 355–364. [CrossRef]
39. Pant, A.; Negi, J.S. Novel Controlled Ionic Gelation Strategy for Chitosan Nanoparticles Preparation Using TPP-β-CD Inclusion Complex. *Eur. J. Pharm. Sci.* **2018**, *112*, 180–185. [CrossRef]
40. Othman, N.; Masarudin, M.; Kuen, C.; Dasuan, N.; Abdullah, L.; Jamil, S.M. Synthesis and Optimization of Chitosan Nanoparticles Loaded with L-Ascorbic Acid and Thymoquinone. *Nanomaterials* **2018**, *8*, 920. [CrossRef]
41. Fan, W.; Yan, W.; Xu, Z.; Ni, H. Formation Mechanism of Monodisperse, Low Molecular Weight Chitosan Nanoparticles by Ionic Gelation Technique. *Colloids Surf. B Biointerfaces* **2012**, *90*, 21–27. [CrossRef] [PubMed]
42. Calvo, P.; Remuñán-López, C.; Vila-Jato, J.L.; Alonso, M.J. Chitosan and Chitosan/Ethylene Oxide-Propylene Oxide Block Copolymer Nanoparticles as Novel Carriers for Proteins and Vaccines. *Pharm. Res.* **1997**, *14*, 1431–1436. [CrossRef]
43. Anand, M.; Sathyapriya, P.; Maruthupandy, M.; Hameedha Beevi, A. Synthesis of Chitosan Nanoparticles by TPP and Their Potential Mosquito Larvicidal Application. *Front. Lab. Med.* **2018**, *2*, 72–78. [CrossRef]
44. Qiu, C.; Hu, Y.; Jin, Z.; McClements, D.J.; Qin, Y.; Xu, X.; Wang, J. A Review of Green Techniques for the Synthesis of Size-Controlled Starch-Based Nanoparticles and Their Applications as Nanodelivery Systems. *Trends Food Sci. Technol.* **2019**, *92*, 138–151. [CrossRef]
45. Ji, N.; Hong, Y.; Gu, Z.; Cheng, L.; Li, Z.; Li, C. Fabrication and Characterization of Complex Nanoparticles Based on Carboxymethyl Short Chain Amylose and Chitosan by Ionic Gelation. *Food Funct.* **2018**, *9*, 2902–2912. [CrossRef] [PubMed]
46. Anita, S.; Ramachandran, T.; Rajendran, R.; Koushik, C.; Mahalakshmi, M. A Study of the Antimicrobial Property of Encapsulated Copper Oxide Nanoparticles on Cotton Fabric. *Text. Res. J.* **2011**, *81*, 1081–1088. [CrossRef]
47. Liu, Q.; Li, M.; Xiong, L.; Qiu, L.; Bian, X.; Sun, C.; Sun, Q. Oxidation Modification of Debranched Starch for the Preparation of Starch Nanoparticles with Calcium Ions. *Food Hydrocoll.* **2018**, *85*, 86–92. [CrossRef]
48. Subramanian, S.B.; Francis, A.P.; Devasena, T. Chitosan–Starch Nanocomposite Particles as a Drug Carrier for the Delivery of Bis-Desmethoxy Curcumin Analog. *Carbohydr. Polym.* **2014**, *114*, 170–178. [CrossRef]
49. Nait Mohamed, F.A.; Laraba-Djebari, F. Development and Characterization of a New Carrier for Vaccine Delivery Based on Calcium-Alginate Nanoparticles: Safe Immunoprotective Approach against Scorpion Envenoming. *Vaccine* **2016**, *34*, 2692–2699. [CrossRef]
50. Wan-Hong, C.; Suk-Fun, C.; Pang, S.-C.; Kok, K.-Y. Synthesis and Characterisation of Piperine-Loaded Starch Nanoparticles. *J. Phys. Sci.* **2020**, *31*, 57–68. [CrossRef]
51. Farrag, Y.; Ide, W.; Montero, B.; Rico, M.; Rodríguez-Llamazares, S.; Barral, L.; Bouza, R. Preparation of Starch Nanoparticles Loaded with Quercetin Using Nanoprecipitation Technique. *Int. J. Biol. Macromol.* **2018**, *114*, 426–433. [CrossRef]
52. El-Naggar, M.E.; El-Rafie, M.H.; El-sheikh, M.A.; El-Feky, G.S.; Hebeish, A. Synthesis, Characterization, Release Kinetics and Toxicity Profile of Drug-Loaded Starch Nanoparticles. *Int. J. Biol. Macromol.* **2015**, *81*, 718–729. [CrossRef]
53. Chin, S.F.; Pang, S.C.; Tay, S.H. Size Controlled Synthesis of Starch Nanoparticles by a Simple Nanoprecipitation Method. *Carbohydr. Polym.* **2011**, *86*, 1817–1819. [CrossRef]
54. Gutiérrez, G.; Morán, D.; Marefati, A.; Purhagen, J.; Rayner, M.; Matos, M. Synthesis of Controlled Size Starch Nanoparticles (SNPs). *Carbohydr. Polym.* **2020**, *250*, 116938. [CrossRef] [PubMed]
55. Pan, X.; Liu, P.; Wang, Y.; Yi, Y.; Zhang, H.; Qian, D.-W.; Xiao, P.; Shang, E.; Duan, J.-A. Synthesis of Starch Nanoparticles with Controlled Morphology and Various Adsorption Rate for Urea. *Food Chem.* **2022**, *369*, 130882. [CrossRef] [PubMed]
56. Wu, X.; Chang, Y.; Fu, Y.; Ren, L.; Tong, J.; Zhou, J. Effects of Non-Solvent and Starch Solution on Formation of Starch Nanoparticles by Nanoprecipitation: Effects of Non-Solvent and Starch Solution on Formation of SNPs. *Starch-Stärke* **2016**, *68*, 258–263. [CrossRef]
57. Sotelo-Boyás, M.E.; Correa-Pacheco, Z.N.; Bautista-Baños, S.; Corona-Rangel, M.L. Physicochemical Characterization of Chitosan Nanoparticles and Nanocapsules Incorporated with Lime Essential Oil and Their Antibacterial Activity against Food-Borne Pathogens. *LWT* **2017**, *77*, 15–20. [CrossRef]

58. Luque-Alcaraz, A.G.; Lizardi-Mendoza, J.; Goycoolea, F.M.; Higuera-Ciapara, I.; Argüelles-Monal, W. Preparation of Chitosan Nanoparticles by Nanoprecipitation and Their Ability as a Drug Nanocarrier. *RSC Adv.* **2016**, *6*, 59250–59256. [CrossRef]
59. Rasooli, I.; Rezaei, M.B.; Allameh, A. Ultrastructural Studies on Antimicrobial Efficacy of Thyme Essential Oils on Listeria Monocytogenes. *Int. J. Infect. Dis.* **2006**, *10*, 236–241. [CrossRef]
60. Sotelo-Boyás, M.; Correa-Pacheco, Z.; Bautista-Baños, S.; Gómez y Gómez, Y. Release Study and Inhibitory Activity of Thyme Essential Oil-Loaded Chitosan Nanoparticles and Nanocapsules against Foodborne Bacteria. *Int. J. Biol. Macromol.* **2017**, *103*, 409–414. [CrossRef]
61. Asgari, S.; Saberi, A.H.; McClements, D.J.; Lin, M. Microemulsions as Nanoreactors for Synthesis of Biopolymer Nanoparticles. *Trends Food Sci. Technol.* **2019**, *86*, 118–130. [CrossRef]
62. Kafshgari, M.H.; Khorram, M.; Mansouri, M.; Samimi, A.; Osfouri, S. Preparation of Alginate and Chitosan Nanoparticles Using a New Reverse Micellar System. *Iran. Polym. J.* **2012**, *21*, 99–107. [CrossRef]
63. Nesamony, J.; Singh, P.R.; Nada, S.E.; Shah, Z.A.; Kolling, W.M. Calcium Alginate Nanoparticles Synthesized through a Novel Interfacial Cross-Linking Method as a Potential Protein Drug Delivery System. *J. Pharm. Sci.* **2012**, *101*, 2177–2184. [CrossRef]
64. Duyen, T.T.M.; Van Hung, P. Morphology, Crystalline Structure and Digestibility of Debranched Starch Nanoparticles Varying in Average Degree of Polymerization and Fabrication Methods. *Carbohydr. Polym.* **2021**, *256*, 117424. [CrossRef]
65. Wang, Y.; Wang, X.; Luo, G.; Dai, Y. Adsorption of Bovin Serum Albumin (BSA) onto the Magnetic Chitosan Nanoparticles Prepared by a Microemulsion System. *Bioresour. Technol.* **2008**, *99*, 3881–3884. [CrossRef] [PubMed]
66. Dubey, R.; Bajpai, J.; Bajpai, A.K. Chitosan-Alginate Nanoparticles (CANPs) as Potential Nanosorbent for Removal of Hg (II) Ions. *Environ. Nanotechnol. Monit. Manag.* **2016**, *6*, 32–44. [CrossRef]
67. Zhang, G.; Zhou, H.; An, C.; Liu, D.; Huang, Z.; Kuang, Y. Bimetallic Palladium–Gold Nanoparticles Synthesized in Ionic Liquid Microemulsion. *Colloid Polym. Sci.* **2012**, *290*, 1435–1441. [CrossRef]
68. Zhou, G.; Luo, Z.; Fu, X. Preparation and Characterization of Starch Nanoparticles in Ionic Liquid-in-Oil Microemulsions System. *Ind. Crops Prod.* **2014**, *52*, 105–110. [CrossRef]
69. Wang, X.; Cheng, J.; Ji, G.; Peng, X.; Luo, Z. Starch Nanoparticles Prepared in a Two Ionic Liquid Based Microemulsion System and Their Drug Loading and Release Properties. *RSC Adv.* **2016**, *6*, 4751–4757. [CrossRef]
70. Kaloti, M.; Bohidar, H.B. Kinetics of Coacervation Transition versus Nanoparticle Formation in Chitosan–Sodium Tripolyphosphate Solutions. *Colloids Surf. B Biointerfaces* **2010**, *81*, 165–173. [CrossRef]
71. Barthold, S.; Kletting, S.; Taffner, J.; de Souza Carvalho-Wodarz, C.; Lepeltier, E.; Loretz, B.; Lehr, C.-M. Preparation of Nanosized Coacervates of Positive and Negative Starch Derivatives Intended for Pulmonary Delivery of Proteins. *J. Mater. Chem. B* **2016**, *4*, 2377–2386. [CrossRef] [PubMed]
72. Saboktakin, M.R.; Tabatabaie, R.M.; Maharramov, A.; Ramazanov, M.A. Synthesis and In Vitro Evaluation of Carboxymethyl Starch–Chitosan Nanoparticles as Drug Delivery System to the Colon. *Int. J. Biol. Macromol.* **2011**, *48*, 381–385. [CrossRef] [PubMed]
73. Tavares, I.S.; Caroni, A.L.P.F.; Neto, A.A.D.; Pereira, M.R.; Fonseca, J.L.C. Surface Charging and Dimensions of Chitosan Coacervated Nanoparticles. *Colloids Surf. B Biointerfaces* **2012**, *90*, 254–258. [CrossRef]
74. Patra, S.; Basak, P.; Tibarewala, D.N. Synthesis of Gelatin Nano/Submicron Particles by Binary Nonsolvent Aided Coacervation (BNAC) Method. *Mater. Sci. Eng. C* **2016**, *59*, 310–318. [CrossRef] [PubMed]
75. Pawar, A.; Thakkar, S.; Misra, M. A Bird's Eye View of Nanoparticles Prepared by Electrospraying: Advancements in Drug Delivery Field. *J. Controlled Release* **2018**, *286*, 179–200. [CrossRef]
76. Bock, N.; Woodruff, M.A.; Hutmacher, D.W.; Dargaville, T.R. Electrospraying, a Reproducible Method for Production of Polymeric Microspheres for Biomedical Applications. *Polymers* **2011**, *3*, 131–149. [CrossRef]
77. Kurakula, M.; Raghavendra Naveen, N. Electrospraying: A Facile Technology Unfolding the Chitosan Based Drug Delivery and Biomedical Applications. *Eur. Polym. J.* **2021**, *147*, 110326. [CrossRef]
78. Wang, Y.; Zhang, R.; Qin, W.; Dai, J.; Zhang, Q.; Lee, K.; Liu, Y. Physicochemical Properties of Gelatin Films Containing Tea Polyphenol-Loaded Chitosan Nanoparticles Generated by Electrospray. *Mater. Des.* **2020**, *185*, 108277. [CrossRef]
79. Thien, D.V.H.; Hsiao, S.W.; Ho, M.H. Synthesis of Electrosprayed Chitosan Nanoparticles for Drug Sustained Release. *Nano. Life* **2012**, *2*, 1250003. [CrossRef]
80. Kempe, K.; Nicolazzo, J.A. Biodegradable Polymeric Nanoparticles for Brain-Targeted Drug Delivery. In *Nanomedicines for Brain Drug Delivery*; Morales, J.O., Gaillard, P.J., Eds.; Springer: New York, NY, USA, 2021; Volume 157, pp. 1–27, ISBN 978-1-07-160837-1.
81. Jana, P.; Shyam, M.; Singh, S.; Jayaprakash, V.; Dev, A. Biodegradable Polymers in Drug Delivery and Oral Vaccination. *Eur. Polym. J.* **2021**, *142*, 110155. [CrossRef]
82. Gholamali, I.; Yadollahi, M. Bio-Nanocomposite Polymer Hydrogels Containing Nanoparticles for Drug Delivery: A Review. *Regen. Eng. Transl. Med.* **2021**, *7*, 129–146. [CrossRef]
83. Rabiee, N.; Ahmadi, S.; Afshari, R.; Khalaji, S.; Rabiee, M.; Bagherzadeh, M.; Fatahi, Y.; Dinarvand, R.; Tahriri, M.; Tayebi, L.; et al. Polymeric Nanoparticles for Nasal Drug Delivery to the Brain: Relevance to Alzheimer's Disease. *Adv. Ther.* **2021**, *4*, 2000076. [CrossRef]
84. Thakuria, A.; Kataria, B.; Gupta, D. Nanoparticle-Based Methodologies for Targeted Drug Delivery—An Insight. *J. Nanoparticle Res.* **2021**, *23*, 87. [CrossRef]

85. Jackson, T.C.; Obiakor, N.M.; Iheanyichukwu, I.N.; Ita, O.O.; Ucheokoro, A.S. Biotechnology and Nanotechnology Drug Delivery: A Review. *J. Pharm. Pharmacol.* **2021**, *9*, 127–132. [CrossRef]
86. Luo, M.-X.; Hua, S.; Shang, Q.-Y. Application of Nanotechnology in Drug Delivery Systems for Respiratory Diseases (Review). *Mol. Med. Rep.* **2021**, *23*, 325. [CrossRef]
87. Ibrahim, M.A.; Abdellatif, A.A.H. Applications of Nanopharmaceuticals in Delivery and Targeting. In *Nanopharmaceuticals: Principles and Applications Vol 1*; Yata, V.K., Ranjan, S., Dasgupta, N., Lichtfouse, E., Eds.; Springer: Cham, Switzeland, 2021; Volume 46, pp. 73–114, ISBN 978-3-030-44924-7.
88. Dahiya, R.; Dahiya, S. Advanced Drug Delivery Applications of Self-Assembled Nanostructures and Polymeric Nanoparticles. In *Handbook on Nanobiomaterials for Therapeutics and Diagnostic Applications*; Elsevier: Amsterdam, the Netherlands, 2021; pp. 297–339, ISBN 978-0-12-821013-0.
89. Hrkach, J.; Von Hoff, D.; Ali, M.M.; Andrianova, E.; Auer, J.; Campbell, T.; De Witt, D.; Figa, M.; Figueiredo, M.; Horhota, A.; et al. Preclinical Development and Clinical Translation of a PSMA-Targeted Docetaxel Nanoparticle with a Differentiated Pharmacological Profile. *Sci. Transl. Med.* **2012**, *4*, 3003651. [CrossRef] [PubMed]
90. Swierczewska, M.; Han, H.S.; Kim, K.; Park, J.H.; Lee, S. Polysaccharide-Based Nanoparticles for Theranostic Nanomedicine. *Adv. Drug Deliv. Rev.* **2016**, *99*, 70–84. [CrossRef] [PubMed]
91. Liu, S.; Yang, S.; Ho, P.C. Intranasal Administration of Carbamazepine-Loaded Carboxymethyl Chitosan Nanoparticles for Drug Delivery to the Brain. *Asian J. Pharm. Sci.* **2018**, *13*, 72–81. [CrossRef]
92. Zu, M.; Ma, Y.; Cannup, B.; Xie, D.; Jung, Y.; Zhang, J.; Yang, C.; Gao, F.; Merlin, D.; Xiao, B. Oral Delivery of Natural Active Small Molecules by Polymeric Nanoparticles for the Treatment of Inflammatory Bowel Diseases. *Adv. Drug Deliv. Rev.* **2021**, *176*, 113887. [CrossRef]
93. Xie, X.; Zhang, Y.; Zhu, Y.; Lan, Y. Preparation and Drug-Loading Properties of Amphoteric Cassava Starch Nanoparticles. *Nanomaterials* **2022**, *12*, 598. [CrossRef]
94. Alp, E.; Damkaci, F.; Guven, E.; Tenniswood, M. Starch Nanoparticles for Delivery of the Histone Deacetylase Inhibitor CG-1521 in Breast Cancer Treatment. *Int. J. Nanomedicine* **2019**, *14*, 1335–1346. [CrossRef]
95. de Oliveira Pedro, R.; Hoffmann, S.; Pereira, S.; Goycoolea, F.M.; Schmitt, C.C.; Neumann, M.G. Self-Assembled Amphiphilic Chitosan Nanoparticles for Quercetin Delivery to Breast Cancer Cells. *Eur. J. Pharm. Biopharm.* **2018**, *131*, 203–210. [CrossRef] [PubMed]
96. Rahbar, M.; Morsali, A.; Bozorgmehr, M.R.; Beyramabadi, S.A. Quantum Chemical Studies of Chitosan Nanoparticles as Effective Drug Delivery Systems for 5-Fluorouracil Anticancer Drug. *J. Mol. Liq.* **2020**, *302*, 112495. [CrossRef]
97. Zhang, C.; Shi, G.; Zhang, J.; Song, H.; Niu, J.; Shi, S.; Huang, P.; Wang, Y.; Wang, W.; Li, C.; et al. Targeted Antigen Delivery to Dendritic Cell via Functionalized Alginate Nanoparticles for Cancer Immunotherapy. *J. Control. Release* **2017**, *256*, 170–181. [CrossRef] [PubMed]
98. Chin, S.F.; Jimmy, F.B.; Pang, S.C. Size Controlled Fabrication of Cellulose Nanoparticles for Drug Delivery Applications. *J. Drug Deliv. Sci. Technol.* **2018**, *43*, 262–266. [CrossRef]
99. Moghaddam, S.V.; Abedi, F.; Alizadeh, E.; Baradaran, B.; Annabi, N.; Akbarzadeh, A.; Davaran, S. Lysine-Embedded Cellulose-Based Nanosystem for Efficient Dual-Delivery of Chemotherapeutics in Combination Cancer Therapy. *Carbohydr. Polym.* **2020**, *250*, 116861. [CrossRef]
100. Sorasitthiyanukarn, F.N.; Muangnoi, C.; Ratnatilaka Na Bhuket, P.; Rojsitthisak, P.; Rojsitthisak, P. Chitosan/Alginate Nanoparticles as a Promising Approach for Oral Delivery of Curcumin Diglutaric Acid for Cancer Treatment. *Mater. Sci. Eng. C* **2018**, *93*, 178–190. [CrossRef] [PubMed]
101. Alkholief, M. Optimization of Lecithin-Chitosan Nanoparticles for Simultaneous Encapsulation of Doxorubicin and Piperine. *J. Drug Deliv. Sci. Technol.* **2019**, *52*, 204–214. [CrossRef]
102. Burhan, A.M.; Klahan, B.; Cummins, W.; Andrés-Guerrero, V.; Byrne, M.E.; O'Reilly, N.J.; Chauhan, A.; Fitzhenry, L.; Hughes, H. Posterior Segment Ophthalmic Drug Delivery: Role of Muco-Adhesion with a Special Focus on Chitosan. *Pharmaceutics* **2021**, *13*, 1685. [CrossRef]
103. Gholizadeh, S.; Wang, Z.; Chen, X.; Dana, R.; Annabi, N. Advanced Nanodelivery Platforms for Topical Ophthalmic Drug Delivery. *Drug Discov. Today* **2021**, *26*, 1437–1449. [CrossRef] [PubMed]
104. Lugoloobi, I.; Maniriho, H.; Jia, L.; Namulinda, T.; Shi, X.; Zhao, Y. Cellulose Nanocrystals in Cancer Diagnostics and Treatment. *J. Control. Release* **2021**, *336*, 207–232. [CrossRef] [PubMed]
105. Lim, C.-K.; Shin, J.; Kwon, I.C.; Jeong, S.Y.; Kim, S. Iodinated Photosensitizing Chitosan: Self-Assembly into Tumor-Homing Nanoparticles with Enhanced Singlet Oxygen Generation. *Bioconjug. Chem.* **2012**, *23*, 1022–1028. [CrossRef]
106. Rayhan, M.A.; Hossen, M.S.; Niloy, M.S.; Bhuiyan, M.H.; Paul, S.; Shakil, M.S. Biopolymer and Biomaterial Conjugated Iron Oxide Nanomaterials as Prostate Cancer Theranostic Agents: A Comprehensive Review. *Symmetry* **2021**, *13*, 974. [CrossRef]
107. Shreffler, J.; Koppelman, M.; Mamnoon, B.; Mallik, S.; Layek, B. Biopolymeric Systems for Diagnostic Applications. In *Tailor-Made and Functionalized Biopolymer Systems*; Elsevier: Amsterdam, the Netherlands, 2021; pp. 705–722, ISBN 978-0-12-821437-4.
108. Bonferoni, M.C.; Gavini, E.; Rassu, G.; Maestri, M.; Giunchedi, P. Chitosan Nanoparticles for Therapy and Theranostics of Hepatocellular Carcinoma (HCC) and Liver-Targeting. *Nanomaterials* **2020**, *10*, 870. [CrossRef]

109. Kim, K.; Kim, J.H.; Park, H.; Kim, Y.-S.; Park, K.; Nam, H.; Lee, S.; Park, J.H.; Park, R.-W.; Kim, I.-S. Tumor-Homing Multifunctional Nanoparticles for Cancer Theragnosis: Simultaneous Diagnosis, Drug Delivery, and Therapeutic Monitoring. *J. Control. Release* **2010**, *146*, 219–227. [CrossRef] [PubMed]
110. Mahmoud, K.A.; Mena, J.A.; Male, K.B.; Hrapovic, S.; Kamen, A.; Luong, J.H.T. Effect of Surface Charge on the Cellular Uptake and Cytotoxicity of Fluorescent Labeled Cellulose Nanocrystals. *ACS Appl. Mater. Interfaces* **2010**, *2*, 2924–2932. [CrossRef]
111. Vodyashkin, A.A.; Rizk, M.G.H.; Kezimana, P.; Kirichuk, A.A.; Stanishevskiy, Y.M. Application of Gold Nanoparticle-Based Materials in Cancer Therapy and Diagnostics. *Chem. Eng.* **2021**, *5*, 69. [CrossRef]
112. Zhang, L.; Gao, S.; Zhang, F.; Yang, K.; Ma, Q.; Zhu, L. Activatable Hyaluronic Acid Nanoparticle as a Theranostic Agent for Optical/Photoacoustic Image-Guided Photothermal Therapy. *ACS Nano* **2014**, *8*, 12250–12258. [CrossRef]
113. Chen, G.; Zhao, Y.; Xu, Y.; Zhu, C.; Liu, T.; Wang, K. Chitosan Nanoparticles for Oral Photothermally Enhanced Photodynamic Therapy of Colon Cancer. *Int. J. Pharm.* **2020**, *589*, 119763. [CrossRef]
114. Wang, F.; Li, J.; Chen, C.; Qi, H.; Huang, K.; Hu, S. Preparation and Synergistic Chemo-Photothermal Therapy of Redox-Responsive Carboxymethyl Cellulose/Chitosan Complex Nanoparticles. *Carbohydr. Polym.* **2022**, *275*, 118714. [CrossRef]
115. Mathews, S.; Gupta, P.K.; Bhonde, R.; Totey, S. Chitosan Enhances Mineralization during Osteoblast Differentiation of Human Bone Marrow-Derived Mesenchymal Stem Cells, by Upregulating the Associated Genes: Chitosan Enhances Mineralization. *Cell Prolif.* **2011**, *44*, 537–549. [CrossRef] [PubMed]
116. Jafary, F.; Vaezifar, S. Immobilization of Alkaline Phosphatase onto Chitosan Nanoparticles: A Novel Therapeutic Approach in Bone Tissue Engineering. *BioNanoScience* **2021**, *11*, 1160–1168. [CrossRef]
117. Herdocia-Lluberes, C.S.; Laboy-López, P.; Morales, S.; Gonzalez-Robles, T.J.; González-Feliciano, J.A.; Nicolau, E. Evaluation of Synthesized Nanohydroxyapatite-Nanocellulose Composites as Biocompatible Scaffolds for Applications in Bone Tissue Engineering. *J. Nanomater.* **2015**, *2015*, 310935. [CrossRef]
118. Chen, X.; Zhou, R.; Chen, B.; Chen, J. Nanohydroxyapatite/Cellulose Nanocrystals/Silk Fibroin Ternary Scaffolds for Rat Calvarial Defect Regeneration. *RSC Adv.* **2016**, *6*, 35684–35691. [CrossRef]
119. HPS, A.K.; Saurabh, C.K.; Adnan, A.S.; Nurul Fazita, M.R.; Syakir, M.I.; Davoudpour, Y.; Rafatullah, M.; Abdullah, C.K.; Haafiz, M.K.; Dungan, R. A Review on Chitosan-Cellulose Blends and Nanocellulose Reinforced Chitosan Biocomposites: Properties and Their Applications. *Carbohydr. Polym.* **2016**, *150*, 216–226. [CrossRef]
120. Fathi-Achachelouei, M.; Knopf-Marques, H.; Ribeiro da Silva, C.E.; Barthès, J.; Bat, E.; Tezcaner, A.; Vrana, N.E. Use of Nanoparticles in Tissue Engineering and Regenerative Medicine. *Front. Bioeng. Biotechnol.* **2019**, *7*, 113. [CrossRef]
121. Tayebi, T.; Baradaran-Rafii, A.; Hajifathali, A.; Rahimpour, A.; Zali, H.; Shaabani, A.; Niknejad, H. Biofabrication of Chitosan/Chitosan Nanoparticles/Polycaprolactone Transparent Membrane for Corneal Endothelial Tissue Engineering. *Sci. Rep.* **2021**, *11*, 7060. [CrossRef]
122. Karagoz, S.; Kiremitler, N.B.; Sarp, G.; Pekdemir, S.; Salem, S.; Goksu, A.G.; Onses, M.S.; Sozdutmaz, I.; Sahmetlioglu, E.; Ozkara, E.S.; et al. Antibacterial, Antiviral, and Self-Cleaning Mats with Sensing Capabilities Based on Electrospun Nanofibers Decorated with ZnO Nanorods and Ag Nanoparticles for Protective Clothing Applications. *ACS Appl. Mater. Interfaces* **2021**, *13*, 5678–5690. [CrossRef] [PubMed]
123. Liu, F.; Cheng, X.; Xiao, L.; Wang, Q.; Yan, K.; Su, Z.; Wang, L.; Ma, C.; Wang, Y. Inside-Outside Ag Nanoparticles-Loaded Polylactic Acid Electrospun Fiber for Long-Term Antibacterial and Bone Regeneration. *Int. J. Biol. Macromol.* **2021**, *167*, 1338–1348. [CrossRef]
124. Singh, T.A.; Sharma, A.; Tejwan, N.; Ghosh, N.; Das, J.; Sil, P.C. A State of the Art Review on the Synthesis, Antibacterial, Antioxidant, Antidiabetic and Tissue Regeneration Activities of Zinc Oxide Nanoparticles. *Adv. Colloid Interface Sci.* **2021**, *295*, 102495. [CrossRef]
125. Horie, M.; Tabei, Y. Role of Oxidative Stress in Nanoparticles Toxicity. *Free Radic. Res.* **2021**, *55*, 331–342. [CrossRef]
126. Kuang, X.; Wang, Z.; Luo, Z.; He, Z.; Liang, L.; Gao, Q.; Li, Y.; Xia, K.; Xie, Z.; Chang, R.; et al. Ag Nanoparticles Enhance Immune Checkpoint Blockade Efficacy by Promoting of Immune Surveillance in Melanoma. *J. Colloid Interface Sci.* **2022**, *616*, 189–200. [CrossRef]
127. Zohri, M.; Alavidjeh, M.S.; Haririan, I.; Ardestani, M.S.; Ebrahimi, S.E.S.; Sani, H.T.; Sadjadi, S.K. A Comparative Study between the Antibacterial Effect of Nisin and Nisin-Loaded Chitosan/Alginate Nanoparticles on the Growth of Staphylococcus Aureus in Raw and Pasteurized Milk Samples. *Probiotics Antimicrob. Proteins* **2010**, *2*, 258–266. [CrossRef] [PubMed]
128. Landriscina, A.; Rosen, J.; Friedman, A.J. Biodegradable Chitosan Nanoparticles in Drug Delivery for Infectious Disease. *Nanomedicine* **2015**, *10*, 1609–1619. [CrossRef] [PubMed]
129. Chien, R.-C.; Yen, M.-T.; Mau, J.-L. Antimicrobial and Antitumor Activities of Chitosan from Shiitake Stipes, Compared to Commercial Chitosan from Crab Shells. *Carbohydr. Polym.* **2016**, *138*, 259–264. [CrossRef] [PubMed]
130. Severino, R.; Ferrari, G.; Vu, K.D.; Donsì, F.; Salmieri, S.; Lacroix, M. Antimicrobial Effects of Modified Chitosan Based Coating Containing Nanoemulsion of Essential Oils, Modified Atmosphere Packaging and Gamma Irradiation against *Escherichia Coli* O157:H7 and Salmonella Typhimurium on Green Beans. *Food Control* **2015**, *50*, 215–222. [CrossRef]
131. Birsoy, K.; Wang, T.; Chen, W.W.; Freinkman, E.; Abu-Remaileh, M.; Sabatini, D.M. An Essential Role of the Mitochondrial Electron Transport Chain in Cell Proliferation Is to Enable Aspartate Synthesis. *Cell* **2015**, *162*, 540–551. [CrossRef] [PubMed]
132. Alqahtani, F.; Aleanizy, F.; El Tahir, E.; Alhabib, H.; Alsaif, R.; Shazly, G.; AlQahtani, H.; Alsarra, I.; Mahdavi, J. Antibacterial Activity of Chitosan Nanoparticles against Pathogenic *N. gonorrhoea*. *Int. J. Nanomed.* **2020**, *15*, 7877–7887. [CrossRef]

133. Aleanizy, F.S.; Alqahtani, F.Y.; Shazly, G.; Alfaraj, R.; Alsarra, I.; Alshamsan, A.; Gareeb Abdulhady, H. Measurement and Evaluation of the Effects of PH Gradients on the Antimicrobial and Antivirulence Activities of Chitosan Nanoparticles in *Pseudomonas aeruginosa*. *Saudi Pharm. J.* **2018**, *26*, 79–83. [CrossRef]
134. Mousavi, S.A.; Ghotaslou, R.; Kordi, S.; Khoramdel, A.; Aeenfar, A.; Kahjough, S.T.; Akbarzadeh, A. Antibacterial and Antifungal Effects of Chitosan Nanoparticles on Tissue Conditioners of Complete Dentures. *Int. J. Biol. Macromol.* **2018**, *118*, 881–885. [CrossRef] [PubMed]
135. Zimet, P.; Mombrú, Á.W.; Faccio, R.; Brugnini, G.; Miraballes, I.; Rufo, C.; Pardo, H. Optimization and Characterization of Nisin-Loaded Alginate-Chitosan Nanoparticles with Antimicrobial Activity in Lean Beef. *LWT* **2018**, *91*, 107–116. [CrossRef]
136. Pawar, V.; Topkar, H.; Srivastava, R. Chitosan Nanoparticles and Povidone Iodine Containing Alginate Gel for Prevention and Treatment of Orthopedic Implant Associated Infections. *Int. J. Biol. Macromol.* **2018**, *115*, 1131–1141. [CrossRef]
137. Kashyap, P.L.; Xiang, X.; Heiden, P. Chitosan Nanoparticle Based Delivery Systems for Sustainable Agriculture. *Int. J. Biol. Macromol.* **2015**, *77*, 36–51. [CrossRef] [PubMed]
138. Ghormade, V.; Deshpande, M.V.; Paknikar, K.M. Perspectives for Nano-Biotechnology Enabled Protection and Nutrition of Plants. *Biotechnol. Adv.* **2011**, *29*, 792–803. [CrossRef]
139. Sathiyabama, M. Biopolymeric Nanoparticles as a Nanocide for Crop Protection. In *Nanoscience for Sustainable Agriculture*; Pudake, R.N., Chauhan, N., Kole, C., Eds.; Springer: Cham, Switzeland, 2019; pp. 139–152, ISBN 978-3-319-97851-2.
140. Pereira, A.E.S.; Silva, P.M.; Oliveira, J.L.; Oliveira, H.C.; Fraceto, L.F. Chitosan Nanoparticles as Carrier Systems for the Plant Growth Hormone Gibberellic Acid. *Colloids Surf. B Biointerfaces* **2017**, *150*, 141–152. [CrossRef] [PubMed]
141. Arsova, B.; Foster, K.J.; Shelden, M.C.; Bramley, H.; Watt, M. Dynamics in Plant Roots and Shoots Minimize Stress, Save Energy and Maintain Water and Nutrient Uptake. *New Phytol.* **2020**, *225*, 1111–1119. [CrossRef] [PubMed]
142. Weemstra, M.; Kiorapostolou, N.; Ruijven, J.; Mommer, L.; Vries, J.; Sterck, F. The Role of Fine-root Mass, Specific Root Length and Life Span in Tree Performance: A Whole-tree Exploration. *Funct. Ecol.* **2020**, *34*, 575–585. [CrossRef]
143. Lynch, J.P.; Brown, K.M. New Roots for Agriculture: Exploiting the Root Phenome. *Philos. Trans. R. Soc. B Biol. Sci.* **2012**, *367*, 1598–1604. [CrossRef] [PubMed]
144. Kochian, L.V. Root Architecture: Editorial. *J. Integr. Plant Biol.* **2016**, *58*, 190–192. [CrossRef]
145. Chakraborty, M.; Hasanuzzaman, M.; Rahman, M.; Khan, M.A.R.; Bhowmik, P.; Mahmud, N.U.; Tanveer, M.; Islam, T. Mechanism of Plant Growth Promotion and Disease Suppression by Chitosan Biopolymer. *Agriculture* **2020**, *10*, 624. [CrossRef]
146. Ing, L.Y.; Zin, N.M.; Sarwar, A.; Katas, H. Antifungal Activity of Chitosan Nanoparticles and Correlation with Their Physical Properties. *Int. J. Biomater.* **2012**, *2012*, 632698. [CrossRef] [PubMed]
147. Chowdappa, P.; Shivakumar, C.; Chethana, C.S. Madhura Antifungal Activity of Chitosan-Silver Nanoparticle Composite against Colletotrichum Gloeosporioides Associated with Mango Anthracnose. *Afr. J. Microbiol. Res.* **2014**, *8*, 1803–1812. [CrossRef]
148. Saharan, V.; Kumaraswamy, R.V.; Choudhary, R.C.; Kumari, S.; Pal, A.; Raliya, R.; Biswas, P. Cu-Chitosan Nanoparticle Mediated Sustainable Approach To Enhance Seedling Growth in Maize by Mobilizing Reserved Food. *J. Agric. Food Chem.* **2016**, *64*, 6148–6155. [CrossRef]
149. Li, R.; He, J.; Xie, H.; Wang, W.; Bose, S.K.; Sun, Y.; Hu, J.; Yin, H. Effects of Chitosan Nanoparticles on Seed Germination and Seedling Growth of Wheat (*Triticum aestivum* L.). *Int. J. Biol. Macromol.* **2019**, *126*, 91–100. [CrossRef]
150. Gomes, D.G.; Pelegrino, M.T.; Ferreira, A.S.; Bazzo, J.H.; Zucareli, C.; Seabra, A.B.; Oliveira, H.C. Seed Priming with Copper-loaded Chitosan Nanoparticles Promotes Early Growth and Enzymatic Antioxidant Defense of Maize (*Zea mays* L.) Seedlings. *J. Chem. Technol. Biotechnol.* **2021**, *96*, 2176–2184. [CrossRef]
151. Kadam, P.M.; Prajapati, D.; Kumaraswamy, R.V.; Kumari, S.; Devi, K.A.; Pal, A.; Harish; Sharma, S.K.; Saharan, V. Physio-Biochemical Responses of Wheat Plant towards Salicylic Acid-Chitosan Nanoparticles. *Plant Physiol. Biochem.* **2021**, *162*, 699–705. [CrossRef]
152. Nguyen Van, S.; Dinh Minh, H.; Nguyen Anh, D. Study on Chitosan Nanoparticles on Biophysical Characteristics and Growth of Robusta Coffee in Green House. *Biocatal. Agric. Biotechnol.* **2013**, *2*, 289–294. [CrossRef]
153. Abdel-Aziz, H.M.M.; Hasaneen, M.N.A.; Omer, A.M. Nano Chitosan-NPK Fertilizer Enhances the Growth and Productivity of Wheat Plants Grown in Sandy Soil. *Span. J. Agric. Res.* **2016**, *14*, e0902. [CrossRef]
154. Verma, M.L.; Dhanya, B.S.; Sukriti; Rani, V.; Thakur, M.; Jeslin, J.; Kushwaha, R. Carbohydrate and Protein Based Biopolymeric Nanoparticles: Current Status and Biotechnological Applications. *Int. J. Biol. Macromol.* **2020**, *154*, 390–412. [CrossRef]
155. Costa, M.M.E.; Cabral-Albuquerque, E.C.M.; Alves, T.L.M.; Pinto, J.C.; Fialho, R.L. Use of Polyhydroxybutyrate and Ethyl Cellulose for Coating of Urea Granules. *J. Agric. Food Chem.* **2013**, *61*, 9984–9991. [CrossRef]
156. Senna, A.M.; Braga do Carmo, J.; Santana da Silva, J.M.; Botaro, V.R. Synthesis, Characterization and Application of Hydrogel Derived from Cellulose Acetate as a Substrate for Slow-Release NPK Fertilizer and Water Retention in Soil. *J. Environ. Chem. Eng.* **2015**, *3*, 996–1002. [CrossRef]
157. Melaj, M.A.; Daraio, M.E. HPMC Layered Tablets Modified with Chitosan and Xanthan as Matrices for Controlled-Release Fertilizers. *J. Appl. Polym. Sci.* **2014**, *131*, 40839. [CrossRef]
158. Servin, A.; Elmer, W.; Mukherjee, A.; De la Torre-Roche, R.; Hamdi, H.; White, J.C.; Bindraban, P.; Dimkpa, C. A Review of the Use of Engineered Nanomaterials to Suppress Plant Disease and Enhance Crop Yield. *J. Nanoparticle Res.* **2015**, *17*, 92. [CrossRef]
159. Chandra, S.; Chakraborty, N.; Dasgupta, A.; Sarkar, J.; Panda, K.; Acharya, K. Chitosan Nanoparticles: A Positive Modulator of Innate Immune Responses in Plants. *Sci. Rep.* **2015**, *5*, 15195. [CrossRef]

160. Chun, S.-C.; Chandrasekaran, M. Chitosan and Chitosan Nanoparticles Induced Expression of Pathogenesis-Related Proteins Genes Enhances Biotic Stress Tolerance in Tomato. *Int. J. Biol. Macromol.* **2019**, *125*, 948–954. [CrossRef] [PubMed]
161. Divya, K.; Jisha, M.S. Chitosan Nanoparticles Preparation and Applications. *Environ. Chem. Lett.* **2018**, *16*, 101–112. [CrossRef]
162. Anusuya, S.; Sathiyabama, M. Preparation of β-d-Glucan Nanoparticles and Its Antifungal Activity. *Int. J. Biol. Macromol.* **2014**, *70*, 440–443. [CrossRef] [PubMed]
163. Anusuya, S.; Sathiyabama, M. Protection of Turmeric Plants from Rhizome Rot Disease under Field Conditions by β-d-Glucan Nanoparticle. *Int. J. Biol. Macromol.* **2015**, *77*, 9–14. [CrossRef]
164. Anusuya, S.; Sathiyabama, M. β-d-Glucan Nanoparticle Pre-Treatment Induce Resistance against Pythium Aphanidermatum Infection in Turmeric. *Int. J. Biol. Macromol.* **2015**, *74*, 278–282. [CrossRef]
165. Manikandan, A.; Sathiyabama, M. Preparation of Chitosan Nanoparticles and Its Effect on Detached Rice Leaves Infected with Pyricularia Grisea. *Int. J. Biol. Macromol.* **2016**, *84*, 58–61. [CrossRef] [PubMed]
166. Divya, K.; Thampi, M.; Vijayan, S.; Varghese, S.; Jisha, M.S. Induction of Defence Response in *Oryza sativa* L. against *Rhizoctonia solani* (Kuhn) by Chitosan Nanoparticles. *Microb. Pathog.* **2020**, *149*, 104525. [CrossRef]
167. Choudhary, R.C.; Kumaraswamy, R.V.; Kumari, S.; Pal, A.; Raliya, R.; Biswas, P.; Saharan, V. Synthesis, Characterization, and Application of Chitosan Nanomaterials Loaded with Zinc and Copper for Plant Growth and Protection. In *Nanotechnology*; Prasad, R., Kumar, M., Kumar, V., Eds.; Springer: Singapore, 2017; pp. 227–247, ISBN 978-981-10-4572-1.
168. Chen, L.-C.; Kung, S.-K.; Chen, H.-H.; Lin, S.-B. Evaluation of Zeta Potential Difference as an Indicator for Antibacterial Strength of Low Molecular Weight Chitosan. *Carbohydr. Polym.* **2010**, *82*, 913–919. [CrossRef]
169. Pinto, R.J.B.; Almeida, A.; Fernandes, S.C.M.; Freire, C.S.R.; Silvestre, A.J.D.; Neto, C.P.; Trindade, T. Antifungal Activity of Transparent Nanocomposite Thin Films of Pullulan and Silver against *Aspergillus niger*. *Colloids Surf. B Biointerfaces* **2013**, *103*, 143–148. [CrossRef]
170. Liang, W.; Yu, A.; Wang, G.; Zheng, F.; Jia, J.; Xu, H. Chitosan-Based Nanoparticles of Avermectin to Control Pine Wood Nematodes. *Int. J. Biol. Macromol.* **2018**, *112*, 258–263. [CrossRef]
171. Liu, Y.; Yan, L.; Heiden, P.; Laks, P. Use of Nanoparticles for Controlled Release of Biocides in Solid Wood. *J. Appl. Polym. Sci.* **2001**, *79*, 458–465. [CrossRef]
172. Wang, Y.; Xiong, X.; Li, T.; Liang, J.; Chen, J. Aqueous Nano Insecticide Suspension and Its Preparation Process. *CN1486606 Chem Abs* **2004**, *142*, 12.
173. Lao, S.-B.; Zhang, Z.-X.; Xu, H.-H.; Jiang, G.-B. Novel Amphiphilic Chitosan Derivatives: Synthesis, Characterization and Micellar Solubilization of Rotenone. *Carbohydr. Polym.* **2010**, *82*, 1136–1142. [CrossRef]
174. Guan, H.; Chi, D.; Yu, J.; Li, X. A Novel Photodegradable Insecticide: Preparation, Characterization and Properties Evaluation of Nano-Imidacloprid. *Pestic. Biochem. Physiol.* **2008**, *92*, 83–91. [CrossRef]
175. Feng, B.-H.; Peng, L.-F. Synthesis and Characterization of Carboxymethyl Chitosan Carrying Ricinoleic Functions as an Emulsifier for Azadirachtin. *Carbohydr. Polym.* **2012**, *88*, 576–582. [CrossRef]
176. Paula, H.C.B.; Sombra, F.M.; de Freitas Cavalcante, R.; Abreu, F.O.M.S.; de Paula, R.C.M. Preparation and Characterization of Chitosan/Cashew Gum Beads Loaded with Lippia Sidoides Essential Oil. *Mater. Sci. Eng. C* **2011**, *31*, 173–178. [CrossRef]
177. dos Santos Silva, M.; Cocenza, D.S.; Grillo, R.; de Melo, N.F.S.; Tonello, P.S.; de Oliveira, L.C.; Cassimiro, D.L.; Rosa, A.H.; Fraceto, L.F. Paraquat-Loaded Alginate/Chitosan Nanoparticles: Preparation, Characterization and Soil Sorption Studies. *J. Hazard. Mater.* **2011**, *190*, 366–374. [CrossRef]
178. Grilloa, R.; Pereira, A.E.S.; Nishisaka, C.S.; De, L.R.; Oehlke, K.; Greiner, R.; Fraceto, L.F.J. Chitosan Nanoparticle Based Delivery Systems for Sustainable Agriculture. *Hazard. Mater* **2014**, *278*, 163–171.
179. Sen, S.K.; Chouhan, D.; Das, D.; Ghosh, R.; Mandal, P. Improvisation of Salinity Stress Response in Mung Bean through Solid Matrix Priming with Normal and Nano-Sized Chitosan. *Int. J. Biol. Macromol.* **2020**, *145*, 108–123. [CrossRef] [PubMed]
180. Mosavikia, A.A.; Mosavi, S.G.; Seghatoleslami, M.; Baradaran, R. Chitosan Nanoparticle and Pyridoxine Seed Priming Improves Tolerance to Salinity in Milk Thistle Seedling. *Not. Bot. Horti Agrobot. Cluj-Napoca* **2020**, *48*, 221–233. [CrossRef]
181. Sheikhalipour, M.; Esmaielpour, B.; Behnamian, M.; Gohari, G.; Giglou, M.T.; Vachova, P.; Rastogi, A.; Brestic, M.; Skalicky, M. Chitosan–Selenium Nanoparticle (Cs–Se NP) Foliar Spray Alleviates Salt Stress in Bitter Melon. *Nanomaterials* **2021**, *11*, 684. [CrossRef]
182. Balusamy, S.R.; Rahimi, S.; Sukweenadhi, J.; Sunderraj, S.; Shanmugam, R.; Thangavelu, L.; Mijakovic, I.; Perumalsamy, H. Chitosan, Chitosan Nanoparticles and Modified Chitosan Biomaterials, a Potential Tool to Combat Salinity Stress in Plants. *Carbohydr. Polym.* **2022**, *284*, 119189. [CrossRef]
183. Ahmad, J.; Qamar, S.; Kausar, N.; Qureshi, M.I. Nanoparticles: The Magic Bullets in Mitigating Drought Stress in Plants. In *Nanobiotechnology in Agriculture: An Approach Towards Sustainability*; Hakeem, K.R., Pirzadah, T.B., Eds.; Springer: Cham, Switzerland, 2020; pp. 145–161, ISBN 978-3-030-39978-8.
184. Behboudi, F.; Tahmasebi-Sarvestani, Z.; Kassaee, M.Z.; Modarres-Sanavy, S.A.M.; Sorooshzadeh, A.; Mokhtassi-Bidgoli, A. Evaluation of Chitosan Nanoparticles Effects with Two Application Methods on Wheat under Drought Stress. *J. Plant Nutr.* **2019**, *42*, 1439–1451. [CrossRef]
185. Ali, E.F.; El-Shehawi, A.M.; Ibrahim, O.H.M.; Abdul-Hafeez, E.Y.; Moussa, M.M.; Hassan, F.A.S. A Vital Role of Chitosan Nanoparticles in Improvisation the Drought Stress Tolerance in *Catharanthus roseus* (L.) through Biochemical and Gene Expression Modulation. *Plant Physiol. Biochem.* **2021**, *161*, 166–175. [CrossRef] [PubMed]

186. Lusser, M.; Parisi, C.; Plan, D.; Rodríguez-Cerezo, E. Deployment of New Biotechnologies in Plant Breeding. *Nat. Biotechnol.* **2012**, *30*, 231–239. [CrossRef]
187. Abdallah, N.A.; Prakash, C.S.; McHughen, A.G. Genome Editing for Crop Improvement: Challenges and Opportunities. *GM Crops Food* **2015**, *6*, 183–205. [CrossRef]
188. Martin-Ortigosa, S.; Valenstein, J.S.; Lin, V.S.-Y.; Trewyn, B.G.; Wang, K. Gold Functionalized Mesoporous Silica Nanoparticle Mediated Protein and DNA Codelivery to Plant Cells Via the Biolistic Method. *Adv. Funct. Mater.* **2012**, *22*, 3576–3582. [CrossRef]
189. Martin-Ortigosa, S.; Peterson, D.J.; Valenstein, J.S.; Lin, V.S.-Y.; Trewyn, B.G.; Lyznik, L.A.; Wang, K. Mesoporous Silica Nanoparticle-Mediated Intracellular Cre Protein Delivery for Maize Genome Editing via LoxP Site Excision. *Plant Physiol.* **2014**, *164*, 537–547. [CrossRef]
190. Torney, F.; Trewyn, B.G.; Lin, V.S.-Y.; Wang, K. Mesoporous Silica Nanoparticles Deliver DNA and Chemicals into Plants. *Nat. Nanotechnol.* **2007**, *2*, 295–300. [CrossRef] [PubMed]
191. Lv, Z.; Jiang, R.; Chen, J.; Chen, W. Nanoparticle-Mediated Gene Transformation Strategies for Plant Genetic Engineering. *Plant J.* **2020**, *104*, 880–891. [CrossRef]
192. Chandrasekaran, R.; Rajiv, P.; Abd-Elsalam, K.A. 14-Carbon Nanotubes: Plant Gene Delivery and Genome Editing. In *Carbon Nanomaterials for Agri-Food and Environmental Applications*; Abd-Elsalam, K.A., Ed.; Micro and Nano Technologies; Elsevier: Amsterdam, the Netherlands, 2020; pp. 279–296, ISBN 978-0-12-819786-8.
193. Demirer, G.S.; Silva, T.N.; Jackson, C.T.; Thomas, J.B.; Ehrhardt, D.W.; Rhee, S.Y.; Mortimer, J.C.; Landry, M.P. Nanotechnology to Advance CRISPR–Cas Genetic Engineering of Plants. *Nat. Nanotechnol.* **2021**, *16*, 243–250. [CrossRef]
194. Hu, P.; An, J.; Faulkner, M.M.; Wu, H.; Li, Z.; Tian, X.; Giraldo, J.P. Nanoparticle Charge and Size Control Foliar Delivery Efficiency to Plant Cells and Organelles. *ACS Nano* **2020**, *14*, 7970–7986. [CrossRef] [PubMed]
195. Li, L.; He, Z.-Y.; Wei, X.-W.; Gao, G.-P.; Wei, Y.-Q. Challenges in CRISPR/CAS9 Delivery: Potential Roles of Nonviral Vectors. *Hum. Gene Ther.* **2015**, *26*, 452–462. [CrossRef] [PubMed]
196. Qiao, J.; Sun, W.; Lin, S.; Jin, R.; Ma, L.; Liu, Y. Cytosolic Delivery of CRISPR/Cas9 Ribonucleoproteins for Genome Editing Using Chitosan-Coated Red Fluorescent Protein. *Chem. Commun.* **2019**, *55*, 4707–4710. [CrossRef] [PubMed]
197. Duceppe, N.; Tabrizian, M. Advances in Using Chitosan-Based Nanoparticles for In Vitro and In Vivo Drug and Gene Delivery. *Expert Opin. Drug Deliv.* **2010**, *7*, 1191–1207. [CrossRef]
198. Borchard, G. Chitosans for Gene Delivery. *Adv. Drug Deliv. Rev.* **2001**, *52*, 145–150. [CrossRef]
199. Kwak, S.-Y.; Lew, T.T.S.; Sweeney, C.J.; Koman, V.B.; Wong, M.H.; Bohmert-Tatarev, K.; Snell, K.D.; Seo, J.S.; Chua, N.-H.; Strano, M.S. Chloroplast-Selective Gene Delivery and Expression in Planta Using Chitosan-Complexed Single-Walled Carbon Nanotube Carriers. *Nat. Nanotechnol.* **2019**, *14*, 447–455. [CrossRef]
200. Maher, M.F.; Nasti, R.A.; Vollbrecht, M.; Starker, C.G.; Clark, M.D.; Voytas, D.F. Plant Gene Editing through de Novo Induction of Meristems. *Nat. Biotechnol.* **2020**, *38*, 84–89. [CrossRef] [PubMed]
201. Lowe, K.; Wu, E.; Wang, N.; Hoerster, G.; Hastings, C.; Cho, M.-J.; Scelonge, C.; Lenderts, B.; Chamberlin, M.; Cushatt, J.; et al. Morphogenic Regulators Baby Boom and Wuschel Improve Monocot Transformation. *Plant Cell* **2016**, *28*, 1998–2015. [CrossRef]

MDPI
St. Alban-Anlage 66
4052 Basel
Switzerland
Tel. +41 61 683 77 34
Fax +41 61 302 89 18
www.mdpi.com

Polymers Editorial Office
E-mail: polymers@mdpi.com
www.mdpi.com/journal/polymers

www.ingramcontent.com/pod-product-compliance
Lightning Source LLC
LaVergne TN
LVHW070048120526
838202LV00101B/1589